HPBooks

HOW TO MAKE YOUR CAR HANDLE

by Fred Puhn

HPBooks

CONTENTS

ANOTHER FACT-FILLED AUTOMOTIVE BOOK FROM HPBooks

NOTICE: The information in this book is true and complete to the best of our knowledge. All recommendations are made without guarantees on the part of the authors or the publisher. The authors and publisher disclaim all liability in connection with use of this information.

HPBooks
are published by The Berkley Publishing Group,
200 Madison Avenue, New York, New York 10016.
ISBN 0-912656-46-8
Library of Congress Catalog Number: 80-8527

Printed in U.S.A.

26 25 24 23 22

The fun of driving is not all due to power or top speed. A major part is the fun of driving a good-handling car. The phrase *man and machine have become one* best describes good handling. It is the confidence-arousing condition where every command of the driver is obeyed precisely by the car.

An evil-handling car will ruin all the fun of driving, as well as cause danger to its occupants and other drivers sharing the highway. For the various types of racing, good handling is essential—even on the drag strip. It is my purpose in this book to show what makes good handling, and give you methods of achieving this goal with your own car.

Handling involves all conditions found in driving—acceleration, straight-line driving, braking and cornering on good surfaces and bad. The good-handling car will be controllable and predictable in all these conditions, and it will provide some degree of riding comfort. Naturally there are compromises, and the use of the car will determine how to make them.

Ideal handling for all types of driving is impossible to obtain, so a multi-purpose vehicle is not the best for any specialized use. A good street car may have a great combination of ride and cornering on street tires, but will not corner as fast as an all-out racing car on racing rubber. A racing machine may wear you out on a 500 mile trip, and would scrape on every driveway you enter. A car set up for oval-track racing will corner poorly on right-hand corners, but a racing sports car will not lap a high-banked track as fast as the car set up for left turns only.

Good handling means superior traction. This is lack of wheelspin or axle hop in acceleration, maximum braking power on all surfaces, and outstanding cornering speed. Here again we are faced with compromises, depending on the use of the car. A car set up for maximum acceleration at a drag race will suffer a loss of traction in cornering. Likewise the car that sticks to the road best under all weather conditions will not have the cornering ability of a car set up for a clean dry racetrack. Always keep in mind the use of the car, and work for best handling in that type of driving.

The fun in driving is a car that handles well. No matter what type of car you drive, from family sedan to grand prix racer, it can be made to handle by methods in this book.

Here's what can be done in an all-out effort to get better handling on the street. The Datsun on the right belongs to George Louis of San Diego. Besides the obvious lowering of ride height, the car has many hidden improvements: Interpart Mulholland springs on Koni shocks, 23mm anti-roll bar front and rear, Goodrich 50-series radial TA tires on 13 x 7 mag wheels. Solid bushings are used in the suspension pivots. Mitchell steering rack in stiffer rack-mounting bushings gives quicker steering response. Stopping is improved with Velvetouch brake linings, steel braided brake hoses. Brake cooling ducts are in the front spoiler. This car is an ultimate street machine, and is only slightly illegal in California. A stock Datsun is shown on the left.

Alan Jones' Formula-One Williams is designed for maximum cornering. Additional downforce is provided by a rear wing and ground-effect venturis located at the sides of the car with spring-loaded skirts which seal against the road surface. The large rear tires make sense on a 500-horsepower car, but these tires on a 100-horsepower Formula Ford would add many seconds to lap times due to the added weight and aerodynamic drag. (Photo courtesy of Goodyear Tire & Rubber Company.)

Perhaps you drive to work every day in an ordinary production car. That dull trip might be made fun if your car could be driven with confidence at speed on banked freeway ramps, and around the occasional tight turn. Driving in traffic can be more enjoyable and safe if you know your car can stop faster and turn more precisely than the other cars in the mob around you.

If you drive for recreation on weekends you surely want better handling. That trip to the country to get away from it all can be an exciting adventure, as you look for interesting new turns. A good-handling car makes pleasure driving much more enjoyable and satisfying. If you take pride in your driving ability you want the utmost in response from your machine. The best driver in the world is frustrated with a bad-handling car.

If racing is your bag, good handling is essential. This book deals with all the intricate problems that are common to racing machines. Here are methods for suspension tuning and modification techniques. You will find that it is possible to convert an everyday sports car into a slalom winner by addition of some new equipment plus suspension adjustments. If the car is set up properly it can be returned to street use after the event. For drag racing, handling and traction are vital, and following chapters will tell you how to achieve that end. A family car can be competitive at the strip and then driven to the grocery store if properly set up.

There are three stages in making your car handle. The first is minor tinkering around the garage on weekends, with minimum money and time invested in the job. Even with this limited approach your car can be improved and this book will show you how. Suspension tuning is the answer here. First you should analyze what the car is and then change it for the better. Every car owner should start with this stage to see how well his car can be made to handle without any major modifications. Suspension tuning is cheap too, as you virtually eliminate that costly trip to the front-end shop. With simple tools it is possible to do most of that work in your own garage.

Chapter 2 deals with the reasons for good and poor handling. Understanding the reasons is vital when working on your car. A great deal of time and money is wasted each year by uninformed people trying to fix handling problems. You may put on $500 worth of new shocks, wheels and tires, when all the car really needed was a proper bump steer and suspension adjustment. Many suspension alignment shops are not aware of the really sophisticated demands of high-speed driving. Many professionals can do a fine job of making the factory settings, but are not trained to *alter* handling with suspension adjustments. You can be your own expert after understanding Chapter 3 and doing some experimenting.

Stage 2 in suspension improvement is modifications, which is the main object of this book. In this stage you have to spend some real money on parts and install them on your car. Chapter 4 explains how to get the right parts the first time. Too many car owners go to the store and come back with wide tires, mag wheels, and heavy-duty shocks—and then are disappointed with the results. The tires may hit the body because the wheels are wrong. The shocks may hurt the ride without improving the handling. Careful *planning* of modifications is a must, and Chapter 4 gives you information on buying each new part. There is a suppliers list in the back of the book. In addition there are home-built modifications that can be of great benefit. If you are a home craftsman, you can build such things as adjustable anti-roll bars which may not even be available for your car. It not only provides great satisfaction to build your own parts, but also can put you ahead of other drivers on the road. It is possible to have a really superior and unique car through modification.

Stage 3 is one that few people should attempt. This is designing your own chassis. There are books on design, but none really emphasize all aspects of the entire job. To put your ideas on paper is one thing, but it is entirely another to fabricate the parts and get the entire package assembled and tested to perfection.

In this book I have supplied sufficient information to attempt a design, but have not covered many of the related technical problems. Structural strength and stiffness is one of these problems. Unless you are a mechanical engineer, most of these problems will be impossible to solve. Another fundamental problem is fabrication techniques. It takes a high degree of skill to build critical parts in your garage, and the results are greatly influenced by workmanship. This part of designing and building a complete chassis is best left to professionals.

Even though you may never design and build a complete car, an appreciation of design procedures and goals is important to you because *any modification to an existing car is a change in the design of that car.*

Very few *production* cars are a design failure in the sense that they don't do what the designer and factory intended. Practically all domestic cars are developed to be very stable for Interstate driving. As a result, they suffer from poor response and feel "mushy." People who are not satisfied with the car they drive are often unusual drivers or those who want the car to do unusual things. Cornering well in a family car is unusual—both in the design of family cars and in the drivers who expect it.

You may rue the day you bought the machine or argue that the factory should have built it better, but that doesn't alter two important facts: To get it the way you want it, you have to change things. When you change something, you have changed the car to a different design. *Your* design.

Changing one part of the design often has far-reaching effects and it is rare that making a single change does anything good. It is necessary to think all the way to the end result and explore all side-effects before you start cutting and welding, or installing racing tires.

This isn't your friendly neighborhood freeway, but it could be. Photo by Goodyear on their proving ground near San Angelo, Texas, shows the difference tires make to wet weather control.

Because the discipline of looking clear to the end is vital in any modification program, the remainder of this chapter is a discussion of things the designer has in mind when he designs a racing car.

It will be useful to you as a way of thinking when you undertake to re-design even a part of your car.

There are a few technical terms used in this discussion which are commonly understood by people interested in automobiles. If you happen not to understand one or two of them, don't worry about it now, because the main purpose here is to give you an overview of design procedure.

All technical terms are carefully explained and defined in later chapters of this book, where they are applied to specific modifications and improvements you can do.

Before putting paper on the drawing board a designer establishes his design *criteria.* This is a list of objectives and specifications for the racing car which will be guide and ground rules all through the design. Criteria are essential to having a finished product that is consistent in all areas and meets the design objectives. How many times have you seen a specially-built car that is lightweight in one area and super-heavy in another? Criteria, if followed faithfully, will prevent that sort of mistake.

The criteria should answer these questions.

Why are you designing the car rather than buying one?
What makes your design better than the ones currently winning races?
What class will you compete in?
What length races will the car be designed for?
What force-loading conditions are expected?
What specifications and performance are you trying to beat?

The first two questions deserve a good answer before starting a design. Many racing cars are designed by one man, and it is human nature for him to think that just because he designed it the car must be superior. There are a number of valid reasons for designing a new car. Some of these are: to take advantage of the latest tires, to use a unique engine in a compatible chassis, or to test out a new suspension idea. These and other reasons are based on sound engineering and must be clearly written in the criteria before the job is started. Once written down, these reasons should be reviewed by the designer frequently to be sure he does not lose sight of his initial objectives.

To start the engineering job the designer should first establish the force-loading conditions for the car. Obviously there is quite a difference between an Indy racer and a dune buggy in the loading conditions. The designer will have to limit his design to withstand those that are important and reasonable. If he doesn't

If you plan on designing a better car than the Chaparral Indy car, first decide what things you intend to do better. It isn't easy, particularly if you consider you have to come up with ideas passed over by designer Jim Hall.

the car will turn out too heavy.

The most severe load imposed on a racing car usually occurs when one wheel hits a bump. The suspension, frame, and other structure are all designed for a bump load condition. Obviously if the driver runs off the road, as happens in the life of any race car, it may hit any sort of bump. The car can always hit a bump larger than it was designed for and break. The reason for selecting a particular bump load is so the entire car is designed to the same set of conditions. Think of the structure as a chain. The strength of the entire chain is determined by the weakest link. The *lightest possible* chain always has links *equal* in strength.

This is supposed to be road racing, but sometimes it becomes off-road. If suspension components are designed for a reasonable bump load, there will be no damage except for the driver's ego. The terrain out there sometimes gets a bit rough, such as the rocky areas next to this track at Willow Springs, California.

The designer may have his own ideas on how strong he wants the car to be, but the following loading conditions seem to give reasonable results on road-racing cars competing in short races:

5 g bump on one wheel (one g is the force of gravity)
5 g bump on all four wheels
Cornering at 2.5 g's
Braking at 2.5 g's
Crash at 10 g's (for design of safety equipment only)
Maximum engine torque times impact factor of 2 (used to design drive train and engine mounts)

These loading conditions are modified for other types of racing cars. Aerodynamic loads from wings and body must be added to the loads listed above.

All during the design process the designer will make decisions which determine the requirements for the car. The following are desired for most racing cars:

Lightest possible overall weight
Lowest unsprung weight
Lowest center of gravity height
Lowest frontal area
Minimum rolling drag
Minimum aerodynamic drag
Maximum traction from the tires
Maximum power delivered to the driving tires

Some of these characteristics are in direct conflict. Minimum rolling drag might be obtained with small tires, but this gives up some traction. Maximum aerodynamic downforce will give good traction but will increase aerodynamic drag. The right decisions on these and other compromises make the difference.

Now the designer begins to select components for the car, keeping in mind the above list of goals. Tires are selected first because they influence almost every aspect of the design. Suspension geometry is determined by size and type of the tires. Aerodynamics are to a very great extent affected by the tires. The weight and center of gravity (CG) height is determined by the tires, as well as the loading on the suspension and the drive train. The car is designed around the tires.

To get tire data the designer asks the manufacturers for help. He submits proposed specifications for his car: Weight, speed, power, class of car, and other pertinent operating conditions. Racing-tire manufacturers respond with recommended tire sizes.

In addition the designer goes to the races and checks the winning cars' tires. If they use several different brands or sizes he must try to get comparison information. One technique is timing the cars through a corner to compare cornering speeds. The best spot is a series of closely spaced corners slow enough to require that the driver reduce speed to make the turn. A long series of right and left turns is best, because it gives a longer distance over which to time the cars and reduces errors in timing. Times are recorded, and the different types of tires are compared. A designer may spot one tire as consistently superior to all others.

After collecting all the information, the designer compares available tires in the following areas:

Manufacturer (possibly related to design of the tire)
Outside diameter of the tire
Tread width
Recommended wheel size
Tire weight
Casing construction
Rubber compound

The size of the tires is the most important in determining the final specifications of the car. The designer tries to use the lowest tires possible, all other factors being equal, to lower the CG and reduce frontal area. This is particularly true on the front tires. At the rear the tire size may be determined by the rear axle height, which in turn may be determined by the engine and gearbox used. One engineering decision is whether to keep the axles straight and run larger tires, or to use small tires and allow the U-joints in the axles to run at an angle. Angled U-joints consume power, which may hurt overall performance.

Selecting the tire tread width is also a hard decision. A wider tire gives higher cornering speeds, but the increased weight and air drag slows the car on the straights. Usually use the largest allowable width the class permits. The ideal compromise may require extensive calculations, as well as observations of the existing cars out on the track. If the car is to race on various tracks the ideal set-up at one track may be wrong for another. Selecting the wheel rim width is a similar decision, with a compromise usually being required for the best overall results.

A series of tight corners in rapid succession is an ideal place to time cornering speeds. Tight turns are the best for this because horsepower will not be much of a factor in cornering speeds. In this corner, the formula Vee shown was nearly as quick as cars with 5 times the power.

Wheel diameter may also be a difficult decision. On some cars the size of the brakes is limited by the space available inside the wheel. On this type of car the selection of a smaller wheel diameter reduces weight and air drag, but may hurt braking. If brakes are mounted inboard on the chassis, their size may be almost unlimited, but overall weight and complexity of the car is increased. Because of the added complexity of half-shafts and U-joints, more servicing is required and there is greater chance for failure. In addition there may be space problems, such as brakes interfering with the gearbox or the driver's feet. All these compromises are hard ones requiring sound engineering judgment and comparisons of alternate designs.

Now the designer begins selecting other components and arranging them in the car. The car is drawn to scale, and alternate designs are drawn up so intelligent comparisons can be made. The choice and arrangement of components determines the following characteristics:

Weight, CG location, and polar moment of inertia
Frontal area
Aerodynamic shape

All of these determine the performance of the car as I will explain in later chapters. The decisions involved in component layout include:

Wheelbase length
Track width
Fuel tank shape and location
Location of engine and gearbox
Driver location and driving position
Cooling system design and radiator location
Miscellaneous system design and location

Each decision interacts with other decisions, so the overall goal must be kept clearly in mind. Here is where the designer can go wrong unless he constantly refers back to his design criteria.

When making these decisions the designer must emphasize the performance characteristics most important to his design. There often is conflict between two characteristics, such as high cornering speed and low aerodynamic drag. It is virtually impossible to have the maximum of both in a single design, so a compromise must be reached.

The type of racetrack largely determines the decision. On a fast track where the car spends most of its time near top speed, low drag is vitally important. It may be possible to get faster lap times by reducing the cornering speed to gain a top-speed advantage. One way of doing this is to use smaller tires. On a tight course where much of the lap is spent in cornering, top speed can be sacrificed to gain cornering speed. One method for achieving this is to use large wings on the car. If the car is used at varied tracks, perhaps the designer can figure out ways to convert the car from one specification to another.

The car itself determines which characteristics are important. A small-engine car spends more time at its top speed, so aerodynamic drag is one of the most important characteristics in determining the car's lap time. On the same track a car with a huge engine never reaches top speed, and has to slow down for virtually every bend in the road. Thus cornering speed and traction are more important in a big-engine car than top speed. The designer must arrive at a judgment for his car as to the important characteristics, and try to emphasize those wherever possible.

Along with mechanical design the aerodynamic shape must be developed.

The ADF Formula Ford went through aerodynamic testing before it was built. A model was used to determine air flow patterns over the body. The design was highly successful, winning the National Championship. Proper attention to aerodynamics is essential in this highly competitive class. The engines have just over 100 hp, making low drag very important.

The low profile front tires on Rich Amick's Quasar allow a smooth air flow over the fenders. Larger front tires would seriously affect the air flow all the way back to the rear wing, plus would add frontal area. This small sports racer—designed by the author—will attain speeds of 140 MPH with only about 100 horsepower, due to the low-drag body. The body design was based entirely on the tire sizes.

If the designer leaves this to the very end he will surely build a loser. The shape and size of the body determine handling as well as the top speed. Compromises with the ideal shape are required, but the designer should be aware of all the facts to make an intelligent decision. The layout of the components, particularly engine and driver, will determine the size and shape of the bodywork.

Extra streamlining increases weight and reduces accessibility. Some designers make tests of streamlining devices by mounting the device on another car and measuring the effect on performance. On really expensive racing cars, a model of the body is placed in a wind tunnel, so drag and downforce can be measured.

If wings are used they must be designed along with the rest of the body. The wing is not an add-on device, and its effect must be included in all the calculations made during the design. The suspension is designed to work under varying aerodynamic loads, and the bodywork is designed to flow air over the wing surface in a desired way. Adjustable features are designed-in to allow the wing to be used for suspension tuning.

Proper design of the suspension is essential to good handling. A great deal of effort is spent by the designer on suspension geometry and making the required compromises. This book presents many features of existing suspensions. You will soon see there are no perfect suspension systems—all involve some sort of compromise. A racing car requires fewer compromises than a road car, but they are still there.

Along with all previously mentioned considerations, the designer must develop a rigid frame to hold the car together. To do this properly he must have a working knowledge of stress analysis or get expert help. Frame design can cause an otherwise excellent car to fail due to insufficient stiffness. Some testing can be done prior to completion, such as torsional rigidity testing. Models can also be used to predict stiffness. There are a number of construction methods the designer can choose from including welded space frames and monocoque construction. He can use materials such as steel, aluminum, magnesium, titanium, and fiberglass. Selection of the right material will depend on many factors including such practical considerations as ease of repair. The frame often requires many compromises, and requires many design layouts before the best can be selected.

After the design is complete, parts are fabricated and assembled into a complete car. This job must go in a big hurry, because a good design can become obsolete if too much time elapses before it is ready to race. After completion the design must be proven in test sessions on the track. With adjustable features on the chassis, many test sessions are required to get the car to its optimum state of chassis tune. This book will assist in testing to develop the best overall handling.

In the following pages are solutions to handling problems on existing cars of all types. Many illustrations show racing car suspensions because they are simple and easy to photograph. However, the discussions of racing car suspensions apply also to the suspension of a family road car. The problems and their solutions are common to both racing cars and road cars alike, even though the cars may appear to be quite different. The fact that all cars run on the road and run on rubber tires makes them very similar in the handling department.

THE ORGANIZATION OF THIS BOOK

Everybody agrees, all the parts of a car must work together to get good performance and fine handling. For exactly that reason, the *design* of all parts must be coordinated and interrelated. When the car is built, the *adjustment or tuning* of all parts must be done so everything works together.

This book is intended for readers with a strong interest in automotive performance but little experience in tuning or modifying. It attempts to first provide the theoretical base which allows understanding how things work and therefore understanding how to tune or adjust them to work better. This information is mainly in Chapter 2, *General Information on Handling.*

Chapter 2 is intended to give you what you need to know *before* you approach your car with tools in hand and determination in your stride. Therefore it stops just short of practical applications. If some part of Chapter 2 leaves you hanging; perhaps saying, "Wow, let's get on with the modification!"; forgive me.

The practical how-to-do-it info is mainly in Chapter 3, *Suspension Tuning,* and Chapter 4, *Chassis Modifications.*

I thought it unwise, in this book, to study some part of the vehicle and then proceed directly to modification instructions — for the same reason it is unwise to do that when working on a car. Very seldom is a single modification useful because of its impact on other parts of the vehicle. It's better to study the entire chassis in Chapter 2, then proceed to a balanced modification program using information in later chapters.

Chapter 5 discusses the application of theory and practice to specialized racers where some aspects of performance greatly outweigh others. For example, if a vehicle must turn only to the left, then it should do that as well as possible no matter how poorly it may turn in the other direction.

Chapter 6 is a brief speculation about the future. I'll be surprised if all my guesses are fulfilled, but anyway it's fun to guess.

At the back of the book is a suppliers list, organized to help you locate vendors who sell things you need. And an index, hopefully organized to help you locate some elusive fact or formula later, when you may be using this as a reference.

That outside front tire is really bending under the cornering load. Tire deflection in a turn is what makes the science of suspension tuning possible. A car with completely rigid tires would be untunable and undriveable at speeds above a walking pace.

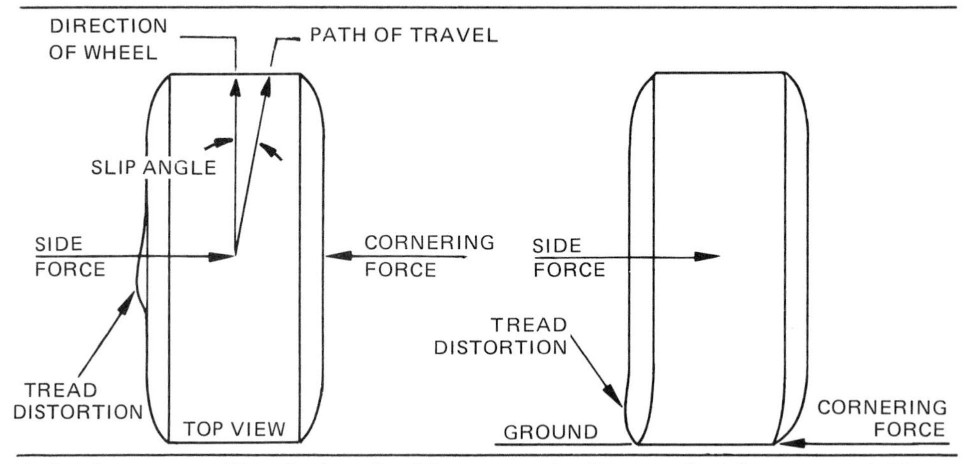

A side force on a rolling tire is resisted by a cornering force against the pavement. The sideways distortion of the tire causes the car to travel in a path at an angle to the direction the wheel is pointing. This angle is the very important *slip angle.*

To set your car up for the best possible handling it is important to understand how the chassis works. The various factors affecting handling are all related to each other, so sometimes the subject seems a bit confusing. Without going into deep mathematics I will show you the physical reasons for certain handling characteristics, and illustrate in later chapters how to change them. Once you know these basic reasons it will then be time to head for the workshop to set your car up for action.

HOW TIRES AFFECT HANDLING

Any discussion of handling must start with the tires. The car's only link with the road are those four *contact patches* of rubber, and their actions can determine how your car will handle and how many races you will win. A chassis expert under stands what is happening at the tires, and learns how to change the situation if needed.

A tire is elastic. A force applied to it in any direction will distort the tire's basic shape. A cornering force on the tire distorts it sideways at the point of contact with the road. This sideways distortion makes the car follow a path at an angle to the direction the wheel is pointing. This angle is called the *slip angle* and it gives the driver a feel of what the car is doing in a turn. A car would be next to impossible to corner fast without slip angles, because there would be no warning to the driver when the tire is about to break loose completely.

A "chicken and the egg" problem arises here. Cornering force causes slip angle, and slip angle causes cornering force. Basically it is the grip of the tire on the pavement which resists sliding sideways when there is side force on the wheel. Because the tire is elastic, it distorts while gripping the surface. Because it is elastic, the tire exerts a force which tends to pull the distorted rubber back to its normal position. *Cornering force* exists between tread and surface due to tire distortion. It is in the opposite direction to *side force* caused by turning. Even if the tire is sliding sideways, the two forces are always equal.

Cornering force measures the ability of a tire to resist sliding sideways in a

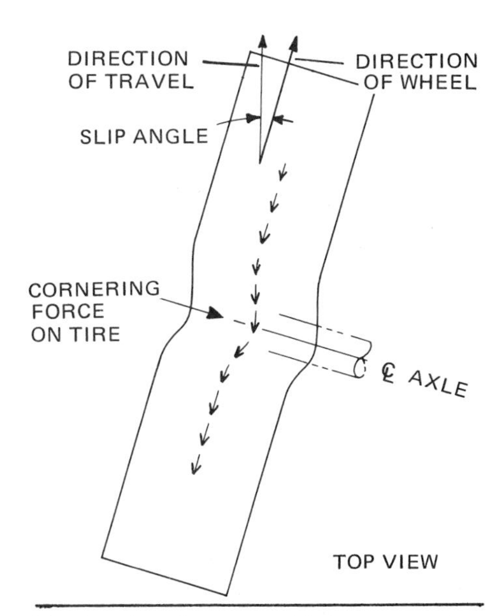

To visualize slip angle, it helps to imagine the path of one small point on the surface of the tire tread as it rotates under the wheel and passes through the tire contact patch. Because the contact patch is distorted by a side force, each point on the tire tread moves to the side as it goes through the contact patch. While it is moving to the side, it is also *in contact* with the surface. The direction the car actually travels is influenced mainly by the path of individual bits of tread as they go through the distorted contact patch.

turn. It would not exist without tire distortion and a resulting slip angle. On the other hand, slip angle would not exist without cornering force.

Both slip angle and cornering force increase as a car is driven into a turn with decreasing radius until a point is reached where the tire slides. When sliding, the tire still has some resistance to sliding—which is a sort of cornering force—but it is no longer caused entirely by tread distortion. It is caused by sliding friction across the surface.

Because slip angle is *defined* as a condition when the tire has not yet lost its grip, slip angle has no meaning when the car is in a skid.

There is no abrupt transition between a tire holding its grip and a tire losing its sideways grip, even though it often feels that way to the driver. Therefore there is no definite point at which the term *slip angle* is no longer relevant to the angle between car travel and the direction the wheel is pointing.

However any driver who has lost it in a turn can testify that *during* a skid there is no relation at all to the direction the car is traveling and the direction the driver is pointing the front wheels. You can turn them any direction you please without much effect on the overall excitement. They will only grip again when the car has slowed down or is making a larger-radius turn so the side force is reduced once more to an amount the tire can withstand without skidding.

To avoid getting trapped by our own words, let's agree that as long as a tire is having some influence on the direction of travel of the car, it has a slip angle even though the angle may be large.

A small cornering force on the tire creates small slip angle, and increasing the cornering force increases the slip angle. A

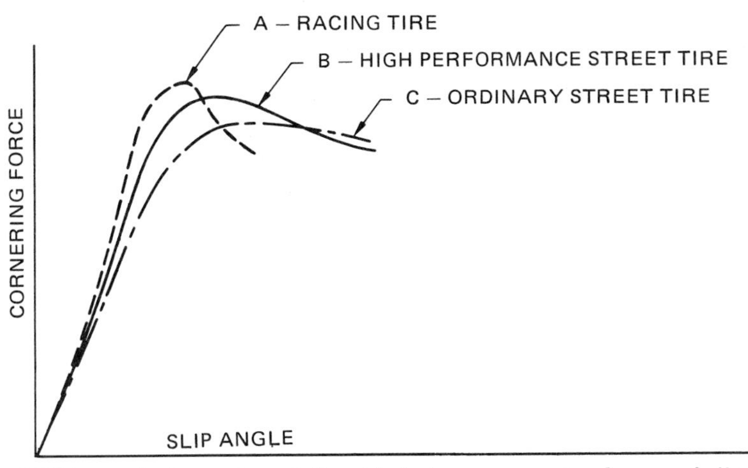

Figure 1/This graph shows the relationship between cornering force and slip angle for three different tires. Curve A has the highest cornering force at the limit of adhesion, but it gives the driver less warning of a skid because of the sharp peak of the curve. Curve C gives a lot of warning as it is relatively flat. The three curves are what you might expect for a racing tire, a high performance street tire, and an ordinary street tire. The street tire never gives as much cornering force as the racing tire.

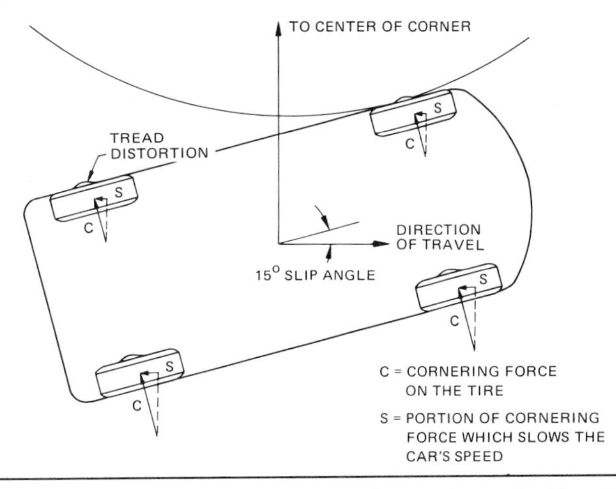

This drawing shows a car cornering with a 15-degree slip angle. For simplicity I am showing neutral steer characteristics—equal slip angles on all tires. It is also assumed that the driver is not applying power, nor is he braking. In this case the cornering forces act sideways on the tires as shown. The force S is the portion of the cornering force which slows the car. The larger the slip angle, the greater this force S becomes and the more quickly the car loses speed.

relationship between slip angle and cornering force is shown in Figure 1. The cornering force at which the driver senses the tire breaking loose is called the *limit of adhesion,* and this is the maximum cornering force that can be obtained from that tire. Notice how the cornering force becomes smaller at slip angles above the limit of adhesion. A sudden light feel in the steering wheel is a signal to the driver that the front tires have exceeded their adhesion limit.

The top of the cornering force curve can be flat or sharp, depending on the characteristics of the tire and its particular operating conditions. A flat curve will give the driver more latitude, due to a large increase in slip angle without any drop in cornering force. Drivers learn to sense the increase in slip angle and interpret it as a warning of imminent loss of adhesion. This is desirable, because it allows the driver a chance to correct excessive cornering speed and keep within the limits of control. A slip angle versus cornering force curve with a sharp peak means sudden breakaway into a skid, with very little warning to the driver. *Unfortunately the tires with the highest limit of adhesion—racing tires—usually give the least warning.*

Luckily for the high-performance driver, a tire operating at a large slip angle has a considerable amount of resistance to forward motion. The larger the slip angle the greater the resistance. Thus if the driver does not apply power in a hard turn, the car will rapidly slow down, much like hitting the brakes.

Using large slip angles to slow the car down is a technique used in racing known as *scrubbing off speed.* This allows the driver to reduce speed while cornering without actually using the brakes. He throws the car very hard into the turn, generating large slip angles. As this slows the car he can tighten the radius of the turn or otherwise use the reduced speed to his advantage. Occasionally a driver of a car without brakes is able to make a reasonable showing in a race by scrubbing off speed entering the turns.

Because of relatively large slip angles many racing cars appear to be sliding sideways through the turns. This attitude in a turn is often called a *drift,* and is looked on by many as the ultimate cornering technique. A car nearing the limit of adhesion will naturally assume this drift if the rear tires operate at large slip angles. However, many racing drivers try to limit slip angles to smaller than maximum values so as not to scrub off speed in fast turns. A smoother driving technique may pay off with higher exit speeds from the turns and faster lap times. Maximum slip angles are used only in extreme conditions or when the driver is deliberately scrubbing off speed.

Now we come to another area where

Rear tires of this Formula Ford have exceeded their limit of adhesion, and they are sliding. You can tell this by the black marks being laid down. The front tires haven't broken loose, and if the driver is lucky and skilled he may have been able to gain control before spinning out. This is not the fastest way around a corner, not to mention it's dangerous.

Two Formula Fords negotiating the same corner as the one shown above. The cars are not exceeding their adhesion limit, and the drivers have reasonably good control. The leading car is in the classic four-wheel drift. All four tires are at an angle to the direction the car is traveling.

an engineering term doesn't serve us well in understanding practical matters. Some books speak of friction between the tire and the road and use the engineering idea of *coefficient of friction.* This coefficient is a number which indicates the difficulty in sliding one surface against another.

Coefficient of friction is the side force required to slide an object, divided by the weight of that object. If a 60-pound box requires a side force of 60 pounds to keep it sliding, the coefficient of friction is one.

Stated another way, the coefficient of friction is the fraction of the weight of the object which is required to slide it. If lubricating the surface reduced the coefficient to 0.5 in the example above, you could slide the box with a side force of only 30 pounds.

This is grip. The photo was snapped the instant the throttle was opened. Note the parachute line still hanging limp. These "wrinkle-wall" drag tires are run at extremely low pressure so the contact patch will be long as well as wide. The wrinkles in the sidewall are due to driving torque being transmitted from the wheel-rim through the tire to the track. (Tom Monroe photo.)

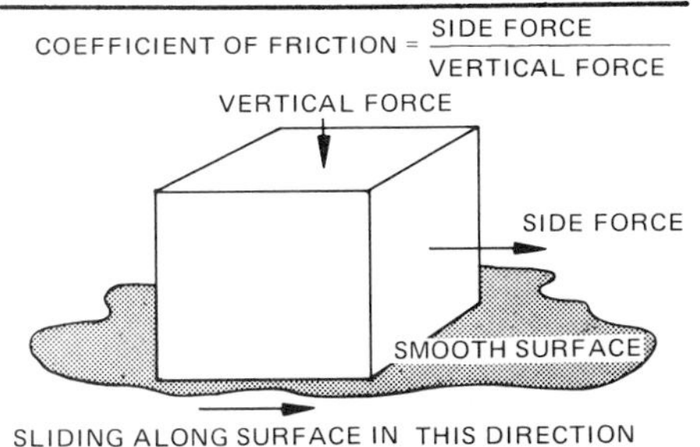

For any object sliding on a surface the coefficient of friction is the force required to slide it sideways divided by the vertical force pressing the object against the surface. This relationship applies to a tire but we call it grip.

Friction is far from simple and it must be subdivided into several categories. The initial force required to move the box first is higher than that required to keep it sliding after you get it into motion. Those are two different kinds of friction: *static and sliding.* Another kind is called rolling friction. When a tire is in a corner, it is doing some sliding and some rolling.

You can make a good case that a tire has a coefficient of friction with a particular surface because there is both side force and downforce due to the weight on the tire, and they have some ratio.

It is the *ratio* of side force to downforce that's important. In the same turn at the same speed, a heavier car will have more centrifugal force tending to make it leave the course. It will also have more downforce on the tires because it is heavier, and it will have more friction because friction is some percent of the downforce.

I prefer to avoid the question of what kind of friction the tire has and what is its theoretical limit and pin the whole responsibility on tire *grip.* Grip is what causes cornering force. It is equal to the side force divided by the vertical force on the tire, just as it would be if we called it *coefficient of friction.*

Grip is also responsible for traction while accelerating and braking, figured in exactly the same way except the directions of the forces are different. A tire with high grip will corner fast, accelerate rapidly without spinning the tire, and brake quickly.

When comparing one car against another or one tire against another we talk about grip at the limit of adhesion. If the driver tries to use more grip than the maximum possible, the car will go beyond the limit and the driver will lose control.

For ordinary street tires, grip is about 0.6 under average conditions on dry pavement. This means that the tire will grip the pavement with a maximum of 600 pounds of force for every 1000 pounds of vertical force on the tire. Racing tires have much higher grips, up to about 1.2 for road racing tires. This type of tire will grip the pavement with a sideways force of 1200 pounds at the limit of adhesion for each 1000 pounds of vertical force on the tire. A car with aerodynamic forces pushing down on it may seem to have even higher grip but in reality the extra cornering force is caused by the added vertical force.

It says in some science books that coefficient of friction can never be larger than 1.0. This "fact" led people to predict that dragsters could never beat a 200 MPH speed or 9 seconds in the quarter mile. Drag racers proved otherwise, as they broke these "impossible" limits and speeds went well above 200 MPH. They did it with *grip.* The drawing shows interlocking of the tire tread with the road surface. The softer the rubber compound, the more the interlocking and the more grip is increased. A rubber tire on pavement interlocks with small holes and bumps in the pavement. Grip between tire and road can be increased by increasing the total area of rubber in contact with the road because there are more interlocking points. Thus the wide and low pressure tire came into use. In general, the larger the contact patch of a tire, the greater will be its grip at the limit of adhesion.

Under a sliding condition, the interlocked rubber is actually ripping off the tire, and this is what leaves skid marks on the road.

The limit of adhesion is not just the maximum cornering force a tire can develop. It is the maximum total traction force in all directions. If you are cornering the car at a speed below the limit of adhesion, there will be some traction remaining for acceleration or for braking. Therefore you can apply the brakes in a corner while driving at a moderate speed. But if the car is cornering at the absolute limit, the slightest touch of brakes will bring instant disaster.

Likewise, if a car is accelerating at the limit of adhesion, say in a drag race, there is very little cornering force available from the rear tires. If the car starts to go sideways, the only way the driver can correct is to back off the throttle.

Both front and rear tires develop a slip angle in a turn. The path followed by the car is determined by the steering of the front tires and the slip angles of both front and rear tires.

If something changes any one of those three, the path followed by the car will change.

Suppose a car is running smoothly around a curve, in a stable condition with some slip angle at front and rear. If the driver increases power, a driving tire has to develop more forward thrust against the pavement, which causes an increase in slip angle of the rear driving tires. This enables the driver to steer with the throttle in a hard corner. By applying more throttle on a rear-drive car the rear slip angle increases and the car turns tighter into the turn. Less throttle does the reverse. This is a way to maintain the delicate balance between cornering on the limit and out of control.

Applying the brakes in a hard corner has a similar effect. The added thrust from the brakes increases the distortion and slip angle of the tires. This can affect both front and rear tires. Braking usually affects the rear tires more than the front, and the result can be a spin-out. It's best to keep your foot off the brake in a hard corner unless you have completely mastered the art of driving at the limit of adhesion. Braking in a corner is an advanced racing technique.

When a tire is cornering there is a sideways force on it. The addition of power adds another force in line with the tire as shown in Figure 2. The total traction force against the pavement is a combination of these two forces. Because they act in different directions, 90 degrees apart, they cannot be combined simply by adding their values together.

Notice that the total traction force in Figure 2 is less than if you added the values of cornering force and acceleration force directly. The value of the total traction force is larger than either the acceleration or the cornering force. This shows in vector form why the tire slip angle increases when acceleration or braking force is added to cornering force.

Imagine what happens to a tire that is cornering at the limit of adhesion. The sideways cornering force is equal to the maximum traction force that the tire can develop in any direction. Thus the addition of *any* acceleration or braking force, no matter how small, will demand a total traction force beyond that possible. The tire will go into a skid.

If you are cornering at the limit of adhesion, any application of accelerator or brake will cause loss of control. In such a case the driver can only steer, keep a constant throttle, and wait for the car to scrub off speed enough to gain traction. Even suddenly backing off the throttle can send the tire into a skid.

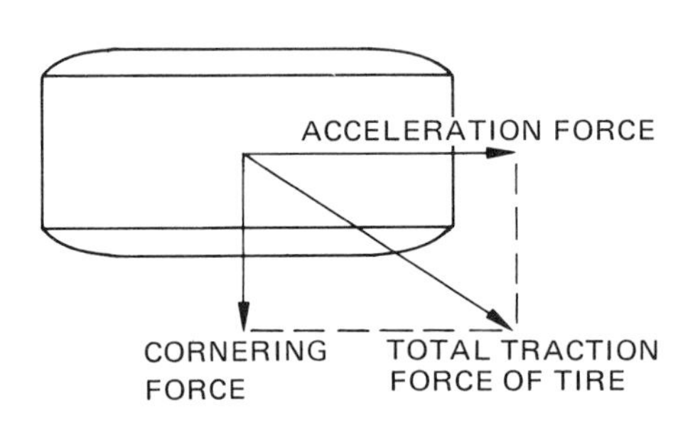

Figure 2/The tire has a maximum amount of traction force available, which can act in any direction. It can be entirely cornering force, entirely acceleration or braking force, or it can be a combination. Two forces acting together on the tire are shown in the drawing. To increase cornering force, acceleration force has to decrease when the tire is at the limit of adhesion. Remember, the total traction force is the limit of adhesion of the tire.

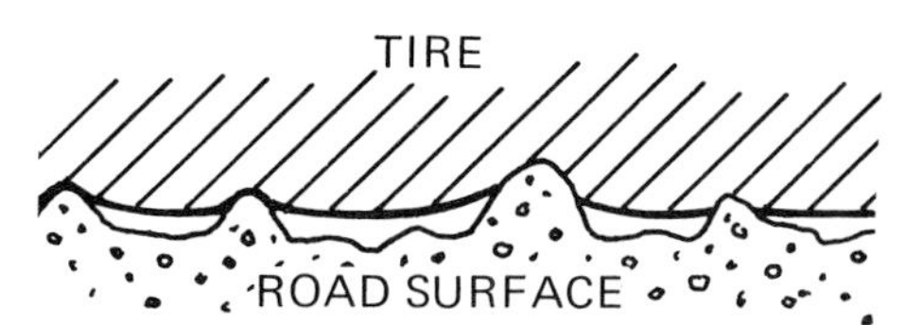

This drawing depicts the interlocking of the tire tread with the road surface.

WHAT'S VECTOR ADDITION?

Combining forces which act in different directions is called *vector addition.* It can be done graphically as illustrated in Figure 2. Each force is represented by an arrow, called a *vector,* because it shows both size and direction of the force. Size or amount of force is indicated by the length of the arrow.

In Figure 2, a cornering force and an acceleration force are drawn 90-degrees apart as they are in real life.

To find the net effect, or *resultant force* due to the action of these two forces at right angles, complete the rectangle as shown by dotted lines on the figure. Then draw the diagonal. The diagonal shows the net force and the direction of the total traction force acting on the tire.

THE DYNAMICS OF CORNERING

A car driving at a constant speed through a constant-radius corner is acted on by a constant force acting away from the center of the turn. This is *centrifugal force,* and it varies with speed and the radius of the turn. Centrifugal force causes the car to lean, slides you across the seat in a turn, and makes the car skid if the limit of adhesion is exceeded.

A law of nature says that anything moving in a straight line will continue in a straight line unless acted on by an external force. With a car the external force is the cornering force of the tires. Centrifugal force represents the *inertia* of the car, or its resistance to traveling in a curved path. If the cornering force is suddenly reduced to zero, say by driving onto a patch of glare ice, the car merely resumes its natural path, a straight line.

The amount of centrifugal force on the car is easily calculated. You need to know the weight of the car, the cornering speed, and the radius of the corner. The formula is:

$$\text{Centrifugal force} = \frac{\text{Car weight} \times (\text{Cornering speed})^2}{14.97 \times \text{Corner radius}}$$

In this formula the car weight is expressed in pounds, cornering speed in miles per hour, and corner radius in feet. The centrifugal force comes out in pounds. This equals total cornering force developed by all four tires, if the car remains in the curved path.

Lateral Acceleration of the car is found by dividing the centrifugal force by the car weight.

$$\text{Lateral acceleration} = \frac{\text{Centrifugal force}}{\text{Car weight}}$$

Lateral acceleration can also be determined without getting weights involved:

$$\text{Lateral acceleration in feet/second}^2 = 39.5\frac{R}{t^2}$$

where **R** is corner radius in feet and **t** is seconds. In terms of g's

$$\text{Lateral acceleration in g's} = \frac{1.226R}{t^2}$$

Lateral acceleration is expressed in g's. If the car has zero aerodynamic lift or downforce, the lateral acceleration equals the average grip of the car's tires. You sometimes hear how many **g's** a car will corner at, and this is a useful relationship for comparing one car against another. The car with the highest *cornering power* is the car which has the highest lateral acceleration in a corner. The actual testing is usually done on a *skid pad,* a flat piece of pavement with a circle painted on it to provide a corner of a known radius. The car is driven around the skid pad as fast as possible without leaving the circle, the cornering speed is measured, and the grip or lateral acceleration is calculated. I will discuss using a skid pad in Chapter 3.

The centrifugal force is distributed all over the car. Each part of the car has a small centrifugal force on it in proportion to its weight. If all these small forces are added up they equal the centrifugal force on the entire car. It can be thought of as a single force acting at the *center of gravity* or *CG* of the car.

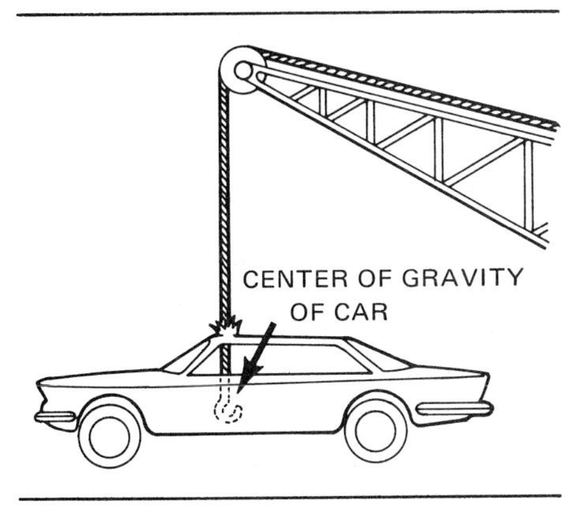

If there were some way to support the car at the center of gravity it would hang balanced. If the car were rotated into another position it would still be balanced. The CG is the *only* support point where it would be balanced no matter to what position the car is rotated.

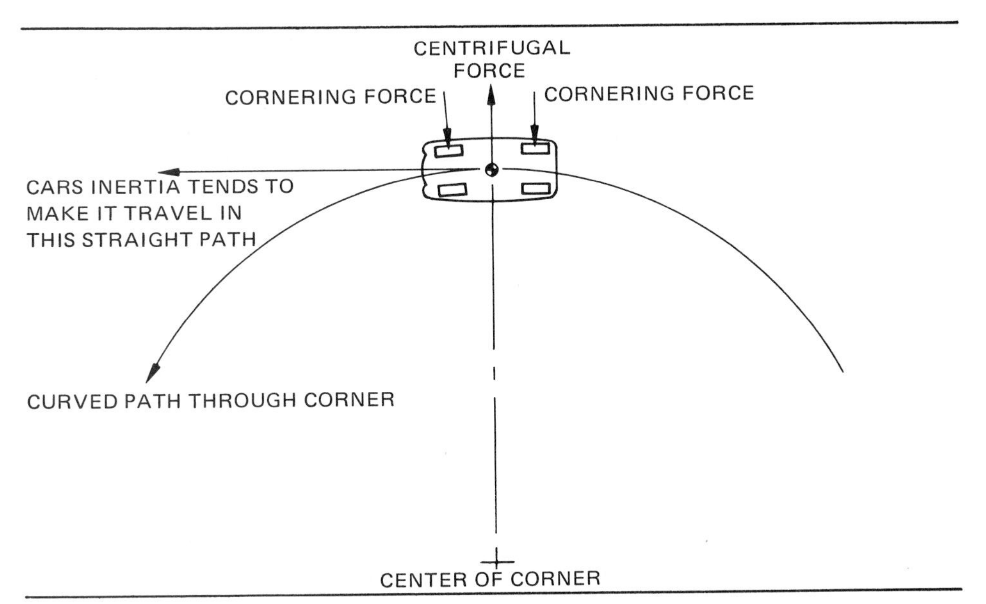

Cornering forces on the tires keep the car in a curved path. Centrifugal force acts away from the center of the corner, and is equal to the cornering forces as long as the car remains on the curved path. If the cornering forces suddenly went to zero, the centrifugal force would disappear and the car would travel in a straight-line as shown.

The CG is the balance point for the entire car. If you could support the car at a single point, the CG would be the only place you could attach the support without the car rotating. The car could be turned on its side, or at any angle and it would still hang balanced if supported at the CG. The location of the CG is always above the road and between the tires. Its exact position can be measured or calculated, and methods are shown later.

The car's *weight distribution* is determined by the location of the CG. Weight distribution can be fore-and-aft or it can be lateral, meaning side to side. Weight distribution is given by a pair of numbers such as 40/60, 50/50, 60/40, etc. For fore-and-aft weight distribution, the numbers are the percentage of weight on the front and rear tires respectively. If a car has 40/60 fore-and-aft weight distribution, 40% of the weight is on the front tires and 60% of the weight is on the rear tires.

Lateral weight distribution can be expressed the same way, but you must also state which side of the car is heavy and which side is light. If a car has 50/50 lateral weight distribution, then the CG is on the centerline. On this car the left tires carry equal weight to the right tires.

Every car has a static weight distribution, whether at rest or traveling in a straight line at a constant speed. This is changed laterally by centrifugal force when the car is in a turn, and it is changed fore and aft when the car is accelerating or braking.

Centrifugal force puts more weight on the outside wheels in a corner. Acceleration puts more weight on the rear wheels and less on the front wheels while the car is accelerating. Braking or deceleration does the reverse—the front tires become more heavily loaded and the rear tires less.

Each of these conditions causes different handling because of different weight distribution among the four tires. This is discussed in more detail later.

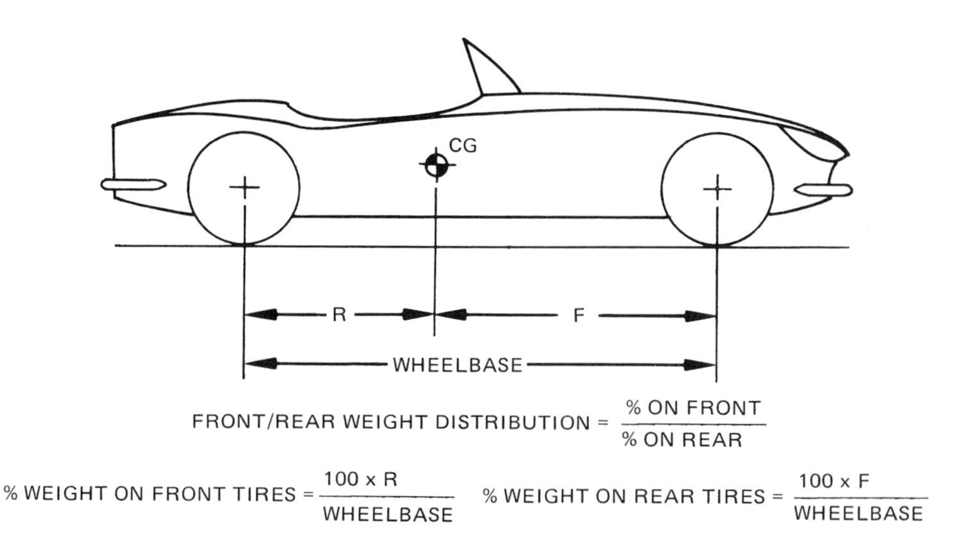

If you know the location of the CG you can compute the front/rear weight distribution. Notice that the percentage of weight at one end is larger if the CG is closer to that end.

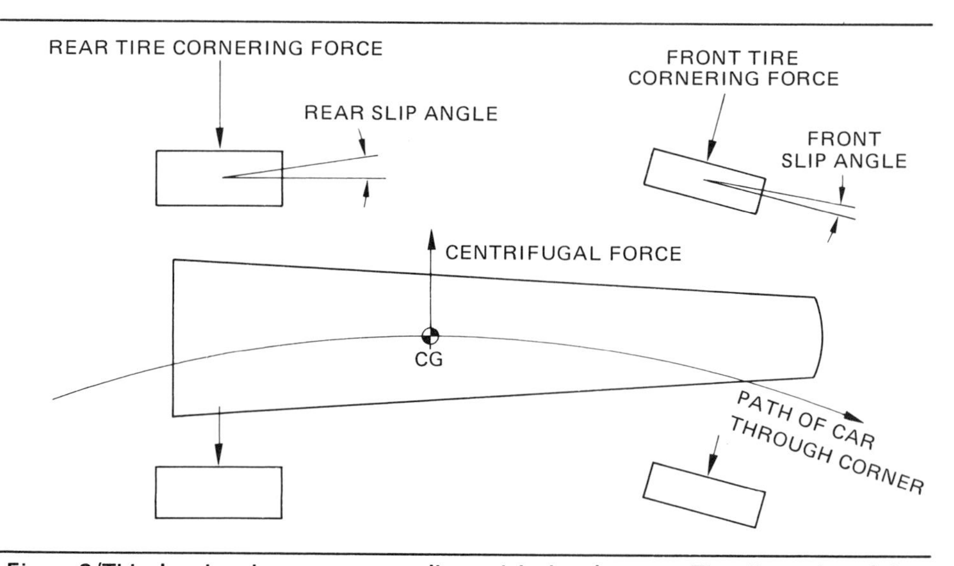

Figure 3/This drawing shows a car rounding a right hand corner. The slip angles of the rear tires are greater than the front in this example—an oversteering condition. We are most concerned about the two outside tires, because they do most of the work in a corner.

UNDERSTEER, OVERSTEER, NEUTRAL STEER

Centrifugal force tries to push the car away from the corner, and it is resisted by the cornering forces of the tires. These two forces act in opposing directions and are equal if the car remains in a constant-radius corner at a steady speed. This is *steady-state cornering.* A diagram of the forces acting on the car during steady-state cornering is in Figure 3.

The tires have slip angles due to the cornering forces acting on them. Because the cornering forces at front and rear of the car are not necessarily equal, the slip angles can be different at front and rear. In this discussion we will only be concerned about the two outside tires—they do most of the work in a corner.

Let's assume the Figure 3 car is tail-heavy. That is, the CG closer to the rear tires, and thus the two rear tires carry more of the car's weight than the front tires. This is typical of a rear engine car. If this car uses the same size tires all around, and all other conditions are equal front and rear, then the rear tires *must* operate at a higher slip angle than the front tires. This is due to the larger side forces acting on them, which in turn results in greater tire distortion. The condition of higher slip angles in the rear of the car causes *oversteer.*

This car is using power coming out of a turn and is in an extreme oversteering attitude. Notice the front wheels are turned to the right even though this is a left hand corner. If the driver backed off on the power the car would straighten out, but he would have to do it very gently to avoid losing control.

The entertainment offered the driver in an oversteering car is something like this: As you approach a corner and turn the steering wheel an amount you think will be just right, the car is very docile about it and goes into the turn. As soon as centrifugal force begins to affect the car, the tires develop slip angles. But the rear tires develop larger slip angles than the front tires, so the car starts turning *into* the turn more than you had planned.

If you don't correct for it, the car begins making a tighter turn and that further increases the centrifugal force. The tires respond again by increasing their slip angles—again more at the rear than the front.

The end result, if not corrected, is the car makes a turn of ever-decreasing radius even though you haven't moved the steering wheel at all since entering the curve. Finally the cornering force of the tires is exceeded by the centrifugal force on the car and you spin out. Guess which tires break loose first.

When the car begins to oversteer, the driver must correct for it by turning the steering wheel away from the direction of the turn. This correction of the steering if done properly will keep the car on its intended path on the road. In a car which has a strong oversteering tendency, the driver may back off on the steering so far that he actually ends up turning the wheels in a direction opposite to the turn. This is *opposite lock cornering,* and is widely used when racing on a dirt track.

Now let's reverse the situation and assume that the car in our example is nose-heavy. For similar reasons, the front tires will end up cornering with larger slip angles than the rear tires, and the car is then said to *understeer.*

Understeer is a common handling characteristic of a stock sedan. It is considered safe by passenger car designers, and so understeer is designed into most road cars.

Stock VW Rabbit comes with severe understeer in tight turns as do most stock sedans, particularly those with front-wheel drive. This is because engine driving and braking power is applied at the heavy front end of the car. Understeer is safe for the general public, but it isn't the quickest way around a race course.

Two racing cars demonstrating different steer characteristics. The car at the top is understeering, as you can see by the larger slip angle of the front tires. The lower car is oversteering, and the driver has to turn the steering wheel away from the corner to maintain control. Neutral steer is generally favorable, depending on the specific track. Oversteer on an Indianapolis car would be suicide, however, it is preferable on a dirt-track car. (Courtesy of Road and Track magazine)

The driver notices understeer by the fact that the car isn't turning as sharply as he wants it to, so he cranks in more steering to compensate. This increases the slip angle on the front tires, and the car stays on the proper path through the corner. The driver always reacts instinctively to an understeering car, and this is why it is considered safe. Also if the driver does not correct soon enough, the car increases its radius of turning, which in turn reduces the centrifugal force on the car. This is a safe and stable condition—if the road is wide enough.

At the limit of adhesion, an understeering car will leave the road nose first on the outside of the corner with the driver vainly turning the wheels as much into the turn as he can. A car out of control in an understeering condition will surely crash if there is a solid object on the outside of the corner, and there is little the driver can do to prevent it.

An *oversteering* car upon exceeding the limit of adhesion will spin out with the driver perhaps turning the steering wheel as far into the skid—away from the turn—as possible in an attempt to save it. The car may swap ends and leave the road on the outside, it may spin off the road to the inside, or if the driver is lucky, it may spin down the center of the road and slide to a safe stop. Thus an oversteering car is less predictable as to where it will end up in an out-of-control situation.

There isn't much difference in the grip of an understeering versus an oversteering car. However, there is quite a large difference in the way the car feels to the driver. It is sometimes said that the condition of understeer is where the driver is scared, and the condition of oversteer is where the passenger is scared. There is some truth to this statement.

For general street driving, understeer is considered safest, because the correction is a natural and instinctive one—turning the steering wheel more into the turn.

Oversteer is usually preferred for racing, because a skilled driver can control the car more easily and come out of the corners faster.

For good handling, the car must not exhibit a *strong* tendency to either understeer or oversteer. The point in between is called *neutral steer,* and you should try to adjust the chassis to get close to this condition. A neutral steering car can be made to either understeer or oversteer by the driver's use of the throttle or by small adjustments to the chassis. Thus the car can be made to handle exactly the way the driver likes it. A strongly understeering or oversteering car is no fun to drive and cannot be finely tuned.

It's worth spending a minute here to review the relationship between grip and steering characteristics.

If a car exceeds the limit of adhesion in a corner, one end or the other will break away first, unless the steer characteristics are exactly neutral. If the rear end slides out first, it had the lowest grip and the car leaves the road in an oversteering condition.

Cars that oversteer when skidding will normally oversteer a smaller amount while still under control. Therefore less grip at the rear means oversteer and less grip at the front means understeer, in normal driving conditions.

We have been considering oversteer and understeer in terms of slip angles, so there must be some relationship between slip angle and grip. At a certain cornering speed a tire has a side force and a vertical force acting on it. Tire distortion under these forces results in a certain slip angle. At this cornering speed perhaps there is a way to reduce the tire distortion, say by using a wider wheel rim to support the tire better. Then the reduced tire distortion results in a lower slip angle at that cornering speed. If you compare the grip of these two arrangements at the limit of adhesion, you will find the tire with the higher slip angle has a lower grip. Thus if you can reduce the slip angle of a tire at a given cornering speed it will *usually* have a higher grip at the limit of adhesion.

This is why a car usually has the same steer characteristics at the limit of adhesion at lower cornering speeds. If a car oversteers due to excess weight on the rear-tires, the car will reach the rear-tire adhesion limit first. Higher slip angles at the rear mean less rear grip. If the car owner can reduce the slip angles on the rear tires, he can expect higher cornering power along with less oversteer.

FACTORS AFFECTING TIRE GRIP

Because of a number of factors which affect the tires, a nose-heavy car is not always doomed to understeer, and a tail-heavy car does not always oversteer. Among the factors that determine grip is tire pressure. The relationship between grip and tire pressure is shown in Figure 4. Notice that there is an optimum pressure for each tire, giving the highest possible grip. Any pressure over this will bulge the tread in the center causing a loss of traction, and any pressure under this will cause excess distortion of the tire and a loss of traction. I will show you how to find these optimum tire pressures in Chapter 3. Obviously tire pressure can be used to fine-tune the chassis when nothing else is possible, but in doing this you are *reducing* grip to get steer characteristics you want.

As an example, let us say you drove your stock Datsun to a slalom, took off the hubcaps, and are running in a stock class. With street suspension and tires, the car understeers so much in tight turns you can't get around them. So you let enough air out of the rear tires to reduce grip at the rear and increase slip angle at the rear. This promotes oversteer, or reduces understeer depending on your point of view. Either way, with less understeer, the car helps you more by wanting to turn tight corners and you win the slalom.

This works unless some crafty competitor has adjusted his suspension to give the *right* amount of oversteer and *still* allow use of proper tire pressures for maximum grip. Then he wins, and you should find the experience instructive.

Changing tire pressures from the values that give best grip is not a normal part of chassis tuning any more and should be done only as a last resort.

If you must do it, here are some rules. Below the pressure for best grip, reducing tire pressure makes the tire more floppy. It will distort more and have a higher slip angle.

Increasing tire pressure makes it less floppy. It resists distortion more and will have smaller slip angles. However, if you increase pressure above the value that gives best grip, the tread bulges out and there is less rubber in contact with the road.

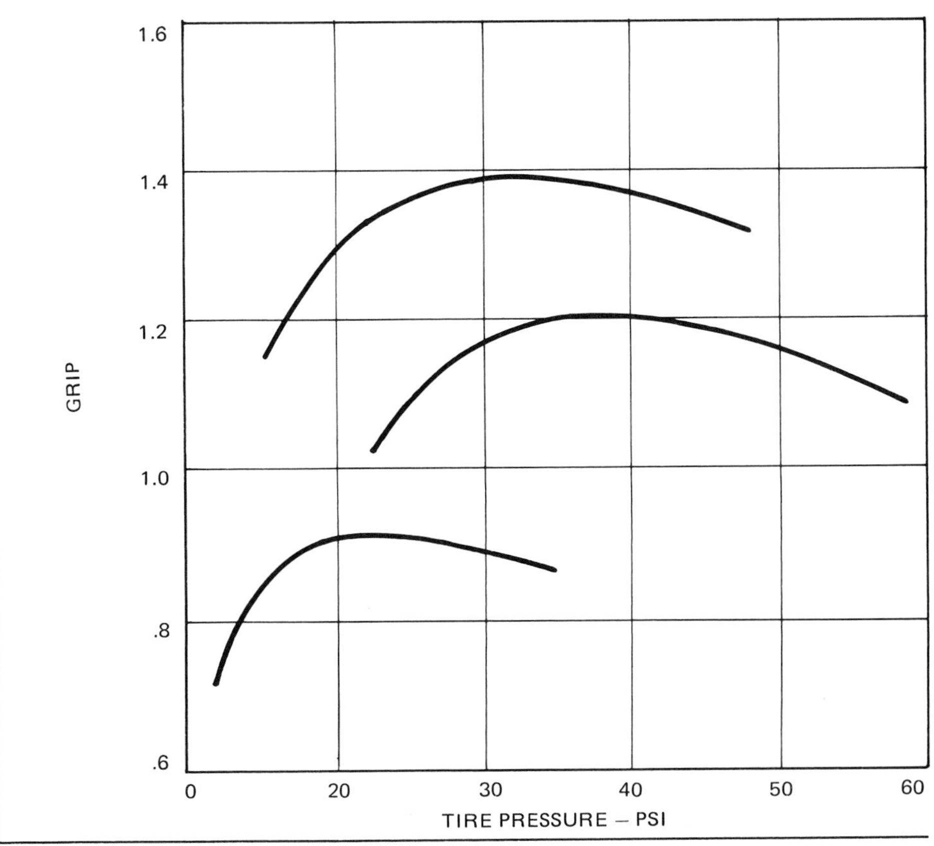

Figure 4/The relationship between tire pressure and grip is shown for three different tires. The optimum pressure will vary with the size of the tire, the wheel rim width, and the sidewall construction. Notice that you lose more by a small amount of underinflation than a small amount of overinflation. When in doubt, increase the pressure before you try decreasing it. You can find the best tire pressures for your car by testing as described in chapter 3.

If you are driving below the limit of adhesion so you don't need all available traction, increasing tire pressure decreases slip angle. If you are driving at the limit with tire pressure higher than optimum, the limit of adhesion is lower and breakaway occurs at a smaller slip angle, so you don't benefit from the higher pressure.

In this case reducing the slip angles at low cornering speed does not necessarily increase the grip at the limit of adhesion. This is an exception to the relationship between grip and slip angle discussed earlier. If you tried to reduce the understeer of your Datsun by putting 50 pounds of air in the front tires the car would oversteer at very low cornering speeds, become neutral at higher speeds, and have understeer at the limit of adhesion. This sort of handling may be fun, but you don't win races with it.

Another factor that affects grip of the tires is *camber.* Camber is the angle the tire makes with the vertical as shown in Figure 5. Note that positive camber tilts the top of the wheel outward; negative camber tilts the top of the tire inward.

Camber is both simple and complicated. From the simple point of view, if you take a bicycle wheel and lean it to the left, it will roll to the left. Also, if you lean the top of a wheel in either direction and put some weight on it, the contact patch between tire and surface will move in the direction the top of the tire is leaning. This distorts the elastic rubber and it will exert a force tending to restore the tread to its normal location on the tire. The resulting force is called *camber thrust.*

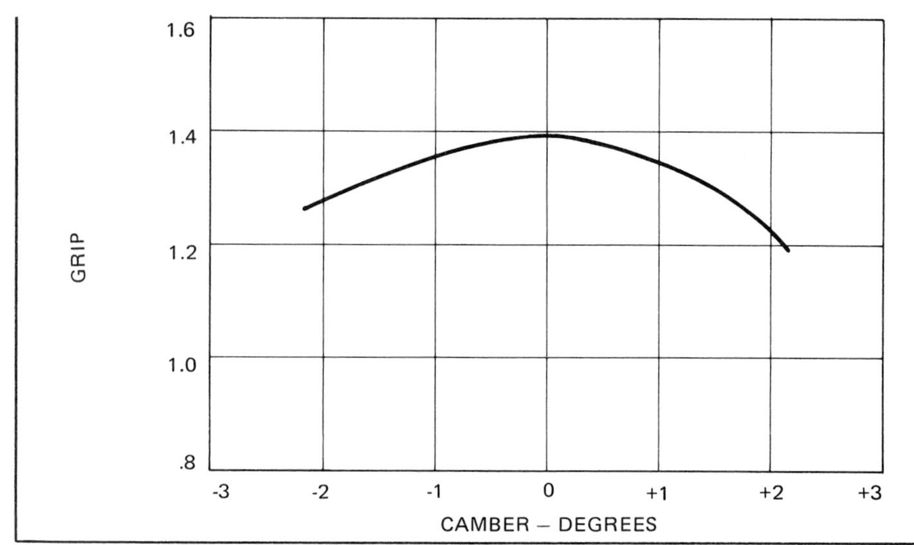

Tires have a relationship between camber and grip as shown in this graph. Note that the best grip occurs with the tire at almost zero camber. This is more critical for wide-tread tires than for narrow ones. Also notice that grip drops more for positive camber than it does for an equal amount of negative camber.

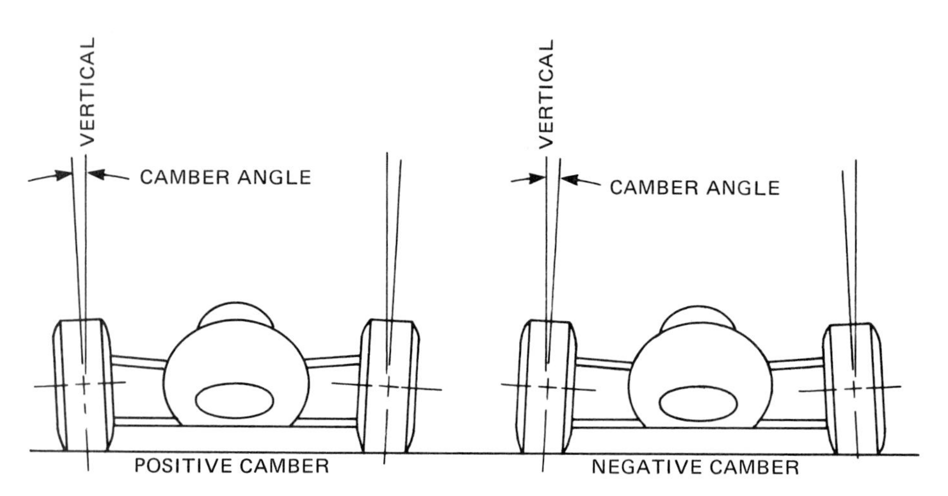

Figure 5/This racing car is shown with the tires set at positive and negative camber. Zero camber is where the tires are exactly vertical. The camber is always measured from vertical. When the car leans in a turn it is important that the tires not lean with the chassis or else camber will change. Keeping camber near zero under all conditions in the primary goal of modern suspension systems.

Camber thrust tends to pull the bottom of the tire in the same direction the top is leaned.

In the early days of chassis design, cars ran with small tires which were round in section—something like a bicycle tire or motorcycle tire. Cambering a wheel with a small tire was interrelated with other factors affecting *steering* with the goal of a balanced design of the front end.

Modern tires, particularly those used on high-performance cars, are wide and flat on the bottom. When inclined or leaned over, such tires do not continue to make a flat contact patch on the surface. They lift up one side of the tread! This has a drastic effect on traction.

The emphasis today is not the effect of camber on steering but its effect on traction. Suspensions are designed and adjusted mainly to keep those wide tires flat on the surface even during suspension movements. This naturally conflicts with other aspects of suspension design but very often "keeping rubber on the road" dominates. Design of the suspension linkage is critical here, and this complex subject is discussed later. Suffice it to say that camber angle is very critical in determining the handling characteristics of the car.

A tire has an optimum camber angle for maximum grip just as it has an optimum tire pressure. Changing camber from the optimum value reduces the maximum grip of the tire, so use this method of adjustment only as a last resort and only in small increments. A method for finding the optimum camber angle is discussed in Chapter 3. In general, a car should have nearly zero camber on the outside tires in cornering, and this can sometimes be checked just by observing the car.

Another factor that affects the grip is the size of the tire. A wide tire will have more grip and a smaller slip angle than a narrow tire. You can adjust steer characteristics of the car by tire width. Racing cars use wider tires at the rear, because the typical racing car is tail-heavy to get maximum straight-line traction. Thus the steer characteristics can be near neutral on a rear-engine race car with two-thirds of the static weight on the rear wheels. Sounds hard to believe, but it's true.

The wheel-rim width also affects grip and slip angle in a way similar to the effect of tire size. A wider rim offers less sideways flexibility and thus reduces tire distortion. Wide rims are used on cars where cornering power is important. The rim can be too wide, however, causing curvature of the tread surface and a subsequent loss of grip. Also wide rims may be limited by racing rules or by the width of the bodywork. As with anything else, there is an optimum condition for the particular use of the car.

Another important characteristic of tires is the effect of temperature on grip. In modern racing tires the rubber is compounded to operate best at a certain temperature, and anything other than this optimum temperature will result in a loss of grip. Tire temperature is determined by both the temperature of the day and also the heating effect of cornering and traction forces applied to the tire. This complex relationship will vary with the race course and the weather, so best performance will vary from one day to the next. A well-equipped racing team will have various tire compounds available and pay attention to tire temperatures. For street tires temperatures are not nearly as important as with racing tires, so the average motorist need not be concerned with this problem.

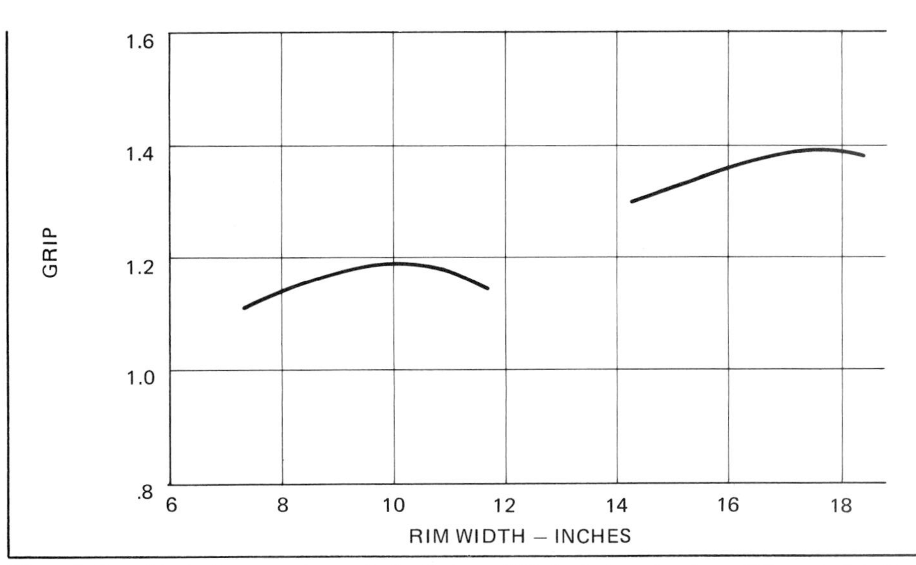

The relationship between grip and wheel rim width is shown for two different size tires. These are wide racing tires as might be used on a Can-Am car. Notice you lose less with too wide a rim than with too narrow a rim. Also note that wider tires have more grip which is the direct result of more rubber on the road.

THE RELATIONSHIP OF GRIP TO WEIGHT

This will be mentioned more than once and used in following discussions because it's important. Grip *is not* a constant percentage of the downforce on the wheel.

Suppose there is 800 pounds on one wheel and the grip is 0.8 or 80 percent. There will be 640 pounds of side force or traction force available at the contact patch of the tire.

Now double the weight on the wheel. Measurements of traction and cornering force show that these forces *do not double.* With a load of 1,600 pounds on the wheel, the side force may be only 1,120 pounds instead of 1,280 as you might expect, based on a grip of 0.8. The grip at a tire loading of 1,600 pounds *decreased* with the higher load and is now only 0.7 Slip angle *increased.*

When a car is in motion, accelerating, braking, turning, going uphill and down, the weight or downforce changes on each tire according to the operating condition of the car at each instant.

The rule is, when a tire is more heavily loaded, grip of that tire is reduced.

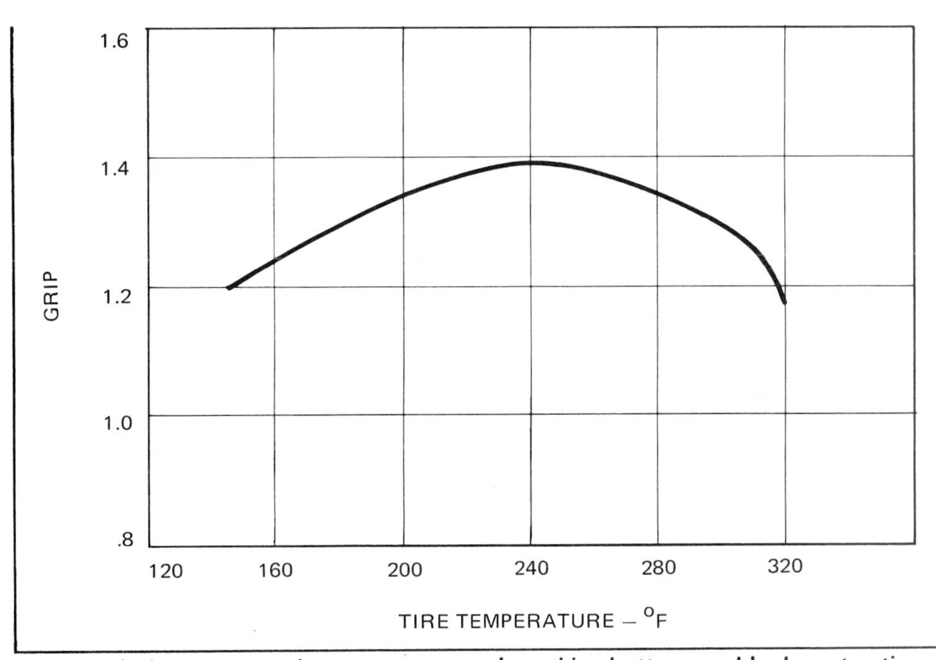

A tire sticks best at a certain temperature, and anything hotter or colder loses traction. The shape of this curve is quite different for various tires, and it is most critical for modern racing tires. Each tire has a maximum allowable temperature beyond which the tire will fail. Thrown treads on street tires are often the result of overheating caused by underinflation or sustained high-speed driving.

TRANSIENT HANDLING CHARACTERISTICS

In previous discussion we talked about steady-state cornering. This condition rarely happens, because most cornering is done under varying conditions of speed, road surface, corner radius and banking. Of particular importance is entering and leaving of a corner, where both speed and cornering radius are changing. This is called *transitional cornering.*

Most critical is entering a corner from the straight. Here the driver turns the steering wheel into the turn and the car reacts. How the car reacts to this initial turn of the wheel will determine what happens in the corner as the car approaches steady-state cornering. Some cars are sluggish and don't want to change directions quickly. Other cars are too sensitive and dart into the turn. Others exhibit too much body roll and delay assuming the steady-state cornering attitude. All these complex characteristics and many more determine transitional or *transient* handling.

One suspension setting that strongly affects transitional cornering is *toe.* Toe setting is the angle measured between the tires, looking down on them from above. This is shown in Figure 6. *Toe-in* brings the front of the tires closer than the rear, and *toe-out* is when the front of the tires are farther apart. *Zero toe* is the setting where the wheels are parallel. Toe can be adjusted at the front of all cars, and at the rear of cars with independent suspension. It is an adjustment that is easy to make and strongly alters the handling characteristics of the car.

It should be obvious that toe setting influences slip angle of the tire. If the toe changes during cornering, say when the body leans or when a tire hits a bump, then this will instantly change the slip angle of that tire. The car will change direction of travel, causing an unexpected event for the driver. Toe change is called *roll steer* if caused by body roll, and *bump steer* if caused by vertical suspension movement. Both are critical in obtaining a stable controllable car, not only in the corners but also on the straight. Adjustment of these characteristics is discussed in Chapter 3. Bump steer is undesirable in any form. Roll steer is seldom found on high-performance cars, but is nearly always designed into production street automobiles to change the steer characteristics, i.e., usually toward more understeer. To reduce oversteer caused by excess weight on the rear tires, larger rear tires can be used instead of roll steer. Roll steer is usually not easily adjustable, but sometimes you can change it with suspension modifications.

Another transient-handling characteristic of a car is high-speed *stability.* This is the term used to describe the handling of a car traveling in a straight line, such as on the freeway, at the drag strip, or on the dry lakes. If a car tends to hold a straight line without regard to wind gusts or changing road conditions it is said to be *stable.* If the car changes direction without the driver turning the wheel it is *unstable*—a scary and sometimes very dangerous condition. If you have ever driven an old VW in a cross wind you know what unstable means—and the faster you go the worse it gets. High-speed instability is a potential problem with all cars, and aerodynamic devices are often used to correct the problem.

A critical part of a turn is the instant the driver first starts to turn the steering wheel. The car's transitional handling characteristics describe what happens at this instant between going straight and turning. This is very complex and sensitive to many factors in chassis design and adjustment. The driver of this vintage Porsche is braking hard and just about to turn into the corner. (Chuck Engberg photo.)

Drag racing tires are brought up to racing temperature with a burnout before a run. Here Don Prudhomme heats the tires of his funny car. The tires are deliberately allowed to break traction by accelerating through liquid. (Photo by Tom Monroe.)

How *quickly* a car reacts to a turn of the steering wheel depends on a characteristic called *polar moment of inertia.* This is the resistance of a car to turning. It is calculated by multiplying the weight of each component of the car by the square of the distance of that component to the CG of the car. Thus the polar moment of inertia not only depends on the car's weight, but also on the distribution of the weight. The farther the weight is from the CG, the greater its polar moment of inertia.

A car with a high polar moment has high resistance to changing direction. Thus it reacts slower to the steering than a car with a low polar moment of inertia. At high speeds a car with a high polar moment tends to be more stable. Forces that tend to turn the car have a slower-acting effect, and the driver has an easier time keeping the car in a straight line.

In a corner, a car with a high polar moment of inertia starts a spin slowly, giving the driver more time to correct. However, once the spin is started it takes longer for a steering correction to stop it. A high polar moment of inertia tends to understeer at the initial turn of the steering wheel.

A large heavy car such as a production sedan has a very high polar moment of inertia. Both size and weight contribute. On the other end of the scale, a go-kart has the lowest polar moment of inertia. Again the tiny dimensions and low weight combine for the low figure. In between these extremes are all the various smaller road cars and race cars.

Another interesting contrast is between old-style racers with a heavy engine up front and the driver and relatively heavy rear axle located at the rear. These cars had high polar moments of inertia and would feel very sluggish to the driver of a modern racer. Current mid-engine cars put most of the engine weight and the driver's weight near the center of the wheelbase, resulting in a low polar moment of inertia and a car that maneuvers very quickly.

The basic size and weight of a car are difficult items to change, so most people make no attempt to change the polar moment of inertia. However, small changes can be made on some cars by relocating various components such as a battery.

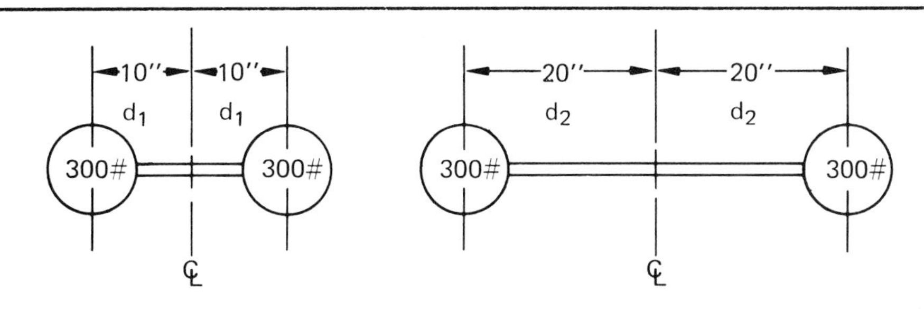

1. LOW POLAR MOMENT
$I_o = 2\ (300)\ (10)^2 = 60{,}000$ lb. in.^{2}

2. HIGH POLAR MOMENT
$I_o = 2\ (300)\ (20)^2 = 240{,}000$ lb. in.^{2}

$I_o = Wd^2$

W = weight of component in lbs.

d = distance of component from center of rotation in inches

I_o = Polar moment of inertia in lb.-inches^2

Polar moment can be described as the "dumb-bell effect." Identical weights can produce different polar moments, depending on the location from the center of rotation.

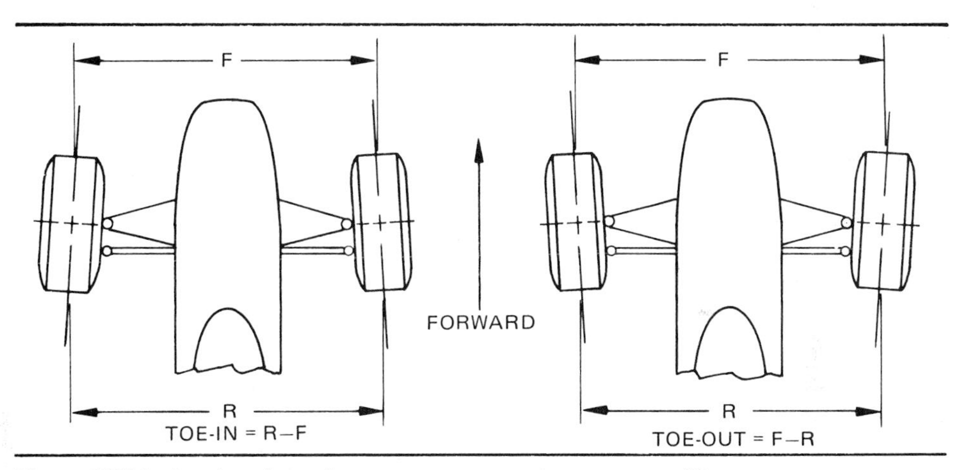

Figure 6/This drawing shows how to measure toe-in or toe-out. The measurements are taken at the tire tread and are given in inches. A small amount of toe-in is the setting for most cars for reasons explained in chapter 3. Toe-out is sometimes used, particularly on front wheel drive cars. The rear suspension can also have a toe setting on independent suspension systems.

TORQUE STEER

This is a transitional instability caused by accelerating or decelerating. A car with torque steer steers in one direction or the other when the throttle opening is changed. This requires the driver to make a correction at the steering wheel. Torque steer is designed-in in some front-wheel cars. For instance, front-wheel-drive cars using transverse-mounted engines and different-length drive shafts usually have torque steer. This is caused by the different windup of different length shafts which causes one wheel to grip the road more than the other. Different axle-shaft-joint angles also affect this "torque distribution."

Thrust on the front tires of a front-wheel-drive car tends to turn the steering in opposite directions, hopefully balancing out the steering effect. But because this is usually not equal from side-to-side, the car turns right or left. My Honda Civic turns to the left under hard throttle in first gear. There is nothing I

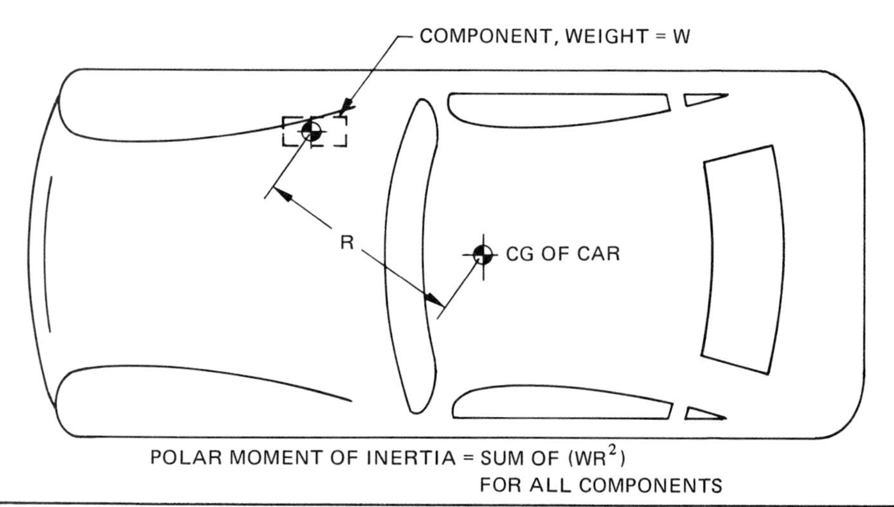

To find polar moment of inertia, add up the sum of (WR^2) for all components of the car. The drawing shows one component, the battery as an example. If its weight is 40 pounds and it is located 30 inches from the CG of the car, its contribution to the total polar moment of inertia is as follows: (WR^2) = (40) $(30)^2$ = 36,000 lb.-in.^{2}
The same calculation is performed for each component of the car, and the results are all added together for the total polar moment of inertia. A high number means a car that reacts slowly to steering. The distance R is measured looking down on the car.

can do about it except correct by steering to the right.

Torque steer usually doesn't affect rear-wheel drive cars unless there is some *unsymmetrical* feature in the suspension system. Semi-trailing-arm independent-rear suspension, as illustrated on page 38, is a notable exception. Rear-tire thrust causes the tires to toe-in during acceleration and toe-out during deceleration because of the control-arm's outer-bushing deflection, or compliance. This can be a bit scary during all-out cornering when "backing-off" the throttle. The heavily-loaded outside-rear tire toes out, steering the car into the turn, or causing oversteer—an unhealthy situation. Porsche corrected this problem with the "Weissach axle" which is used in the 928. It incorporates an additional bushing located directly behind the conventional bushing. This bushing is designed to deflect so the rear tires toe-in during deceleration so it will counteract the normal toe-out and oversteer of the semi-trailing-arm rear suspension during power-off situations.

When the driving wheels also steer, as in the case of front-wheel-drive cars, the designer has to be very careful with driveline design and front-suspension and steering geometry to minimize torque steer. To do this the steering-pivot axis is designed so it intersects the exact center of the tire contact patch, or as close to it as possible. This minimizes the tendency for either wheel to steer during acceleration or braking. However, it is very difficult to make the turning forces zero under all conditions, so some problems remain. If you change front-wheel offsets on a front-wheel-drive car, be prepared for drastic changes in torque steer. I discuss wheel offset and how to change wheel-width without affecting offset in Chapter 4.

FORE-AND-AFT WEIGHT TRANSFER

When a car accelerates, the tires move the car forward by traction force against the road. The car is subjected to an *inertia force* which acts opposite to the force accelerating the car. The inertia force is what pushes you back into the seat during acceleration. Inertia force acts on every part of the car, but it can be considered

WHAT IS A g?

If all this business about g's puzzles you, let's spend a minute getting unpuzzled.

The symbol *g* means the force due to gravity and is also used to mean the acceleration of a falling body due to gravity.

Use of the symbol or expression in race-car talk is mainly as a standard of comparison between what theoretically happens to falling bodies due to gravity and what happens to a race car when it is moving horizontally along the surface. When we say a car is accelerating at one g, we don't mean gravity is doing it. The engine is doing it, but we are proud of the fact that the car is accelerating as much as if you had driven it out of an airplane and it was headed for the finish line. The word *finish* was carefully chosen in the sentence above.

If you pick up something and hold it in your hand, the force of gravity is pulling it toward the earth. What we call *weight* is the force of gravity, g.

If you drop it, it will gain speed as it falls due to acceleration by the force of gravity—until it hits you on the foot. Because it is being accelerated by gravity, the amount of acceleration is standard and is called one g. How much acceleration is that?

If there were no air resistance, a free falling body close to earth will have the following speeds, starting at zero when you drop it.

Time	Speed
0	0
1 second	32 feet per second
2 seconds	64 feet per second
3 seconds	96 feet per second
4 seconds	128 feet per second
5 seconds	160 feet per second

We can stop there, but the speed will increase without limit if there is no air to hold it back—until it hits the ground.

Notice that during each second, the speed increases by the same amount—32 feet per second. Acceleration is the *rate of change* of speed and in this case acceleration is 32 feet per second per second. That means the speed increases by 32 feet per second in every second, while it is accelerating at one g.

We can convert the speed in feet per second shown above into miles per hour. One-hundred-sixty feet per second is about 109 miles per hour. In five seconds, that's romping along pretty good.

One g is one g whether horizontally at the race track or vertically due to falling.

The main reason g is handy is that whenever a car is doing something at one g, we know the force on the car has to be equal to the weight of the car.

There are one-g turns at a certain speed which depends on radius. If your car can make a one-g turn, the centrifugal force on the car is equal to the weight of the car. There is also a centrifugal force on you equal to your weight.

When a car accelerates at one-g, the traction force between driving tires and surface must equal the weight of the car. Same for braking. Same for you.

It happens that the coefficient of friction of tires, which we are calling grip, is the same as the maximum g's the tires can produce. If the grip of the tires is 1.2, the car can brake or turn at 1.2 g's and the forces will be 1.2 times the weight of the car.

In acceleration the traction force is 1.2 times the vertical force on the *driving* tires, so acceleration can only be 1.2 g's if 100% of the car's weight is on the driving tires. This is possible in a wheelie, or with four-wheel drive.

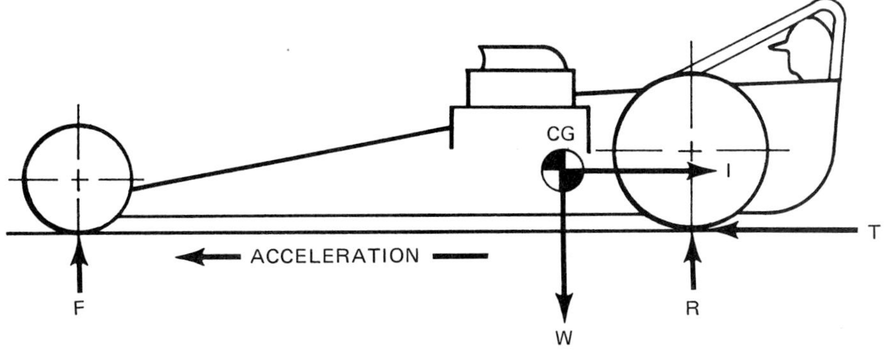

The forces acting on an accelerating car are shown. Both the weight and the inertia force can be thought of as acting at the CG, even though they are distributed on every part of the car. The forces on the tires are reactions against the pavement. Car weight is resisted by upward forces on the tires, and torque on the rear axles results in a forward thrust at the pavement, which accelerates the car. Air drag is not shown in this drawing.

as acting on the CG of the car which is some distance above the surface. The accelerating force between rear tires and the surface is of course at ground level. The effect of these two forces, one pushing forward at ground level and the other pushing backward at the higher CG, is to lift the front of the car. At the drag races if the front tires leave the ground this is called a *wheelstand.* Motorcyclists call it a *wheelie.*

This pickup truck has exceeded the limit of adhesion of the rear tires by accelerating too hard. Burning rubber is not the way to win a drag race, as maximum traction is obtained just before the point where the tires break loose. Nose-heavy weight distribution and powerful engine of this pickup make smoking the rear tires very easy.

Just as centrifugal force in a turn causes the outside tires to be more heavily loaded, during acceleration weight is reduced at the front and increased by the same amount at the rear of the car.

By increasing the weight or vertical force on the rear tires during acceleration, *weight transfer* increases the traction force which can be produced by the rear tires. Grip does not change much and should decrease slightly, but the result is more available traction force at the rear and better acceleration of a rear-drive car.

To calculate weight transfer requires knowing the inertia force, which in turn requires measuring acceleration which is difficult to do.

The formula is

$$\text{Inertia force} = \frac{\text{Car weight x acceleration}}{g}$$

where g is the acceleration due to gravity which is 32 feet per second per second. In this formula, acceleration of the car is expressed in feet per second per second.

If the acceleration of the car is one g, common among powerful cars with racing tires, then the inertia force is equal to the weight of the car.

The maximum possible acceleration of a car is at a g-rate which is the same as the grip of the tires. This is *only* possible if 100% of the car's weight is on the driving

Chuck Neal has this sand dragster set up so that there is just enough weight on the front tires to equal the weight transfer during acceleration. The front tires are off the ground, which means that 100% of the car's weight is on the driving tires. This results in the maximum possible acceleration. (Photo by Shirley DePasse, Off Road Action News.)

wheels. The car would also be on the verge of doing a wheelstand. Cars other than drag racers have some weight on the front tires during acceleration, so the maximum possible acceleration in g's never equals the grip of the rear tires.

Let's assume an acceleration of 1/2 g and show the effect of weight transfer from the front tires to the rear.

Weight transfer =

$$\frac{\text{Inertia force x CG height}}{\text{Wheelbase}}$$

Assume the following specifications for the car:

Wheelbase = 100 inches
CG height = 20 inches above the ground
Car weight = 2,000 pounds
Static weight on front tires = 1,000 pounds
Static weight on rear tires = 1,000 pounds

First calculate the weight transfer by substituting numbers in the formula above.

Weight transfer =

$$\frac{1}{2} \times \frac{2{,}000 \times 20}{100} = 200 \text{ pounds}$$

During a 1/2 g acceleration, 200 pounds would be lifted off the front tires and added to the rear-tire loading. Weight *distribution* while accelerating would be 800 pounds at the front and 1,200 pounds at the rear. With less acceleration the weight transfer is less and when the vehicle stops accelerating and runs at a constant speed there is no inertia force and no weight transfer due to acceleration.

If you make a fraction, dividing CG height by wheelbase—20/100 or 1/5 in this example—that fraction of the total car weight will be transferred from the front tires to the rear tires during a one g acceleration. If you want more weight transfer, make the ratio of CG height to wheelbase a larger number.

Drag racers often try to transfer more weight than exists on the front wheels and they obligingly lift up in the air.

Weight transfer works just the opposite way during braking. The inertia force acts in a forward direction, and the traction forces at the tires act towards the rear, slowing the car. This causes the car to have *negative acceleration*—that is, the speed is reduced. The method of calculation is the same. Weight transfer during braking reduces the vertical force on the rear tires and increases the vertical force on the front. This forward weight transfer is why the front brakes generally wear

Triumph braking hard for a corner. You can see the nosedive resulting from forward weight transfer. The front brakes take most of the heat because the front tires do most of the stopping.

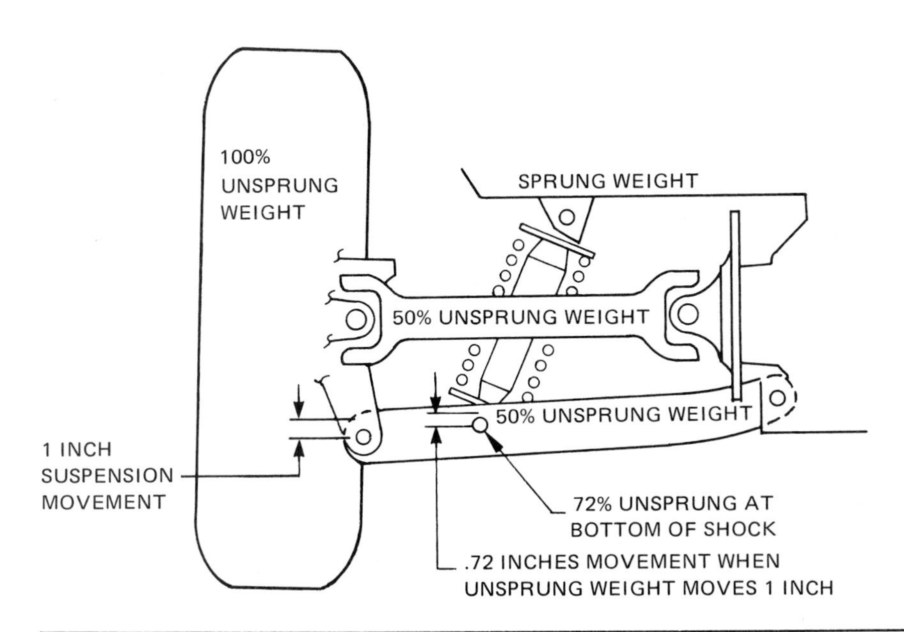

Wheel, tire, stub axle, and upright all move together as a rigid unit, and thus are all 100% unsprung weight. The axle shaft and the lower suspension arm are both 50% unsprung, an average between the two ends. The shock is 36% unsprung because the lower end is moving 72% of the unsprung weight's motion. The 36% is an average of the two ends of the shock. The coil spring mounted on the shock is also 36% unsprung weight.

out quicker than the rear, and also why disk brakes are used on the front of the car. Because the front tires have larger forces on them during heavy braking, the heat put into the front brakes is also greater.

SUSPENSION SYSTEMS

The suspension is the link between the tires and the frame of a car, and includes the springs and shock absorbers. If all roads were smooth, suspension would not be necessary. Specialized racing cars have been built without any suspension such as go-karts, which are very small and light compared to other vehicles. In addition to providing comfort, the suspension is used to tune the chassis for the best possible handling qualities. It is also to blame for most of the poor handling qualities you may be trying to get rid of.

The chassis supports the engine, body and occupants. It rests on springs which insulate the chassis from road irregularities, and from the driver's point of view the chassis bounces up and down on the springs. The weight of the chassis and all parts mounted on the chassis is considered to be *sprung weight.*

The tires, wheels and suspension parts that move up and down with the wheels are effectively underneath the springs and are not insulated from bumps in the road by the cushioning effect of the springs. The moving suspension parts are considered *unsprung weight.*

Some parts are attached to the frame at one end and to the wheel at the other. The frame end is sprung and the wheel end is unsprung. As noted in the accompanying drawing, there are ways to calculate the sprung and the unsprung portions of any moving suspension part.

Every component of a car is sprung, unsprung, or a combination of the two.

The ratio of sprung-to-unsprung weight has an important effect on the handling of a car on bumpy surfaces, and is a factor in automobile design. It can be calculated for the overall car by dividing the total sprung weight by the total unsprung weight of all four wheels. It can also be calculated separately for each end of the car—that is, the sprung weight of the front end divided by the unsprung weight at the front.

A ratio of 5 is good and a ratio of 2 is very poor. Sprung-to-unsprung weight ratio takes different values because of different designs of chassis and suspension. In following discussions, this ratio will be mentioned as a good or bad feature of some designs.

A car with a high sprung-to-unsprung weight ratio will stick better on bumpy roads than a car with a lower ratio. The car will also ride better. The reason is, the heavy body of the car is not disturbed much by the relatively light wheels bouncing up and down. The inertia of the body resists the movement of the tires, and thus the tire is pressed into reasonably firm contact with a bumpy road surface.

You can get this effect by adding a heavy load to your car. The ride will usually be noticeably smoother, particularly on a washboard surface. You may not notice it as much, but the cornering and acceleration or braking traction on bumps is improved too. However, the added weight of the entire car will cancel out some of this gain, so the better ride is the most noticeable effect.

A heavy luxury sedan rides better than a lighter stripped-down model. There isn't much difference in the *unsprung* weight of a Cadillac and a Chevy, but the sprung-weight difference usually changes the ride a noticeable amount.

To get better handling on bumps, it is much better to *reduce* the unsprung weight and keep the sprung weight as light as possible. This is done on all high-performance cars with good suspension designs. The usual reason for buying a car with expensive independent rear suspension is to reduce the unsprung weight and thus improve the handling.

The amount of suspension movement is important in extreme driving conditions. The total movement of the suspension is called the *suspension travel.* A car with a large amount of travel can be operated on rougher roads without *bottoming.* Bottoming is the condition where the suspension runs into the *bump stops* at the end of its travel. The driver can hear bottoming as a loud thump, and sometimes can feel a jolt. The extreme position of the suspension against the bump stops is called *full bump.* At full bump, the body of the car is closest to the road but with proper suspension design the car does not hit the road surface.

Full downward extension of the suspension is called *full droop.* The motion in the droop direction is usually limited by the shocks, but on some cars separate *droop stops* are provided. The body of the car is at its highest position above the road when the suspension is at full droop. The suspension will move to full droop if you jack up the car and put jack stands under the frame.

Motion of the suspension is a combination of vertical motion of the sprung weight, rolling of the car sideways, and pitching in a fore-and-aft direction. In addition, a single tire can hit a bump and move upwards without any change in the position of the sprung weight.

The position of the suspension with the car at rest is called the *static position.* There is no body roll, pitch, or vertical motion on the suspension. All suspension measurements are taken at the static position when tuning the chassis.

Cars have been built with many types of suspension through the years, but they can all be divided into two basic types, solid axle and independent. The solid axle is the earliest type, starting with the horse and buggy. It was used on both the front and rear of all early cars, but today its use is mostly limited to the rear suspension. A solid front axle is still used on oval-track and drag-racing cars, and on heavy-duty trucks. A solid axle is low in cost and simple, and it keeps the wheels upright under all normal road conditions. There are disadvantages in comfort and performance on bumps, due to high unsprung weight.

With a conventional solid rear axle, the relatively heavy axle, axle housing, differential and part of the drive shaft move up and down with the wheels.

A design called De Dion has been used on racing cars and expensive road cars. It mounts the differential on the chassis at the rear of the car and drives the rear wheels through U-jointed drive shafts. Sometimes the rear brakes are also chassis-mounted on the inboard side of the drive shafts. In that case, the wheels and part of the springs move up and down over bumps. The wheels are held vertical and parallel to each other by a stiff tube which runs from wheel to wheel across the car. This *De Dion tube* also moves up and down with the wheels. It reduces unsprung weight because only the De Dion tube moves with the wheels—the differential is bolted to the frame. De Dion is as complex as a fully independent rear suspension, but it does not provide all of the advantages. And, it wastes space.

To reduce unsprung weight and improve the ride most car manufacturers have gone to independent suspension at

Here is the classic solid front axle suspension used on a dirt track racer. It is located by radius rods, one on each side, which run back to pivots near the mid-point of the chassis. These suspensions have been replaced with independent front suspension on almost all cars, but the old solid axle system still is used for dirt oval racing. It is a rugged and simple suspension and is highly developed for this type of racing.

A modern A-arm front suspension on the ADF Formula Ford. Every effort is made to save weight and reduce drag. The suspension is even streamlined by the use of airfoil shaped tubes. Every little bit helps in Formula Ford racing!

least on the front of the car. These take many forms, with the geometry being the major difference. All types tend to be lower in unsprung weight than a solid axle, and they also allow large objects such as the engine to be placed between the wheels.

The most popular type of independent suspension is the double A-arm type. This is commonly used on the front suspension and has been increasing in use at the rear of high-performance cars. It is almost universally used on racing cars.

With double A-arm suspension the upright supporting the wheel is attached to the frame by a pair of links, usually in the form of an A-arm or *wishbone.* The links are connected to the frame by bearings called *suspension pivots,* either metal or rubber. The links are not always parallel, and are usually unequal length. They provide camber characteristics that change with both vertical suspension movement and body roll. These camber characteristics can be designed to be almost an ideal compromise for cornering.

VW, Porsche and other makes have in the past used a pure trailing-arm suspension at the front. The wheels move straight up and down with this suspension system, having zero camber change with vertical motion. However, the wheels camber with body roll, and thus tend to less than optimum for maximum cornering power. This type of suspension is also rather heavy, and is losing popularity with car manufacturers.

Another ancient independent suspension still used by Morgan is the famous sliding pillar. In this one the wheel is guided up and down a nearly vertical track on bushings so it has no camber change with vertical motion. It is simple but suffers the bad feature of cambering as much as the body rolls. In addition, friction against the guides tends to bind the suspension in a corner resulting in very little movement on a rough road. It is probably good that the Morgan looks like an old-fashioned sports car, because it certainly handles like one.

A current independent suspension is the *MacPherson strut.* This is known also as a *Chapman strut* when used at the rear of the car. The wheel is located by a

The slalom car uses the trailing-arm front suspension from a VW. Even though this suspension is used on a lot of race cars its best features are simplicity and strength. It is not very rigid and it does not offer good geometry for cornering. The only reason it works at all in a slalom car is that the springs are very stiff allowing very little lean in the turns.

This early Spitfire is one of the poorest handling sports cars in stock form due to the swing-axle rear suspension. Jacking effect in a turn causes the car to thrust its tail upwards like a stink bug. The result is a large amount of positive camber on both rear tires and lost traction. Late model Spitfires have greatly improved rear suspension.

Simplified drawings of three suspension types courtesy of Road and Track magazine. From top to bottom: Swing axles, De Dion, unequal-length A-arms. Modern design is the A-arms.

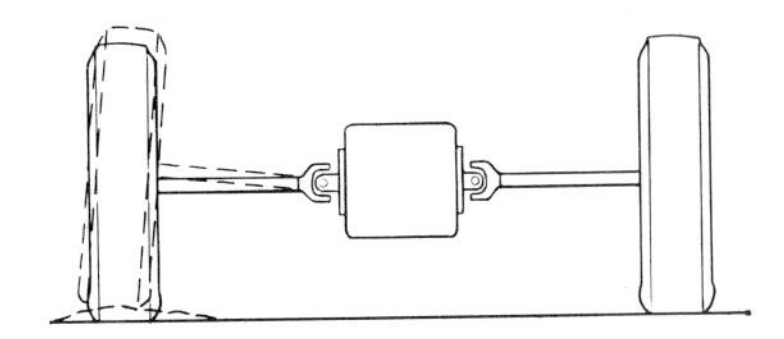

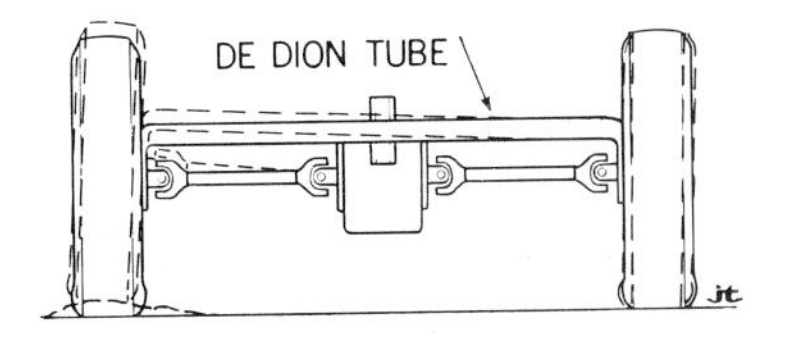

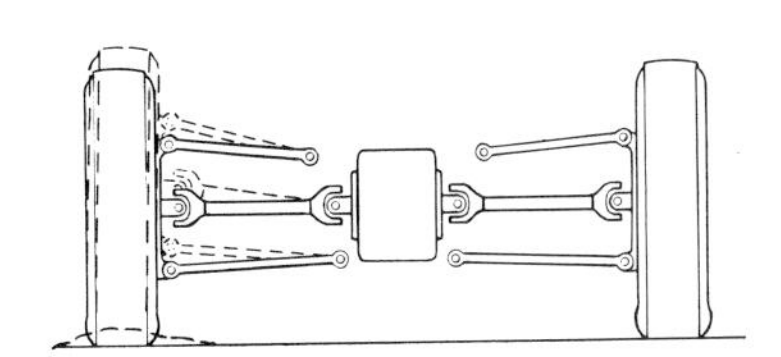

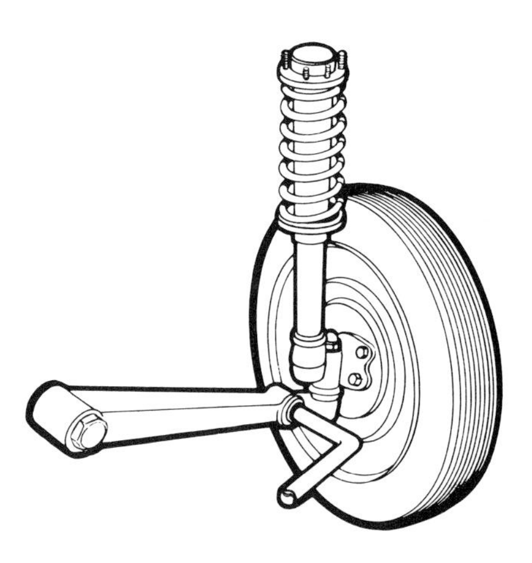

MacPherson or Chapman strut uses a single control arm at bottom of wheel suspension, which may be an A-arm or some other arrangement as shown in this drawing. Remainder of suspension is the vertical or near-vertical strut composed of shock absorber and coil spring. This requires a rather high mounting point for the top of the strut which makes the suspension more suitable for passenger cars than racing cars.

This early VW suspension is the swing-axle type. Each axle pivots about the single U-joint next to the transmission. The axle housing provides a rigid connection with the bearing carrier at the wheel. This autocross car (a motified Formula Vee) uses coil-spring suspension instead of VW torsion bars.

lower A-arm and a near-vertical strut which is integral with the shock absorber. This is a very simple and compact suspension, but it is quite tall. Its vertical height limits its use mainly to road cars, but the geometry can be very favorable for good handling. Some of the best-handling road cars made today use the MacPherson strut, and it is gaining in popularity.

In years gone by a popular type of independent rear suspension was the swing axle. In this design each axle pivots about a U-joint next to the chassis-mounted rear-end housing. It has several nasty characteristics including a tendency to lift the car when acted on by a cornering force—called *jacking effect.* This suspension also has a very large amount of camber change with vertical motion, making it poor for cars with a large variation in load. Except for simplicity and reduced unsprung weight the swing axle has little to recommend it, and let's hope Ralph Nader and others have nailed the lid on its coffin once and for all.

In modern times Mercedes developed a great improvement on the traditional swing-axle suspension. This is the single-pivot swing axle design, and it is used only at the rear of the car. To reduce the camber change and the jacking effect of the swing axle, Mercedes used a single pivot point under the differential as the pivot for both wheels. The effect is a reasonable compromise between the simple swing axle and more complex types of independent suspension.

Another variation on the swing axle design has been used by Ford on their light trucks. They call this *Twin I-Beam* front suspension, but it is nothing more than a pair of elongated swing axles, each pivoting on the opposite side of the vehicle. This still has the bad features of the swing axle, but to a lesser degree. It is my opinion that Ford "had a better idea" with the solid front axle previously used on their light trucks. At least the old solid axle didn't suffer from bump steer.

Another variation on the swing-axle rear suspension is the semi-trailing arm. With this design the wheel is located by a single large A-arm with the pivoting axis at an angle to the longitudinal axis of the car. This semi-trailing arm suspen-

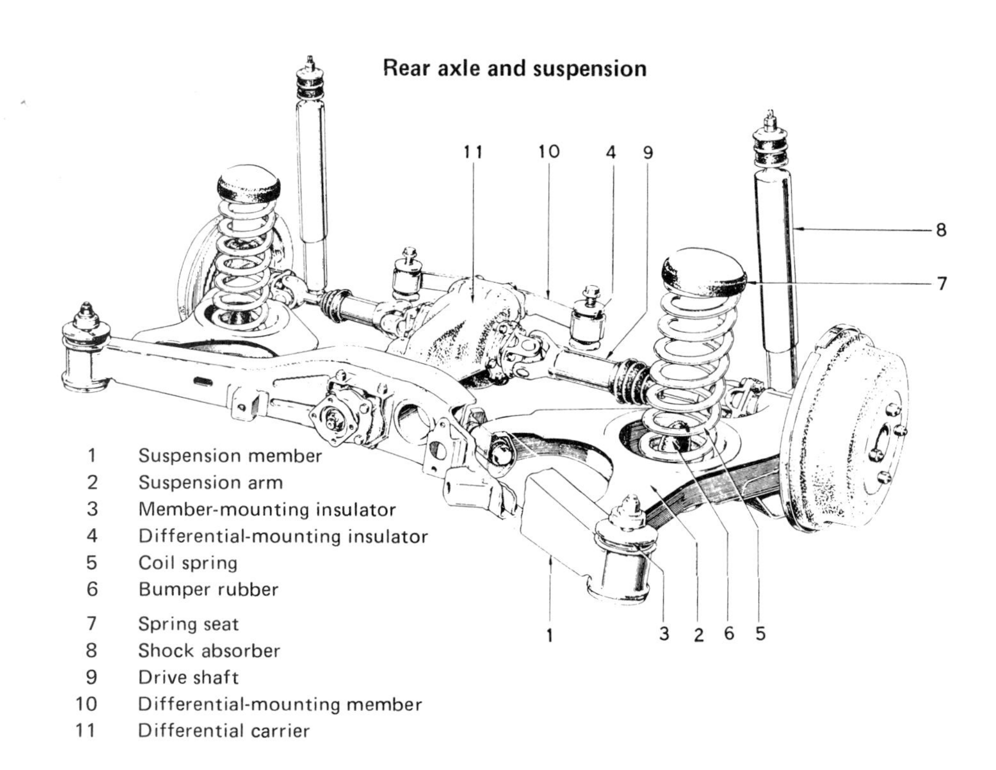

The Datsun 510 rear suspension is typical of the semi-trailing arm type. The arm pivots make an angle with the centerline of the car. The result is a considerable amount of toe-in change with suspension movement. This is not ideal for racing, and works reasonably well only if the suspension is ultra-stiff.

sion is used on the rear of many production cars. Semi-trailing arm suspension almost always exhibits some roll steer, and this is used to alter the handling of the car. Thus this type of suspension is seldom used on racing cars, as the designer usually does not like roll steer for racing.

ROLL CENTERS

When a car is in a turn, centrifugal force causes the body to lean toward the outside of the turn. This is called *roll.* The mechanical facts are, roll tends to compress the springs on the outside of the turn and allow the springs on the inside to extend.

In several aspects of car design, we look for an equivalent, such as CG, which can simplify our thinking—particularly during design of the vehicle.

Such an equivalent is roll center. Viewing a car which is rolling in a turn, it is evident that the body is no longer horizontal with respect to the road. There must be some point in space about which the car body could have rotated to assume that same angle. This point can be real

SPRUNG—TO—UNSPRUNG WEIGHT RATIOS FOR VARIOUS REAR—SUSPENSION TYPES

TYPICAL ROAD CAR WITH 1000 POUNDS ON THE REAR TIRES

Suspension Type	Sprung Weight Per Wheel (LBS)	Unsprung Weight Per Wheel (LBS)	Sprung Weight / Unsprung Weight
Solid Axle with leaf springs	370	130	2.85
Solid Axle with coil springs	390	110	3.54
De Dion with coil springs	400	100	4.00
Independent with brakes at wheel	410	90	4.55
De Dion with inboard brakes, coil springs	425	75	5.66
Independent with inboard brakes	435	65	6.69

Notice the great improvement in the ratio of sprung to unsprung weight between the early solid-axle designs and a modern independent suspension with inboard brakes.

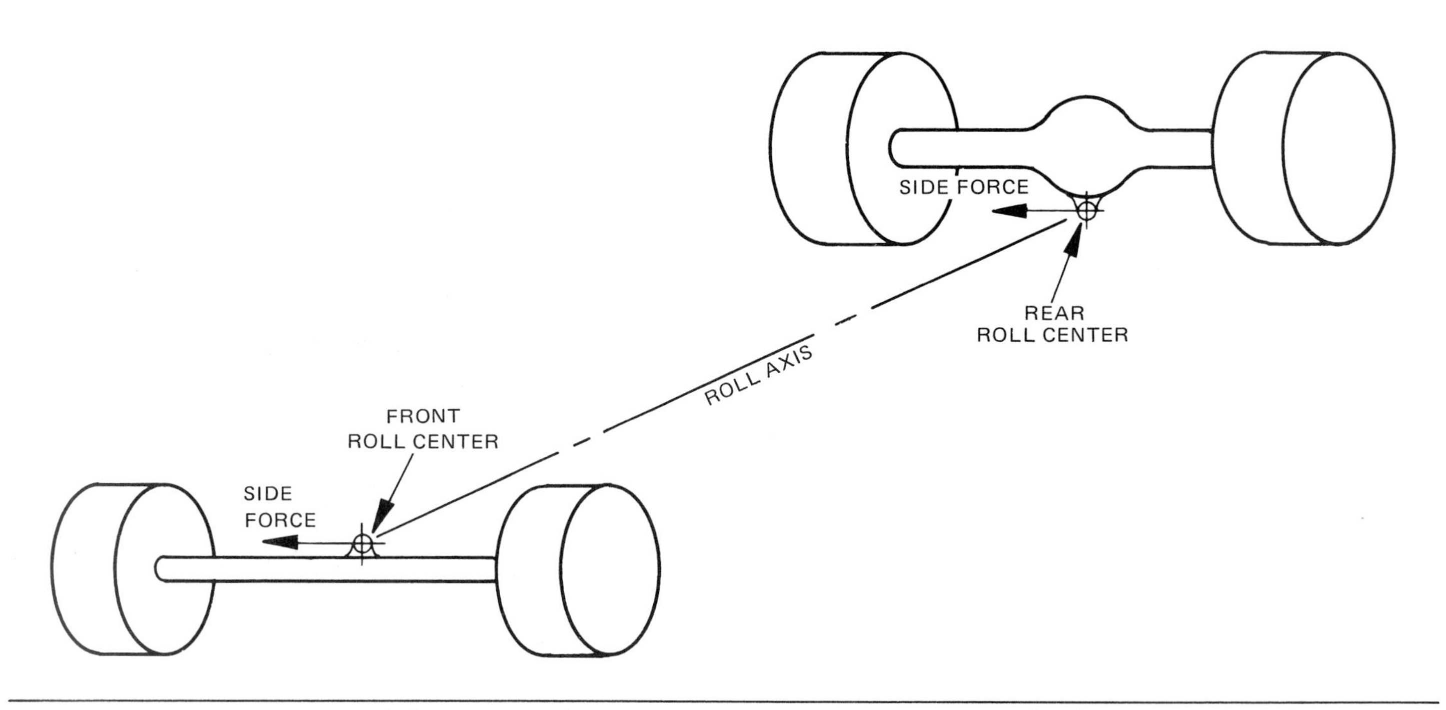

This drawing shows how the side force is applied to the unsprung weight at the roll centers. Solid axle suspension is shown for simplicity, but roll centers serve the same purpose with independent suspension too. When the body rolls in a turn it pivots about the roll axis.

or imaginary, depending on the design of the car. It is called the *roll center.* Height of the roll center above the road is important and it need not be the same at the front as the back of a car.

The discussion of roll centers in this section may be difficult for you, depending on your ability to visualize imaginary points in space or even a car body pivoting around a real point on the car. If so, it should not concern you greatly and you are welcome to read this section for the general information it contains without worrying about the details.

Roll centers on paper are mainly important to designers. The idea of roll-center height is important to car buyers and drivers because it influences handling; however, all the buyer really needs to know is which type of suspensions tend to give good handling due to good roll-center locations. This section discusses that and it comes up again later in the book. Also, the idea of roll-center height is important to anyone who decides to modify the suspension of a car because some modifications change roll center. This is also covered in specific terms later in the book.

Also, the treatment of roll centers in later chapters is more practical—the effects are discussed in terms of handling or performance that you can test and observe. If it doesn't test right, there are practical modifications in some cases that can improve test results or performance on the road.

The point I am circling around here is that there is just no way to talk about roll centers without actually doing it and some people don't communicate on that wavelength. If you don't, don't worry about it. Grab as much of the idea as you can and bring it with you to the testing and modification chapters which follow.

The exact point about which the body rolls is important. The front suspension has a roll center and the rear suspension has a roll center. A line between the front and rear roll centers is called the *roll axis.*

The roll center is an actual pivot point on some suspension system, but on others it is a point in space. The roll center is located near the center of the car, usually at a height somewhere between the road surface and the center of the wheels. The roll center height above the ground is very important, and on some suspensions it changes as the body rolls. This gets quite complicated but it can be figured out and used to help the handling.

In addition to being the pivot point for body roll, the roll center has another important characteristic. The side force due to centrifugal force is transmitted from the sprung weight to the unsprung weight through the roll center. This is very simple to understand if the roll center is an actual pivot point between the suspension and the body. However, even on complicated independent suspension systems the suspension links act in a way that is *equivalent* to a single roll center between the sprung weight and the unsprung weight.

The roll-center height on a solid-axle suspension depends on the type of linkage locating the axle. The linkage locates the axle fore-and-aft, and it prevents sideways motion between axle and frame. The roll-center height is usually found by drawing the linkage to scale and making some geometrical constructions on the drawing.

Figures 7A through 7F show how to find the roll center on various types of solid-axle suspensions. In each case point A is located, and another point B is found. A line is drawn through points A and B. The roll center is located where this line crosses a vertical plane through the axle centerline. The captions under each figure show how to find points A and B, and what they mean.

The simplest type of solid-axle suspension is the Hotchkiss drive, where the axle is located by two leaf springs. Hotchkiss drive is used on most low-cost production cars. Its roll center is found as shown in Figure 7A. If an additional locating device is used between the axle and frame, this may change the roll-center height. Such a locating device, to be effective, must be stiffer than the leaf springs in resisting sideways motion between axle and frame. Thus the leaf springs no longer serve to locate the axle sideways, and they only locate in a fore-and-aft direction. The new locating device must always be made with stiff bearings or bushings, not the rubber mounts commonly used in leaf spring eyes. More detail on axle-locating devices is presented in Chapter 4.

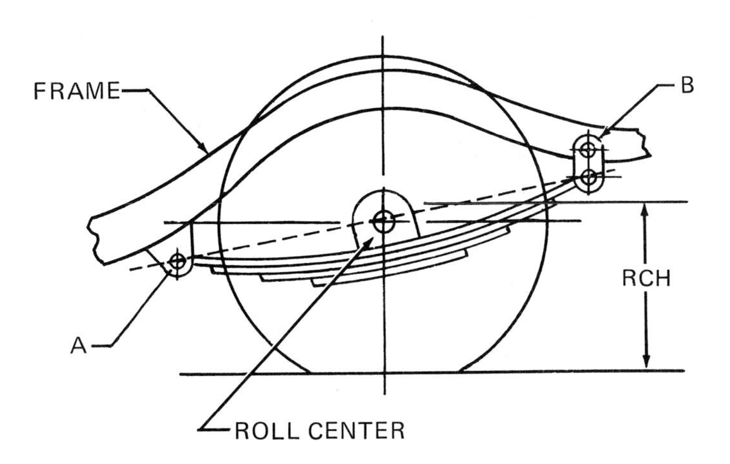

Figure 7A/If the rear axle is located only by two leaf springs it is a Hotchkiss drive. The roll center is found by drawing a line between points A and B, the spring mounting pivots at the ends of the spring. The roll center is located where the line crosses a vertical plane through the axle centerline. Line A–B is the roll axis. Adding a Panhard rod or other axle locating device to a Hotchkiss drive suspension may change the roll-center height.

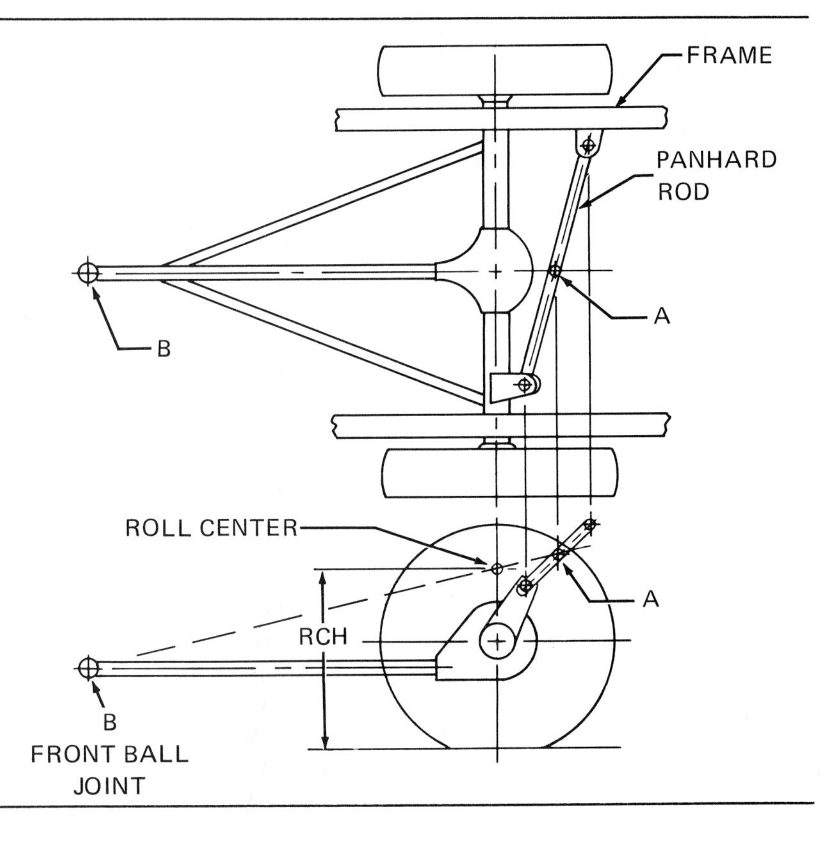

Figure 7B/On a suspension using a Panhard rod for lateral location of the axle first find point *A*. This is where the Panhard rod crosses the centerline of the car as viewed from the top. On the torque tube suspension shown, point *B* is the front ball joint on the torque tube. Point *B* is the point about which the axle would rotate if the Panhard rod were disconnected, viewed from the top. Join points *A* and *B* with a line. The roll center is where this line crosses a vertical plane through the axle centerline.

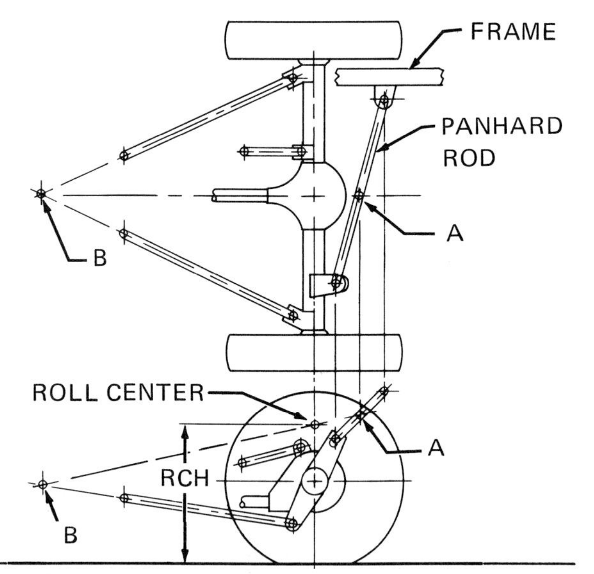

Figure 7C/The three-link suspension with the Panhard rod is similar to the torque tube suspension shown in Figure 7B. Point *A* is on the Panhard rod where it crosses the centerline of the car, viewed from the top. Point *B* is where the lower links intersect. Point *B* is the point about which the axle would rotate if the Panhard rod were disconnected, viewed from the top. The roll center is where a line from *A* to *B* crosses a vertical plane through the axle centerline. This arrangement provides roll understeer.

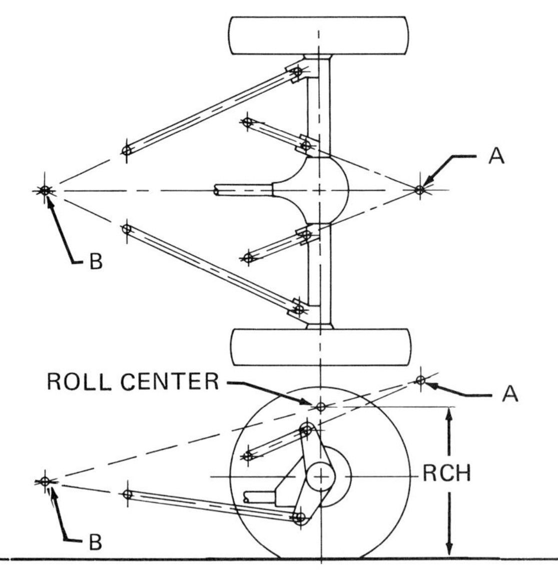

Figure 7D/The four-link suspension using non-parallel links is shown here. Point *A* is the intersection of the top links, and point *B* is the intersection of the lower links. The roll center is located where the line from *A* to *B* crosses a vertical plane through the axle centerline. This arrangement provides roll understeer.

A roll axis which points down towards the front of the car is an understeering roll axis.

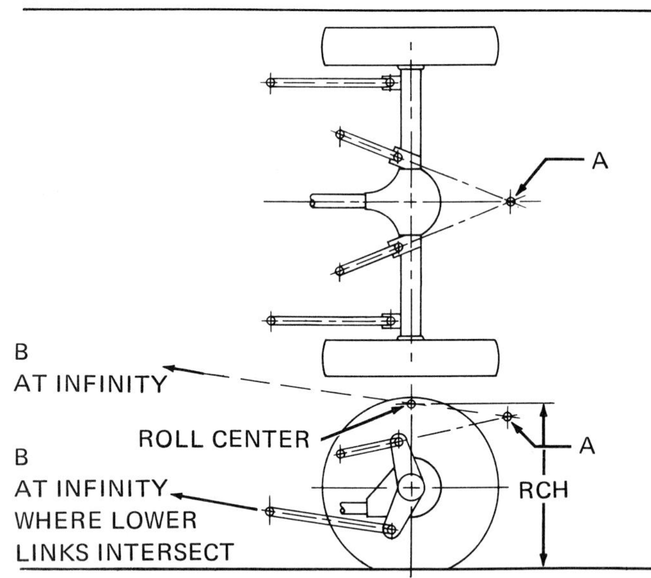

Figure 7E/This is a four-link suspension with parallel lower links. Point *A* is the intersection of the upper links. Point *B* is the point where the lower links would intersect at infinity. The line from *A* to *B* is parallel to the lower links because it too intersects the lower links at infinity. The roll center is where this line crosses a vertical plane through the axle centerline. This arrangement, or any other where the roll axle points up toward the front of the car provides roll oversteer.

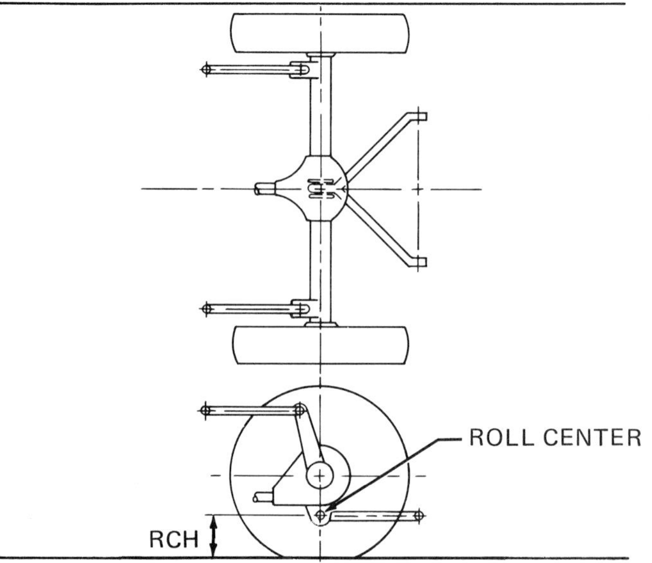

Figure 7F/If the sideways location of the axle is provided by a pivot point on the center of the axle, that pivot point is the roll center. In this case the attachment of the A-arm to the axle housing is the roll center. If the locating device is a Watts linkage or a Panhard rod mounted to the axle in the same spot, the roll center is still at that spot. The type of locating linkage makes no difference; only the height of its attachment to the axle. This arrangement provides neutral steer because the roll axis is parallel to the ground.

Locating the roll center for an independent suspension is usually done by constructing lines on a scale drawing. This requires identifying some additional points and links which may be real or imaginary.

When a wheel has been tilted from vertical as a result of a bump or suspension movement, we can find some point about which the wheel could have rotated to assume the tilted position, and imagine that the wheel is connected to that distant point by a radius arm of some sort. If the wheel assumes a different angle, it may appear to be connected to a different center of rotation, however at any instant there is always some center about which the wheel appears to rotate. This is called the *instantaneous center.* As you will see, the designer can put the instantaneous center nearly anywhere he chooses and it may not even be between the wheels of the car.

Wherever the instantaneous center is, we also imagine the wheel to be connected to it by a radius arm or a swing arm. Because this swing arm is not necessarily real, it is called the *virtual swing arm.*

The only way you can predict such behavior and draw the position of the wheel as it comes to full bump or full droop is to recognize the instantaneous center and the virtual swing arm, real or not.

One of the simplest cases is a swing axle because both are real and made of metal which you can see and touch. Figure 8 shows this suspension. Each axle swings, pivoting around the U-joint next to the differential. The instantaneous center is the U-joint, the virtual swing arm is the length of the axle between the U-joint and the wheel. Both are real and the instantaneous center doesn't move around when the wheel changes its attitude.

To find the roll center for this suspension, draw lines between the center of the tire contact patch on the ground and the instantaneous center for that wheel. Do it on both sides of the car and extend the two lines until they cross each other. The roll center is at the point where the two lines intersect. As you can see in Figure 8A, the roll center is always very high with conventional swing axles.

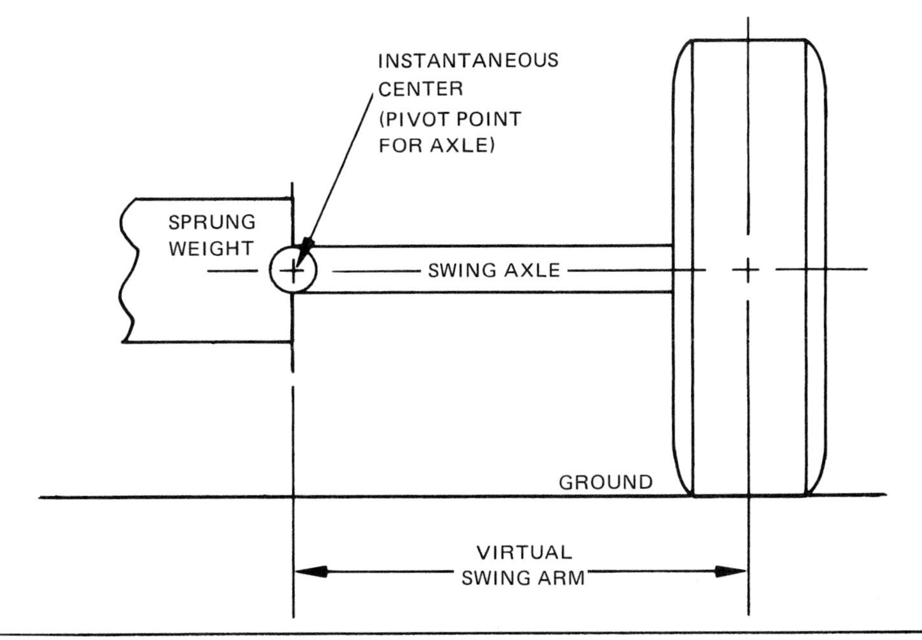

Figure 8/The simple swing axle pivots about a point next to the differential which is mounted on the chassis. This pivot point is the instantaneous center for the suspension. The length of the swing axle is the virtual swing arm length as shown.

SWING AXLE

In this chapter, I have defined a swing-axle suspension as one in which the axle-shaft acts as one part of the suspension linkage which locates the wheel. To keep things simple in the examples shown here, I assume the swing axle always pivots about a line parallel to the centerline of the car. If a swing-axle suspension pivots about a line not parallel to the car centerline, as on an early VW rear suspension, find the roll center using the same method as for a semi-trailing-arm type. You can see from this that there is very little difference in suspension geometry between a semi-trailing-arm suspension and many swing-axle suspensions systems.

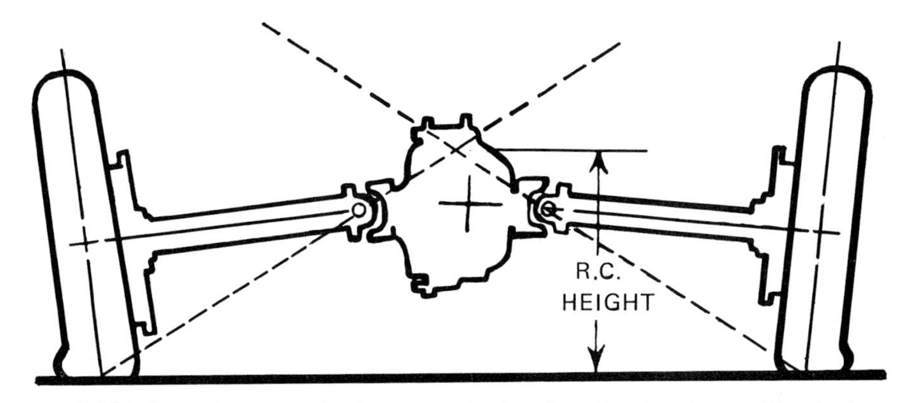

Figure 8A/Swing-axle suspension has a very high roll center. It is located by the intersection of the two lines connecting the tire contact patch with the axle pivot point on each side. The high roll center results in the infamous jacking effect, which tends to raise the car under the action of cornering forces.

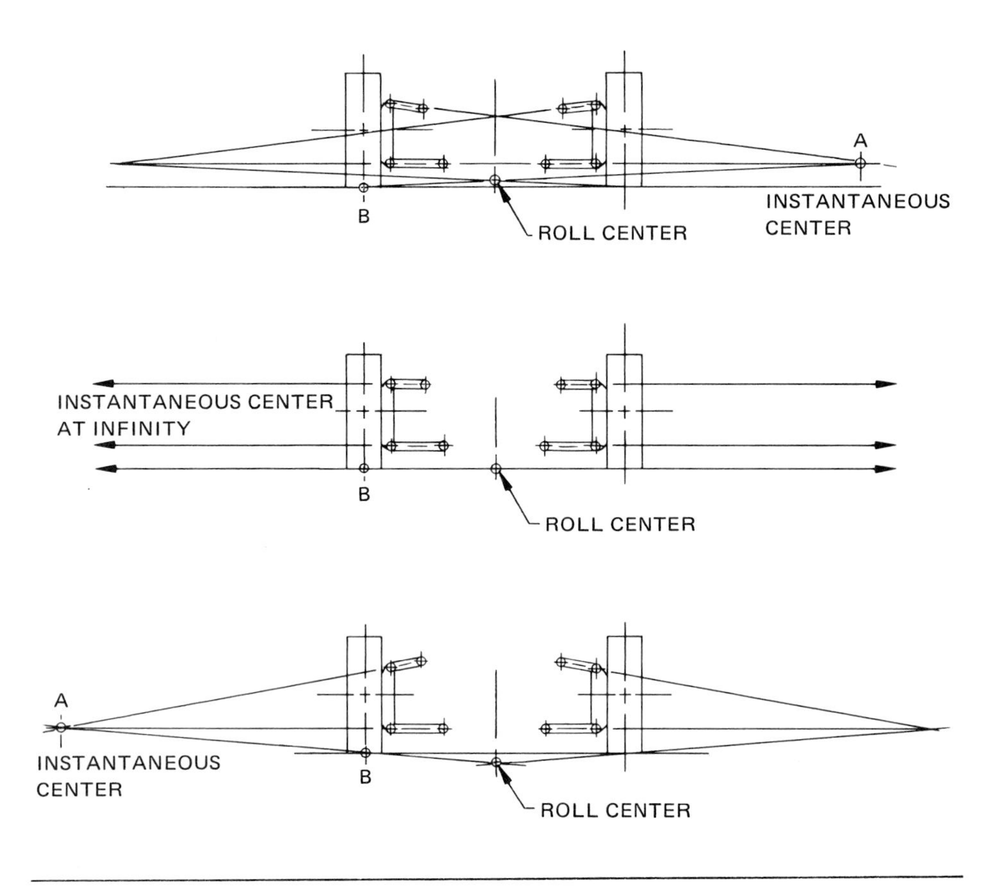

Figure 9/The roll center for double A-arm suspension can be at almost any height, depending on the angles of the A-arms. The three designs shown all have the same length A-arms and the only difference is in the height of the upper inboard pivot point. On conventional cars, the roll center with this type suspension is usually close to the ground. It can even be below the ground.

Other types of independent suspensions are analyzed in the same way. The procedure is: Find the instantaneous center and the virtual swing arm. Construct a line through the tire contact patch of each wheel and its instantaneous center. Locate the roll center at the intersection of these two lines.

The virtual swing arm and instantaneous center are easy to find on a double A-arm suspension. Draw a front or rear view of the suspension. Extend lines down the center of each A-arm until they cross. This is the instantaneous center, and is shown as point A in Figure 9. The tire moves the same as if it had a swing axle in place of the virtual swing arm. However, the virtual swing arm changes length as the suspension moves, and this is why double A-arm suspension offers advantages for high-performance driving. The design can be so the instantaneous center moves to give proper camber changes as the body rolls.

The roll center for the double A-arm suspension is found by connecting the instantaneous center point A with the tire contact patch point B as shown in Figure 9. Due to the long virtual swing arm, the roll center is usually quite close to the ground. It is on the ground if the A-arms are parallel, because the virtual swing arm is infinitely long. Lines which intersect at infinity are parallel to each other.

With suitable A-arm angles the roll center can be below the ground. This is only possible with independent suspension, because it would be very hard to build a solid axle with a lateral locating device below the ground. A roll center below the ground is not usually used on high performance cars, but it does exist on some sedans.

The roll centers shown in Figure 9 are with the car at zero roll angle. The roll centers can move around as the car rolls, and this sometimes hurts handling. A drawing must be made at various angles of body roll to accurately determine how the roll center moves. When making the drawing the body must be pivoted about the roll center at each angle of roll, so the process has to be done in small steps for accuracy. I suggest drawing the car at one degree roll, two degrees roll, and so forth.

The roll center location of a Chapman strut or MacPherson strut suspension is shown in Figure 10. This suspension has similar geometry to the double A-arm type, except that the strut replaces the upper A-arm. To find the instantaneous center draw a line through the pivot at the top of the strut at 90 degrees to the strut centerline. Where this line intersects the line through the center of the lower A-arm is the instantaneous center. The roll center is determined the same way as for the double A-arm suspension by drawing lines between the tire contact patch and the instantaneous center of each wheel.

Because location of the strut limits the geometry, the roll center cannot have as many different locations as with double A-arm suspension. With MacPherson strut suspension, the roll center is usually above the ground, but it is never very high. This is typical of roll-center locations on high-performance cars, no matter what type of suspension system is used.

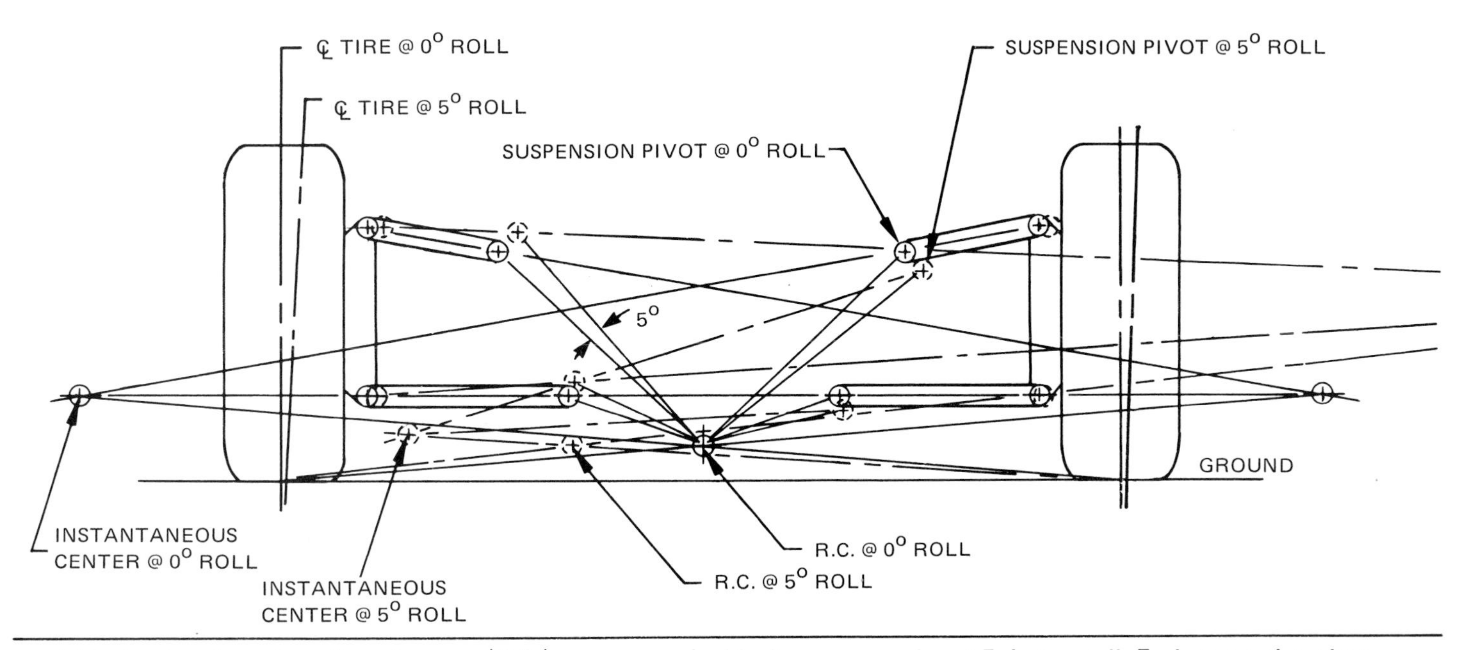

This drawing shows how the roll center (R.C.) moves on a double A-arm suspension at 5 degrees roll. Each suspension pivot on the chassis is drawn in the 5 degree roll position, using the roll center at zero degrees as a pivot point. The links are drawn in the new positions with the tires on the ground and the new roll center at 5 degrees is plotted. With this suspension the roll center remained at about the same height but moved to the left considerably. When designing or analyzing a suspension it is usual to take one degree increments for roll angle and check the roll center position at a number of angles up to a maximum of about 5 degrees.

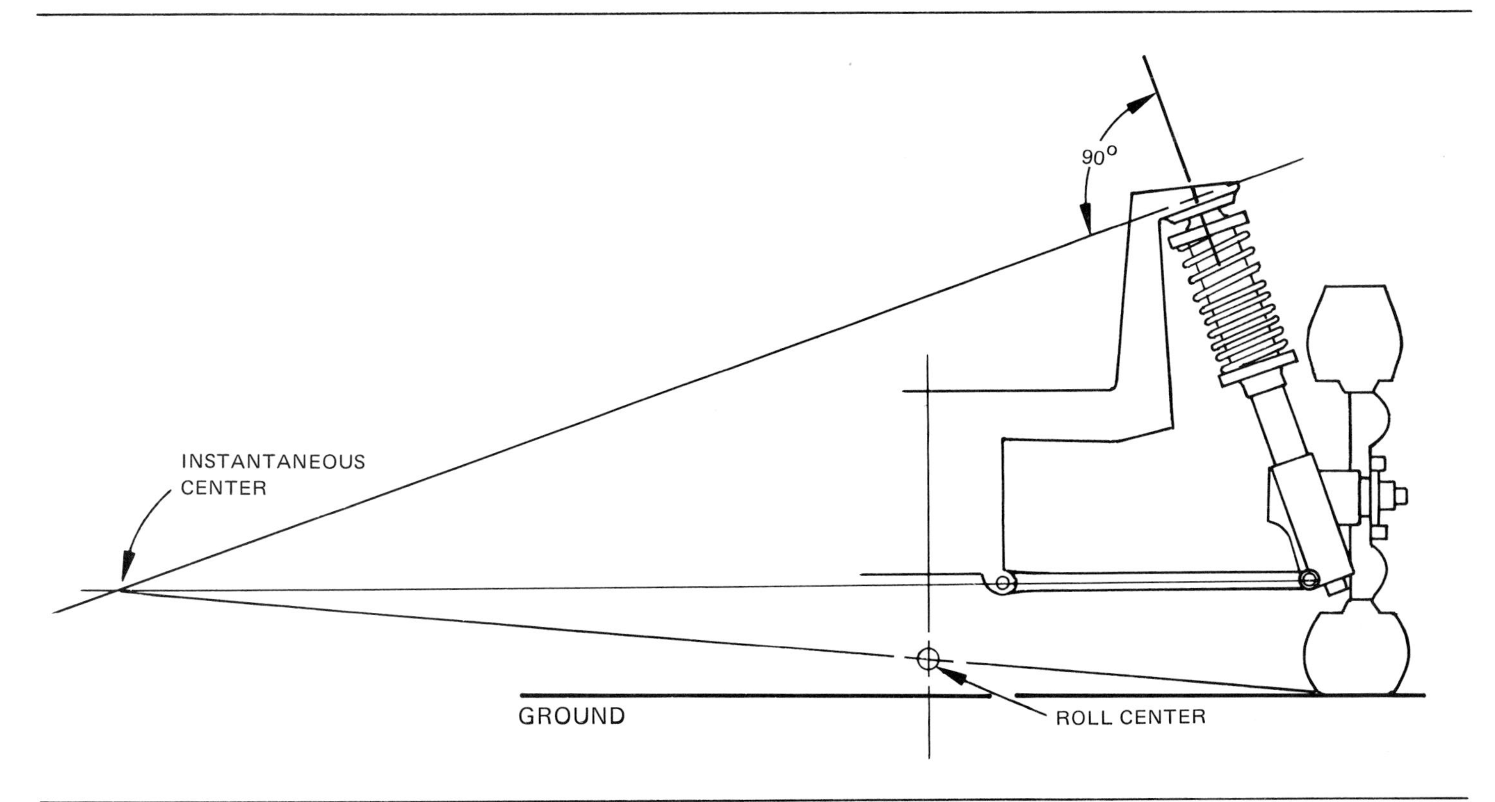

Figure 10/The Chapman strut suspension of the Lotus Elan has the same geometry as the popular MacPherson strut. The only difference is that Chapman strut suspension is used at the rear of the car where MacPherson strut refers to front suspension. The roll center is low, similar to a double A-arm suspension.

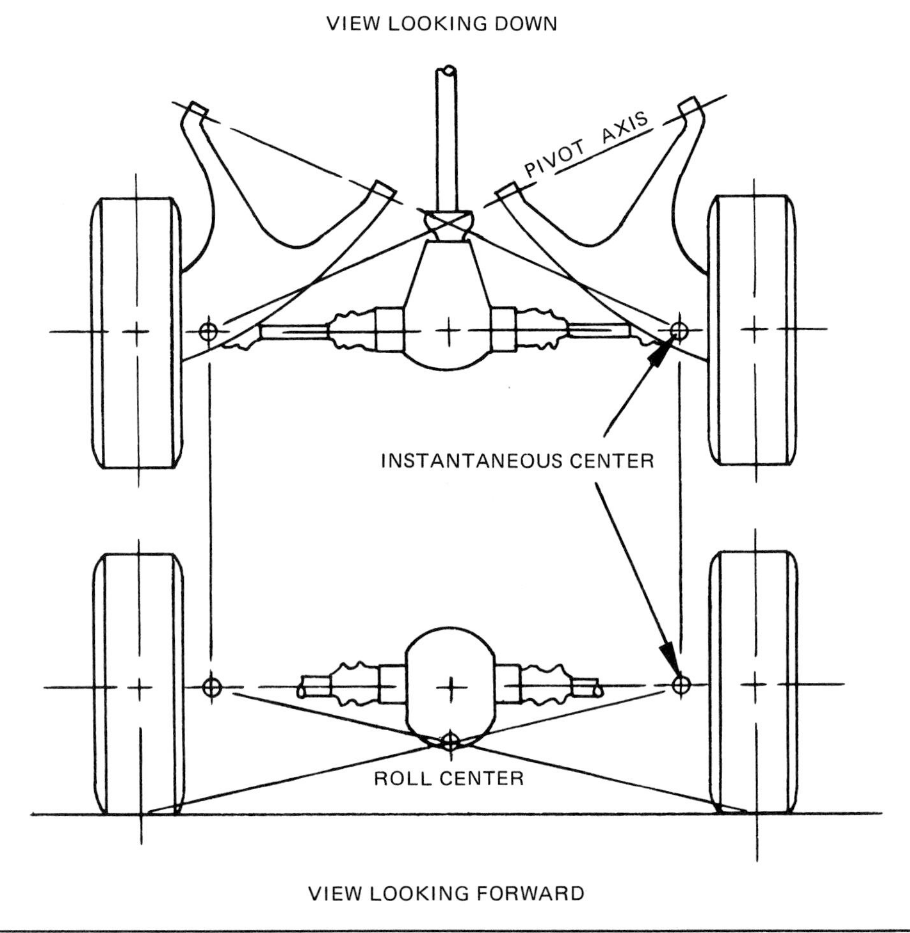

The semi-trailing-arm suspension has a virtual swing arm longer than a pure swing axle. This lowers the roll center and makes this type of suspension more acceptable for high performance cars.

The old-fashioned swing axle is very similar in concept to the modern semi-trailing-arm suspension. The only difference is that the semi-trailing-arm suspension has its pivot axis at an angle to the centerline of the car. To find the instantaneous center of this suspension draw a view looking down on the suspension. Draw a line between the pivot points of the A-arm and extend it until it crosses the axle centerline. This is the instantaneous center. Notice that the virtual swing arm is quite long with this type of suspension—much longer than the actual A-arm length. This results in a roll center much lower than the swing axle. The roll center is found the same way as for other independent suspensions.

Length of the virtual swing arm gives an indication of the amount of camber change which will occur as a wheel moves up and down. Obviously longer virtual swing arms will allow less camber change for the same amount of vertical wheel movement. By inference, a suspension which allows zero camber change must have a virtual swing arm which is infinitely long.

If the semi-trailing arm has its pivot axis moved to 90 degrees from the centerline of the car it has the geometry of a pure trailing arm, or leading arm suspension. This type, familiar as the old VW front suspension, has no camber change with suspension movement. The wheels are always the same camber as the angle of body roll. This type of suspension has its roll center at ground level and the virtual swing arm is infinitely long.

The sliding pillar suspension also has its roll center at ground level. It guides the wheel vertically along a straight line at the frame, thus forcing the wheel always to have a camber angle equal to the angle of body roll. Except for the linkage the sliding pillar has the same geometry as trailing arm suspension.

CENTRIFUGAL FORCE AND MAXIMUM CORNERING SPEED

When a car rounds a bend in the road, the car and its occupants are acted on by centrifugal force. As all of us know from experience, this force increases if the weight of the car is larger, the speed higher, or the radius of the turn shorter.

For a turn of constant radius, centrifugal force can be calculated as follows:

$$\text{Centrifugal force} = \frac{\text{Weight} \times (\text{speed})^2}{14.97 \times \text{radius}}$$

where weight is the total weight of the car and occupants expressed in pounds, radius of the curve is in feet, and speed is in MPH.

It is sometimes interesting to calculate the maximum theoretical speed a car can travel in a turn without tipping over:

$$\text{Max speed} = 2.74 \sqrt{\frac{\text{Radius} \times \text{Track}}{\text{CG Height}}}$$

where maximum speed is in MPH, and all other units are in feet. This is theoretical only in that it requires the car not to spin out or slide off the road. If the tires hold the car on the radius of the turn used in the calculation, then the indicated max speed can be experimentally verified.

Notice that weight of the car doesn't have anything to do with tip-over speed under the stated assumptions and the further assumption that the car isn't heavier on one side than the other. It is all determined by the height of the CG and the track or distance between the wheel centers. A wide car with a low CG is less likely to tip over than a tall narrow car.

As an example, on a turn with a 200-foot radius, a car with a 5-foot track and a CG height of 2 feet can corner at 61 MPH if the tires allow it.

Raise the CG height to 4 feet and maximum cornering speed in that same turn becomes 43 MPH.

LATERAL WEIGHT TRANSFER

Lateral weight transfer occurs in cornering, similar to fore-and aft weight transfer during acceleration or braking. The centrifugal force acts sideways at the CG of the car in a direction away from the center of the corner. Because the CG is some distance above the ground, weight is removed from the inside tires and added to the outside tires. The total weight on all four tires remains the same during this process.

The amount of lateral weight transfer can be easily calculated by the following formula:

$$\text{Total lateral weight transfer} = \frac{\text{Centrifugal force} \times \text{CG height}}{\text{Track width}}$$

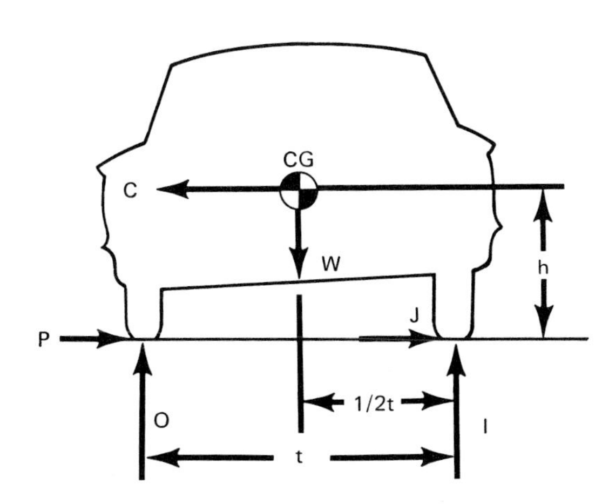

This drawing shows all the forces acting on a car in a corner except aerodynamic forces. The CG is shown in the center of the car, this is, the car is assumed to have 50/50 lateral distribution. The weight on the outside tires is greater than the inside tires because of lateral weight transfer.

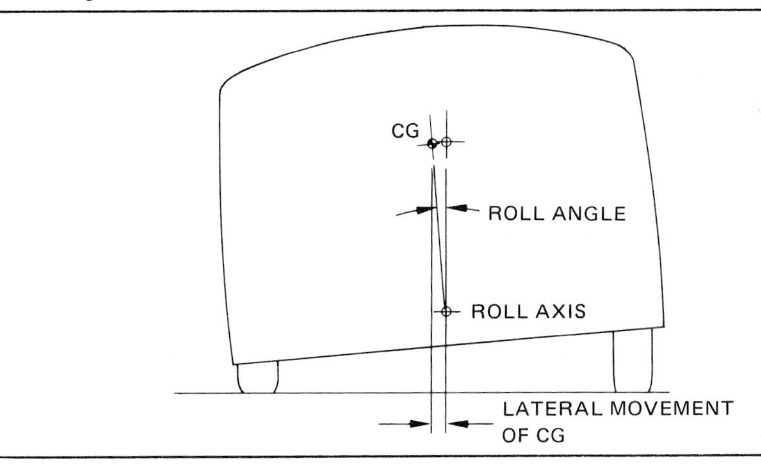

This shows how the CG moves laterally a small amount due to body roll. The car in this example has a very high CG, making the lateral movement quite large. Most high-performance cars have low CG height and a small roll angle, resulting in a very small CG shift.

In this formula the lateral weight transfer and the centrifugal force are in pounds, and the CG height and the track width are in inches. The total lateral weight transfer is the amount of weight removed from *both* inside tires. It gives no information on the distribution of this weight transfer between the front and the rear of the car.

In addition there is a small additional weight transfer due to the sideways movement of the CG as the body rolls. This effect is very small on modern high-performance cars due to the low CG height and small roll angles. I only mention it here so the discussion of weight transfer will be complete. Unless you have a very tall car with a huge amount of lean, the weight transfer due to CG movement will be less than 3% of the car's weight.

In the following discussion I am going to talk about weight transfer associated with body roll and other apparent sources or reasons for weight transfer. It is convenient to assign *some part* of the total weight transfer to a particular part of the car or to its behavior during cornering. However, this can lead to the assumption that something such as body roll *causes* weight transfer, which is not true.

Even though the point has just been made in the preceding formula for total weight transfer, it is worth restating. The total weight transfer from the inside wheels to the outside wheels in a corner is determined entirely by the centrifugal force acting on the CG of the car, assuming that any lateral movement of the CG can be ignored.

No arrangement or design of a car can reduce total weight transfer in a corner unless the height of the CG is lowered, the track increased, or the car weighs less. The driver can reduce weight transfer by slowing down or using a larger radius through the turn, but that's not part of the car design.

A car with no suspension—all parts solidly welded to each other—will have a certain amount of weight transfer in a corner. The addition of a suspension does not change the total weight transfer, but it allows it to *appear* or manifest itself in different ways. Because of springs in suspension the body can roll or lean. The weight transfer changes the forces on the springs, causing the outside springs to compress and the inside springs to extend. Thus *weight transfer causes* body roll, not the other way around.

The outside tires are more heavily laden as a result, and the inside tires are less heavily loaded. We say this part of the total weight transfer is associated with or *due to* body roll, which is OK as long as we remember that the original cause of it is centrifugal force acting on the CG of the car.

The total lateral weight transfer as calculated in the preceding formula is made up of three portions as follows:

Total lateral weight transfer =

Weight transfer due to body roll +
Weight transfer due to roll center heights +
Weight transfer due to unsprung weight.

Each of these portions must be understood, as they are adjusted separately when tuning the chassis. The art of chassis tuning is much concerned with the adjustment of weight transfer—changing *the proportion* between front and rear of the chassis.

To understand lateral weight transfer, you must look at three different parts of the car—the sprung weight, the front suspension unsprung weight, and the rear suspension unsprung weight. The weight of the car is the total weight of these three parts. Each part can be considered to have its own separate CG, and the centrifugal force acts on the CG of each part in proportion to its weight. The effect on the car is exactly the same as considering the total centrifugal force acting at the CG of the entire car. A drawing of a car with weight and centrifugal forces divided among the three parts is shown in Figure 11.

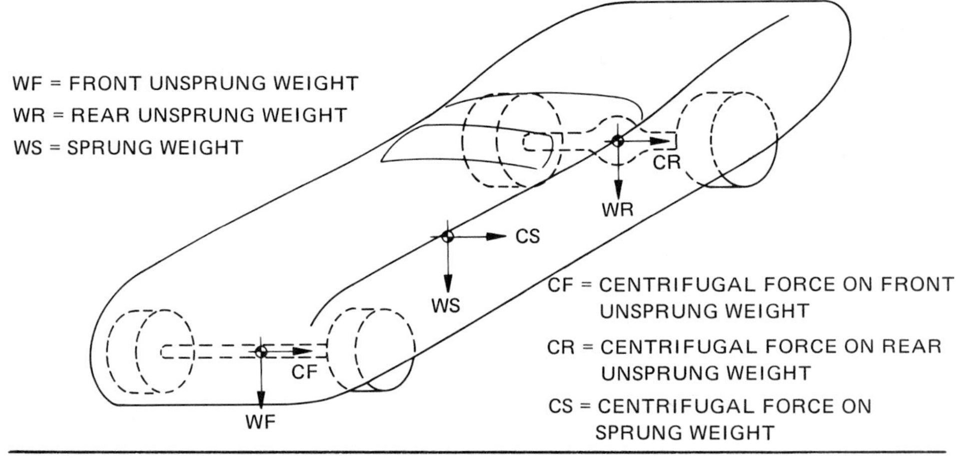

Figure 11/A car can be considered as the three parts shown, each with its own CG. The weight of each part and the centrifugal force on that part are shown acting at the CG of each part. The result on the car is the same as if all weight and centrifugal force were acting at the CG of the entire car.

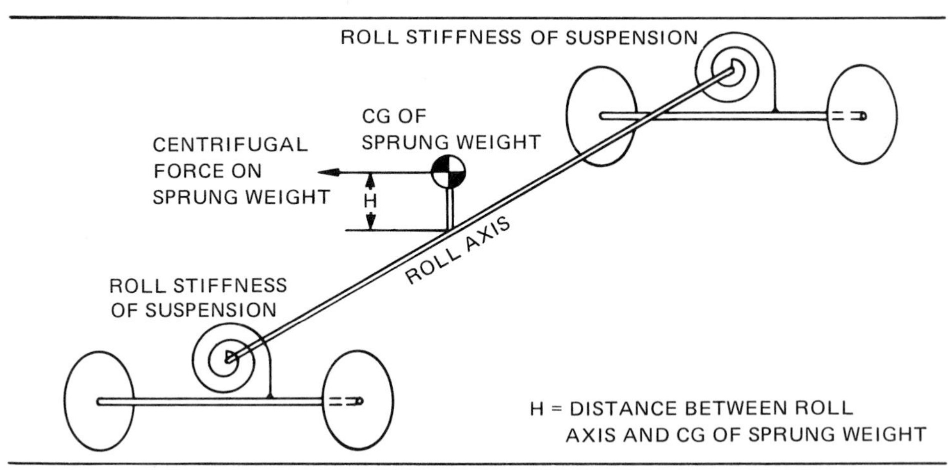

The roll stiffness of each suspension can be represented as a clock spring. The centrifugal force at the CG rolls the sprung weight about the roll axis, twisting each clock spring the same amount and meeting resistance according to the stiffness of each spring. The resistance of both springs is the total roll stiffness of the car.

If for example the front unsprung weight is 10% of the total car weight, then 10% of the total centrifugal force acts on the front unsprung weight. If the rear unsprung weight is 20% of the total car weight, it receives 20% of the total centrifugal force. The sprung weight would receive the remaining 70% of the total centrifugal force, which will cuase body roll.

Weight Transfer Due To Body Roll—*Roll stiffness* of a suspension is its resistance to body roll. The part of the car that rolls in a turn is the sprung weight, and it rotates around the roll axis. As mentioned earlier the roll axis is a line joining the roll centers of the front and rear suspensions. Think of the roll axis as an axle about which the sprung weight of the car pivots.

The stiffness of the springs at each end of the roll axis affects the roll stiffness of the front and the rear suspension respectively. Roll stiffness is expressed in inch-pounds per degree of roll.

If a suspension has a roll stiffness of 5000 inch-pounds per degree, this means it takes 5000 inch-pounds of torque about the roll axis to cause one degree of body roll. The total roll stiffness of the car is equal to the roll stiffness of the front suspension plus the roll stiffness of the rear suspension.

Roll stiffness of the suspension is determined by the suspension springs. The position of the springs as well as their stiffness determines roll stiffness of the suspension. For example, look at a typical solid-axle suspension with longitudinal leaf springs. The distance between the springs is called *spring base* and is applied only to solid axles. As you can see in Figure 12, distance between the springs

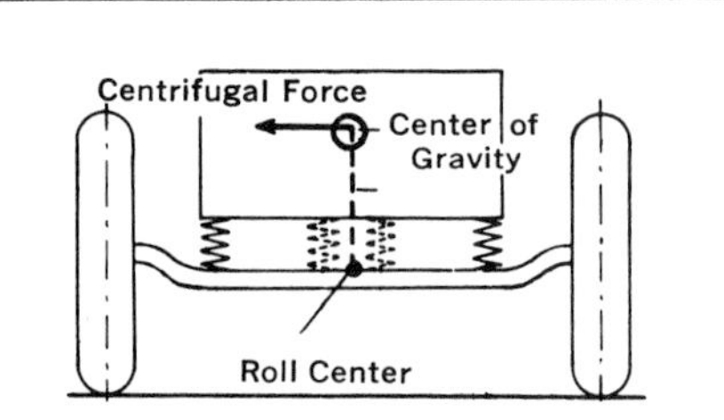

Figure 12/In a solid-axle suspension, notice how the position of the springs on the axle affects the roll stiffness. If the springs are spaced wide apart the roll stiffness is high. If the springs are closely spaced the roll stiffness is low. The vertical stiffness of the springs could be the same in either case.

Rear anti-roll bar installed on the rear bulkhead of this Formula Vee. The anti-roll-bar links are attached to each axle housing and to each end of the bar. Body roll causes one axle to rise and the other to fall relative to the frame. This puts a twist in the anti-roll bar which resists body roll.

determines roll stiffness. The suspension can be designed with stiff springs but with low roll stiffness just by placing the springs close to the center of the car.

The most common device used to increase roll stiffness of the basic suspension is an *anti-roll bar.* This is sometimes called a *sway bar,* a *roll stabilizer bar,* or an *anti-sway bar.* Whatever you call it, it's a simple and highly effective way of limiting body roll. An anti-roll bar is a torsion bar mounted across the frame, with links driving its cranked ends from the two sides of the suspension. The bar is not twisted during vertical suspension movement of both wheels, but is twisted when unequal suspension movements occur on the two sides of the car. When the body rolls in a turn, one side of the suspension travels up and the other down, thus putting a twist in the anti-roll bar. Its torsional stiffness resists twist and thus reduces the amount of rolling movement that occurs.

An anti-roll bar can be built so its stiffness can be adjusted to change the weight-transfer distribution. It is one of the most useful items for chassis tuning.

The CG of the car is above the roll axis on all conventional cars. This means that centrifugal force causes a torque about the roll axis equal to the centrifugal force *on the sprung weight* multiplied by the distance from the roll axis to the CG of the sprung weight. This torque is called *total roll couple.*

The total roll couple acts to rotate the sprung weight around the roll axis, and this motion is resisted by the roll stiffness of the suspension. Notice that the total roll couple can be resisted by either the front suspension, the rear suspension, or a combination of both.

Here is the secret of chassis tuning: **THE SUSPENSION WITH THE HIGHEST ROLL STIFFNESS WILL RECEIVE THE LARGEST PORTION OF WEIGHT TRANSFER CAUSED BY BODY ROLL.**

Because roll stiffness can be adjusted by means of the anti-roll bars, this weight transfer distribution can be adjusted to suit the needs of the chassis tuner. It can even be done without changing the roll angle of the car by increasing the roll stiffness at one end of the car and reducing it the same amount at the other end. The only thing changed by this adjustment is the weight-transfer distribution.

The portion of the total roll couple taken by one end of the car is in direct proportion to the roll stiffness at that end of the car. It can be expressed as a formula as follows:

$$\text{Front roll couple} = \frac{\text{Front roll stiffness}}{\text{Total roll stiffness}} \times \text{Total roll couple}$$

$$\text{Rear roll couple} = \frac{\text{Rear roll stiffness}}{\text{Total roll stiffness}} \times \text{Total roll couple}$$

In this formula *front roll couple* means the roll couple resisted by the front suspension, and the *rear roll couple* is the roll couple resisted by the rear suspension. Obviously the total roll couple is the sum of the front and rear roll couples.

C = CENTRIFUGAL FORCE ON SPRUNG WEIGHT
H = DISTANCE BETWEEN ROLL AXIS AND CG OF SPRUNG WEIGHT
TOTAL ROLL COUPLE = C x H

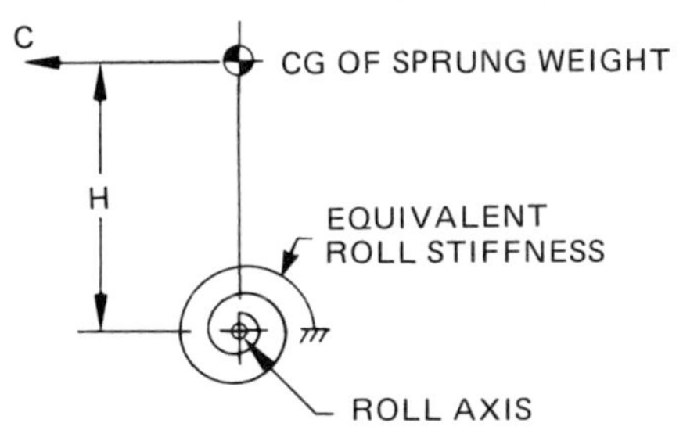

VIEW LOOKING DOWN ROLL AXIS

This is a view looking along the roll axis. The centrifugal force acting laterally on the CG of the sprung weight rolls the sprung weight around the roll axis as shown. Roll stiffness resists. The torque about the roll axis is the total roll couple.

NOTE:
This entire discussion assumes the frame is completely rigid and does not twist. A "flexy" frame will make this sort of suspension tuning impossible. See Chapter 3.

WHAT IS TORQUE AND COUPLE?

Engine people sometimes get a little vague about the meaning of the word *torque* whereas chassis people are always very specific about it. Engine folk sometimes leave the impression that torque inhabits an engine with more of it in larger-displacement engines, and it comes out mainly at low engine speeds.

If you happen to be just now getting interested in chassis tuning as an aid to laying some of that torque on the ground, it may help to say precisely what it is.

Torque is the effect of a force applied in such a way as to cause rotation. This implies a lever arm as shown in this sketch:

FORCE
H
TORQUE

The amount of torque is calculated by multiplying the amount of force by the length of the lever arm. Increasing either makes the turning effort greater. Units of torque are pound-feet, or pound-inches, sometimes written in reverse order such as inch-pounds.

An engine person might offer the opinion that a couple should drive around in a two-passenger car and then go on to tell you some of his experiences.

Chassis people, being more high-minded, define a *couple* as two equal forces in the same plane, acting in opposite directions, as in the sketch below.

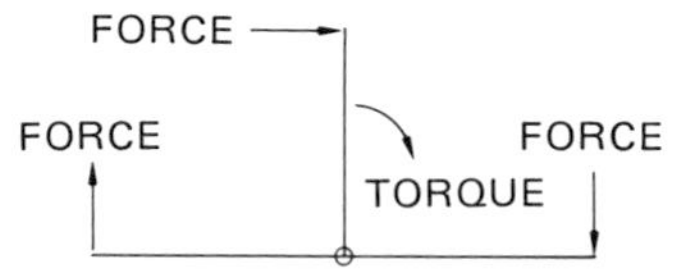

This is applicable to weight transfer of a car in a corner. Centrifugal force acting on the lever arm **h**, which is the height of the CG, causes a torque on the car. The effect is increased downforce on the outside tire and the same amount of force removed from the inside tire—in other words, a couple. Torque applied on the roll axis of a car causes a couple of forces at the tires—one up and one down.

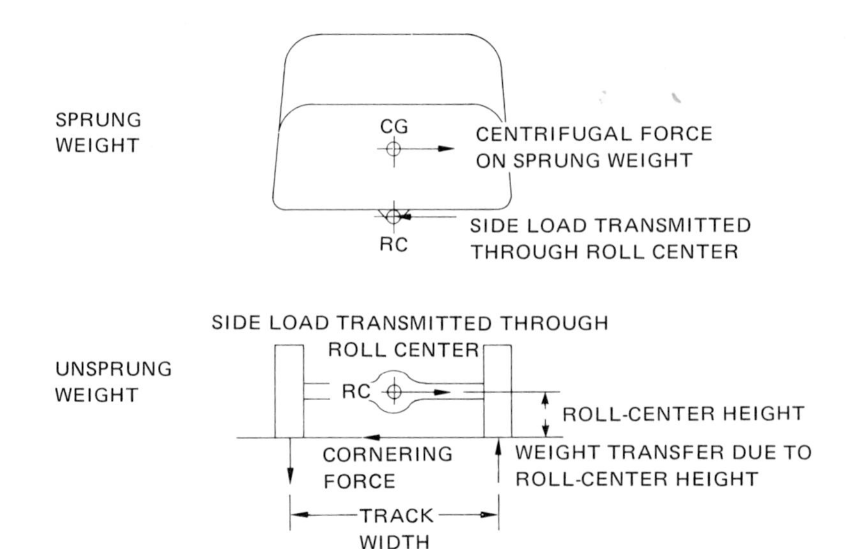

Figure 13/Side loads are transmitted from the sprung weight to the unsprung weight through the roll centers. Only one roll center is shown here, that of the rear suspension. Notice in the lower half of the drawing the side load tends to tip over the axle, causing weight transfer from one side to the other. This is weight transfer due to roll-center height.

It is possible to design a suspension with zero roll stiffness at one end of the car. This has been tried on several racing cars to get an absolute minimum of weight transfer on the rear driving wheels. In such a case the rear roll couple is zero and the total roll couple is resisted by the front suspension. A solid rear axle with a spring in the center of it accomplishes this trick, or a complex linkage design with independent suspension. If you want to try a zero roll stiffness suspension remember that it only eliminates the weight transfer *due to body roll,* and not necessarily the other portions of the total weight transfer.

The weight transfer due to body roll can be calculated for each end of the car as follows:

Front weight transfer due to body roll =

$$\frac{\textbf{Front roll couple}}{\textbf{Front track width}}$$

Rear weight transfer due to body roll =

$$\frac{\textbf{Rear roll couple}}{\textbf{Rear track width}}$$

That's all there is to it. The total weight transfer due to body roll can be found by adding the front and rear weight transfers together. As with other formulas, the roll couple is expressed in inch-pounds, the weight transfer in pounds, and the track width in inches.

Weight transfer Due To Roll-Center Heights—The roll-center heights are the second major item causing weight transfer in a corner. The roll centers can be located anywhere, but the normal height is somewhere between the axle center and the road. Remember I said that one characteristic of the roll center is that it can be considered as the point where all the side load is transferred from the sprung weight to the unsprung weight. To visualize this look at Figure 13. Here the roll center is drawn as a pivot point, as it would be on some solid-axle suspension systems. You can see how the side force acting on the unsprung weight causes weight transfer from the inside tire to the outside tire. The higher the roll center the larger this weight transfer.

The side load acting at the roll centers is a part of the centrifugal force *on the sprung weight.* It is proportioned between the front roll center and the rear roll center according to the location of the CG of the sprung weight. If the sprung weight CG is midway between the front and rear axles, the centrifugal force is divided 50/50. If the CG of the sprung weight is closer to the front axle, then the front roll center will receive the majority of the side load.

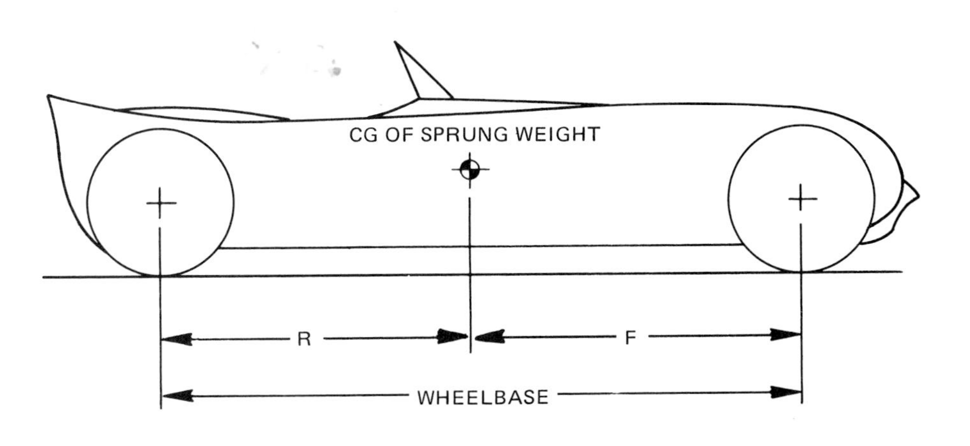

Figure 14/The formulas given in the text calculate the centrifugal force acting on both the front and rear roll centers. The roll centers are located on the vertical lines passing through the axle centers. F and R are the distances from the CG to the front and rear roll centers respectively.

To compute the portion of the centrifugal force carried by each end of the car look at the drawing in Figure 14. The location of the CG of the sprung weight must be known. The technique of finding CG location is discussed in Chapter 4. For now let's assume you already know where it is located for the sprung weight. Determine the distance from CG to the front and to the rear axles, lengths F and R in Figure 14. The formula for the centrifugal force carried at each end of the car is as follows:

Centrifugal force on front roll center =

$$\text{Sprung weight centrifugal force} \times \frac{R}{\text{Wheelbase}}$$

Centrifugal force on rear roll center =

$$\text{Sprung weight centrifugal force} \times \frac{F}{\text{Wheelbase}}$$

Because you now know the side load acting at both roll centers, the final step is to compute weight transfer due to roll-center height.

Front weight transfer due to roll-center height =

$$\frac{\text{Front CF} \times \text{front RC height}}{\text{Front track width}}$$

Rear weight transfer due to roll-center height =

$$\frac{\text{Rear CF} \times \text{rear RC height}}{\text{Rear track width}}$$

Where CF is Centrifugal Force and RC is Roll Center.

Again, the total weight transfer due to roll-center height is the sum of the front and the rear-weight transfer.

In the old days it was common to use roll-center height as an adjustment to tune the chassis. The theory is, by raising the roll-center height at one end of the car you get more weight transfer at that end. There are two problems with this and the method is no longer used.

First, raising the roll center does increase weight transfer due to roll-center height, but it *decreases* the weight transfer due to body roll. This introduces complitions into the chassis tuning process that are hard to figure out. Whenever you make two changes at once with a chassis adjustment you are in for trouble, and if the changes act in opposite directions, then you make a bad situation even worse!

The second and by far the worst problem is the almost impossible task of changing roll-center height without changing the suspension geometry radically. People used to move the upper A-arm of the double A-arm suspension, thinking it would only alter the roll-center height, but they were changing the camber characteristics and bump steer of the suspension at the same time. These multiple changes make adjusting the roll center a very risky and confusing guessing game with independent suspension. To me it seems so much easier to use adjustable anti-roll bars. They are not only easier to adjust, but the anti-roll bars give you an infinitely fine adjustment.

Weight Transfer Due To Unsprung Weight—The third portion of weight transfer in a corner is the weight transfer due to the unsprung weight. This is almost completely non-adjustable, except by a major re-design of the suspension. Even then the results would be small.

The unsprung weight has its CG roughly at the center of the axle. The centrifugal force acting at this CG causes weight transfer, and all of it caused at one end of the car acts on those two tires. Thus on a car with conventional solid-rear-axle suspension, that huge heavy axle is causing quite a bit of weight transfer to exist on the rear tires. There is no way to get rid of this except maybe to go to independent rear suspension. If this were done, the weight of the heavy differential would then be part of the sprung weight, and its weight transfer could be proportioned between the front and rear suspensions as described previously.

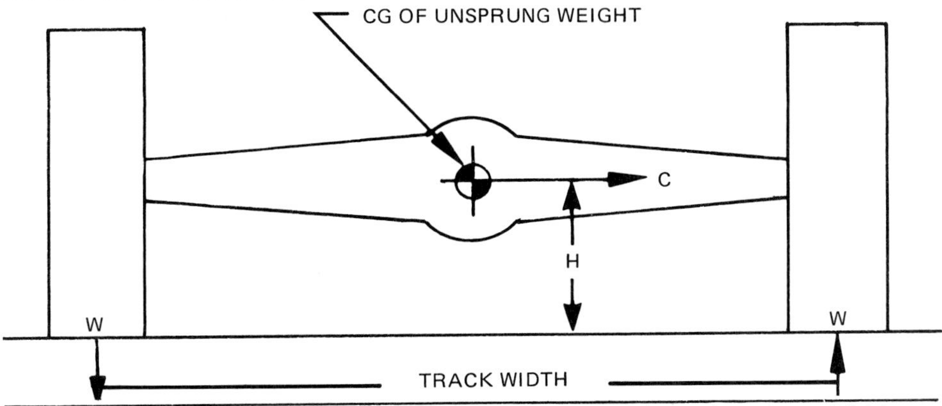

Height of the CG of the unsprung weight above the ground determines the weight transfer caused by the unsprung weight. In this drawing "C" is the centrifugal force on the unsprung weight, "H" is the CG height, and "W" is the weight transfer.

A clever zero roll-stiffness rear suspension on a Vee. The horizontal spring is compressed only by the car's weight. It moves sideways when the car rolls in a corner. Only that portion of weight transfer due to body roll is affected by this device.

This Porsche has been set up with too much roll stiffness in the front suspension, and the front tire is lifting off the road. The car would corner flatter if more roll stiffness were added to the rear suspension—and cornering speed would be improved.

Weight transfer due to the unsprung weight is computed as follows:

$$\text{Front weight transfer due to unsprung weight} = \frac{\text{Front unsprung CF} \times \text{front unsprung CG height}}{\text{Front track width}}$$

$$\text{Rear weight transfer due to unsprung weight} = \frac{\text{Rear unsprung CF} \times \text{rear unsprung CG height}}{\text{Rear track width}}$$

Now the total weight transfer at each end of the car may be found.

Total front weight transfer =
Front weight transfer due to body roll + Front weight transfer due to roll-center height + Front weight transfer due to unsprung weight.

Total rear weight transfer =
Rear weight transfer due to body roll + Rear weight transfer due to roll-center height + Rear weight transfer due to unsprung weight.

When using weight transfer distribution as a chassis-tuning method you will reach a limit. The maximum weight transfer on one end of the car can never exceed the static weight on one of the tires at that end of the car. When you try to exceed this limit of weight transfer, the inside tire transfers as much weight as it had on it without any cornering, then lifts off the road surface. Any additional weight transfer must then be carried on the opposite end of the car *no matter how that car is set up.*

Wheel lifting must be avoided for good handling. A car with one wheel off the ground is a three-wheeled vehicle, and its weight-transfer distribution cannot be adjusted in any way. It also leans an excessive amount, which usually has adverse effects on the camber. With a solid-axle suspension, lifting one side of the axle causes positive camber on the opposite tire with a resulting loss of traction.

The maximum limit to total weight transfer of the car is the total static weight carried by both inside tires. If the total weight transfer exceeds this amount both inside tires lift off the road and the car tips over. The resulting 90-degree camber angle causes total loss of control plus great fear and expense for the driver. The only solution for a car that tends to overturn is to lower the CG, widen the track, or slow down!

WEIGHT TRANSFER IN CHASSIS TUNING

The helpful feature of lateral weight transfer is its use in tuning the chassis. By adjusting the ratio of front weight transfer to rear weight transfer the car's steer characteristics can be adjusted to suit the driver.

The secret behind this chassis tuning is the fact that grip of a tire changes with vertical force on the tire. If you increase the vertical force on a tire, it distorts more and its grip is reduced.

It does have the ability to transmit more side force against the pavement, but *not in proportion to the increase in vertical force.*

The inside tire in a corner has a reduced vertical force on it due to weight transfer, and it gains some grip. However, the increased grip of the inside tire is less than the reduced grip of the outside tire. The result is a *net loss* of grip. Thus increasing weight transfer at one end of the car reduces the grip at that end of the car.

If you ignore the effect of aerodynamic forces, the following rules hold true:

1. All other factors being equal, a light car will corner faster than a heavy car.

2. A car with a small amount of lateral weight transfer will corner faster than a car with a large amount of lateral weight transfer, again all other factors being equal.

3. Increased weight transfer means reduced grip. Thus the steer characteristics can be adjusted by increasing roll stiffness of one end of the car and reducing roll stiffness at the other end.

The effect of changing the vertical force on a tire is shown in Figure 15. This graph shows the relationship between slip angle and vertical force on a tire while cornering at a lateral acceleration of 0.6 g's. At this lateral acceleration the tire is not at the limit of adhesion, so slip angle indicates distortion of the tire. The

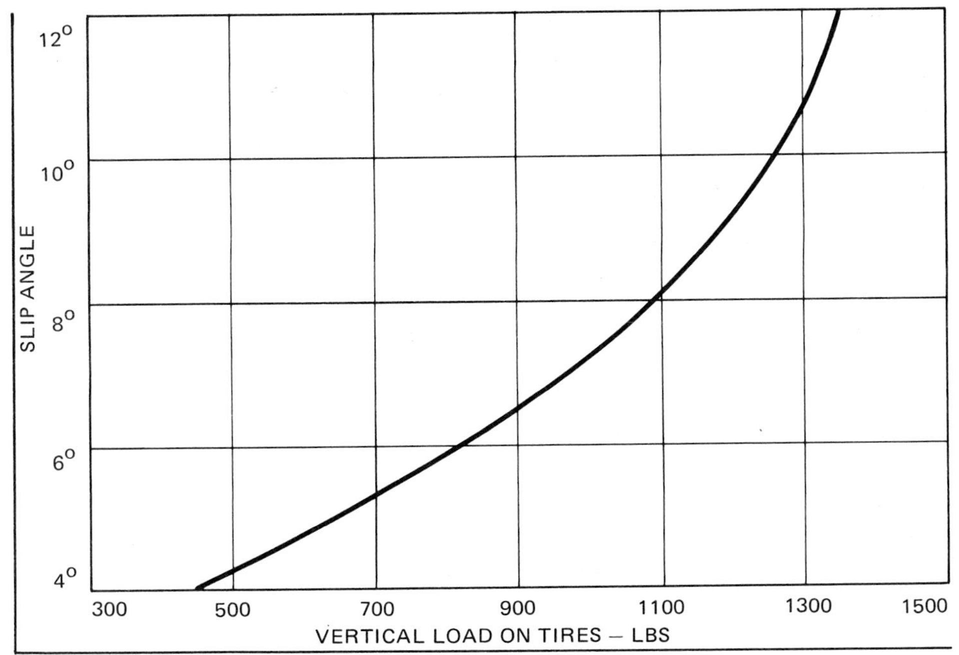

Figure 15/Here is the relationship between slip angle and vertical force on a particular tire for a lateral acceleration of 0.6 g. This same general relationship holds true at any lateral acceleration. Notice that a tire has a low slip angle if the vertical force is low, and it has a high slip angle for a high vertical force. Because low slip angles imply that the tire has more traction available, this curve shows why light cars tend to corner faster than heavy cars. The curve indicates that with lateral weight transfer on a pair of tires, the outside tire tends to lose more grip than the inside tire gains. Thus a pair of tires will have an overall loss of cornering power with increasing weight transfer. This is the basis for chassis tuning using the proportioning of weight transfer between front and rear suspensions.

low slip angles indicate small tire distortion, which to the driver means more control. If the car were pushed to higher cornering speeds, the low slip angle at 0.6 g's means the tire would allow considerably greater cornering speed before reaching the limit of adhesion.

To illustrate what weight transfer does to a pair of tires let's take an example. Assume a pair of tires support a vertical force of 2200 pounds, this means each tire supports 1100 pounds when there is no lateral weight transfer. If the car was somehow designed to have zero weight transfer at that end of the car, both tires would operate at an 8° slip angle at a lateral acceleration of 0.6 g's. This can be seen from Figure 15. Now let's assume we adjust the car to a lateral weight transfer of 200 pounds. This means that the inside tire supports a 900-pound force in the turn, and the outside tire supports 1300 pounds. Again looking at Figure 15 we find that the inside tire would operate at a slip angle of $6\frac{1}{2}°$ and the outside tire at $10\frac{3}{4}°$. This averages to a slip angle of approximately $8\frac{1}{2}°$ for the pair of tires. This is a greater average slip angle than the 8° angle corresponding to zero weight transfer. Thus increasing the lateral weight transfer increases the average slip angle for the pair of tires. This means a total increase in tire distortion, or a net loss of grip for that pair of tires.

THE EFFECT OF BODY ROLL

An excessive amount of body roll feels bad to the driver, but does it affect the cornering power of the car? The answer is a qualified *yes.*

The angle of roll of the sprung weight can affect camber on a car with independent suspension, and in this way it can affect grip of the tires. It is difficult to keep camber angles within reasonable limits under all conditions on a car that rolls a lot. For racing, roll is reduced as much as possible, usually by use of stiff anti-roll bars. For street use, a car that rolls a very small amount may handle well, but the ride usually suffers.

Another undesirable property of large angle of roll is that it uses up most of the suspension travel, and bottoming can occur easier on bumpy or banked corners.

This stock Triumph shows a lot of body roll in this slalom turn. Notice the large amount of positive camber on both outside tires, particularly the front. This car is strongly understeering, and just from looking at the photo it seems mostly due to this camber change. This car would corner faster and probably understeer less if roll stiffness were increased.

Thus you may wish to limit the roll in a turn to some minimum value.

Also there is a small amount of lateral weight transfer due to lateral motion of the CG during body roll. A large amount of body roll may add noticeably to the weight transfer, particularly on a car with a high CG.

For best handling and fun in driving, body roll should be small. Most production cars roll too much, and will handle better at reduced roll angles.

FRAME STRUCTURE

The frame is a large bracket connecting the various components of the car into one unit. It sometimes also includes the body structure. The frame must be strong enough to support the weight of the car and its contents on the worst bumps encountered. In addition the frame must be sufficiently rigid to allow the suspension to act as the designer intended.

For good handling the most important characteristic of the frame is torsional stiffness. If the suspension roll stiffness is different on the two ends of the car, the frame is subjected to torsion loading from front to rear. The torque can be as large as the total roll couple. If the frame is not stiff it will twist under the application of this torque.

Twisting of the frame modifies the lateral weight distribution between the front and rear suspensions. All efforts to use lateral weight transfer for chassis tuning will be useless if the frame does not have sufficient torsional stiffness.

A good frame should be at least 10 times as stiff as the roll stiffness of the suspension. This is measured in inch-pounds per degree of twist, just as roll stiffness is measured. Frame torsional stiffness is measured by clamping one end of the frame and applying a known torque to the opposite end. The angle of twist between front and rear of the frame is measured. The frame torsional stiffness equals the applied torque divided by the measured angle of twist.

There are three basic types of frame structure, twin rail, space frame, and monocoque. The frame materials can be steel, aluminum, reinforced plastic, wood, or exotic metals such as *nowatium* and *unobtanium.* Most production and racing cars use a steel frame because it is a low cost and very stiff material. Other materials are rarely used on production car frames, but more often on racing cars.

The twin-rail frame is the oldest type, dating back to the horse and buggy. It consists of two longitudinal members, either tubes or open sections, usually of steel. These carry the bending loads between the front and rear suspension. Large brackets tie on all the necessary components, and cross members are used for additional strength at several points along the frame. Torsional stiffness is very poor with this type of frame, as only the twisting or bending of the individual frame rails can resist torsion. This type of frame is very simple and easy to build, but it suffers from lack of stiffness, and should not be seriously considered for a track or road racing car. It works reasonably well in drag racing where cornering is not too important. Many road cars use this type of frame, but most derive some extra stiffness from the heavy body structure.

This racing car illustrates the twin-rail frame—old-fashioned but still common on many production cars. It has very low torsional stiffness, and should be beefed up for better handling. On production cars the body can be used to stiffen the frame.

This space frame is fully triangulated on the top. There are a large number of tubes in this frame, which adds weight and cost. The best design is one with the minimum number of tubes, but still fully triangulated.

The space frame is seldom used on road cars due to high manufacturing cost. It has been the popular design for road racing cars for the past 20 years. It consists of a network of relatively slender tubes, arranged in a large three-dimensional framework surrounding the majority of the components of the car. In a good space frame, the basic geometrical form is a triangle. There are no rectangular or trapazoidal planes in the structure along the outside surfaces of the frame. This is said to be fully *triangulated.* Loads in a space frame are carried down the center of the tubes, providing the centerlines of tubes always meet at a point where they are joined. If the centerlines do not meet at a point, bending in the tubes is introduced, with a resulting loss of both strength and stiffness.

All loads from components attached to a space frame must be introduced at the joints. This is a very important aspect of design. A bracket in the middle of a frame tube is a sure way to promote frame failure.

Space frames are welded or brazed together, and have been made of steel, aluminum, or magnesium. Various alloys are used, depending on the application. Most racing cars, particularly those built in England, are made of low-carbon mild steel tubing, brazed together. This material is the lowest in cost, but it also is the least efficient on a strength-to-weight basis. Strength is the ability of a material to withstand a load without damage. Damage can be either fracture or permanent deformation. Mild steel is a good material for a car with a minimum weight restriction in the class.

The stiffness-to-weight ratio for mild steel is as good as any other material, and it is cheap. It also has the desirable property of being very ductile, so repairs in the field can sometimes be made by bending damaged tubes back into shape.

If high strength-to-weight ratio is desired of a space frame, other materials are used. Alloy steels such as 4130 chrome moly offer much higher strength for equal weight and stiffness compared to mild steel. It is more expensive and requires expert welding technique to get good joints. Heliarc TIG welding should be used with chrome moly and other exotic materials to get the best possible welds. Alloy steels must be stress-relieved after welding, and they can also be heat treated to a desired strength level greater than the as-welded condition.

Higher strength implies less ductility, so the compromise is between lightest weight and the ability to absorb overloading in the event of an accident. A material with low ductility will fracture if overstressed, whereas a material with high ductility will stretch, bend, or crumple without fracturing.

Aluminum or magnesium tubing is sometimes used in space frames. These have application on very light cars, where the theoretically optimum steel tubing would be so thin it can't be successfully welded. Then aluminum or magnesium offers a weight savings. On loaded members where the steel is not at minimum practical wall thickness, alloy steel offers

This is a monocoque frame on a racing car, made of sheet aluminum riveted to tubular welded bulkheads. On this car the lower portion of the frame serves as the body and fuel tanks. Notice the anti-roll bar (arrow) on the front suspension.

a higher strength-to-weight ratio.

A combination of materials provides the lightest possible design. Say aluminum for the frame and alloy steel for the highly stressed suspension parts. Some of the best racing cars in the world are designed this way. Aluminum requires expert welding, and magnesium is even more difficult. Fabrication of these materials is only for experts.

Monocoque frames were developed in the aircraft industry. The loads are carried in a stressed skin, forming the outside of the structure. Where the basic geometrical unit of a space frame is a triangle, the monocoque frame uses a sheet, usually rectangular or trapezoidal in shape. While the loads in a space frame are tension or compression, the loads in a sheet are carried in shear in the plane of the sheet. To form a stiff unit, the sheets must form a closed box. This is vital if any stiffness is to be achieved. You can easily visualize how a monocoque frame works by folding up a square box out of paper. Twist the box in torsion from end to end as you are building it, and see how the stiffness dramatically increases as you

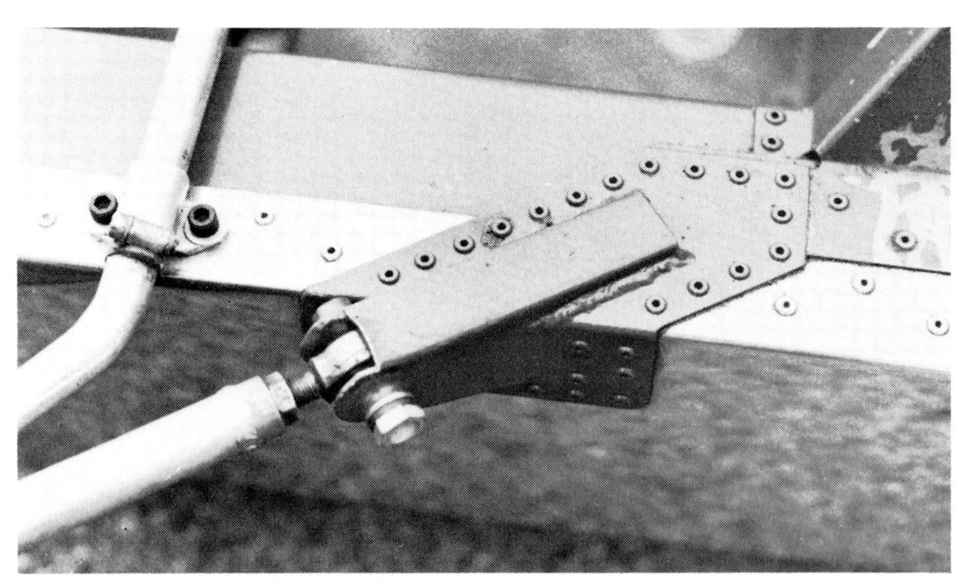

This suspension bracket ties into the corner of the monocoque frame. Loads must be introduced into a monocoque structure at corners where the panels join, and a large number of small rivets must be used to spread the load over a large area. The design of brackets is one of the more troublesome parts of monocoque frame design.

tape the final end, forming a closed box. In a car, any opening in the box structure must be heavily reinforced to be as stiff as the missing panel. This is usually done with welded bulkheads or by edge members surrounding a cut-out.

The monocoque frame is very common on production cars. Here it is known as *unit construction.* Frame and body is a single welded sheet-steel unit. This is a relatively light and stiff frame for a mass-produced car. The disadvantages are difficult repairs and noise transmission from the running gear. Rather than monocoque construction, rail frames are used on some sedans to silence the car. By mounting the separate body on the frame rails with rubber pads, noise is isolated from the passengers. The monocoque frame is superior for good handling due to its stiffness.

On all frames a loss of stiffness occurs in the center of the car where the doors are. Open cars are much worse than sedans, because the body top adds considerable torsional stiffness. On a convertible with a rail frame it is common for the frame to be beefed up with an X-member in the center. Thus convertible frames are preferred for stock car racing classes which require a rail frame.

AERODYNAMICS

The handling of a car can be greatly affected by body aerodynamics. This has a small effect at legal road speeds and becomes increasingly important at higher speeds. It should be noted that high head winds and cross winds encountered during highway driving can be quite similar to those encountered at racing speeds. On racing cars aerodynamics is a major factor in determining the steer characteristics and the cornering speed of the car.

Aerodynamic forces on a car are *drag, downforce, yaw* and *lift.* Drag acts toward the rear and tends to slow the car. Drag exists on all car bodies. It increases with the square of the speed. Drag affects handling. The amount of the effect is largely determined by the center of pressure. The higher the center of pressure, the more the front wheels are unloaded and understeer results. An exaggerated example of this is a Top Fuel Dragster using a wing, as shown in the accompanying sketch. The tradeoff between drag and downforce is important in tuning a racing-car suspension. Drag determines the car's top speed, and can be as important as the engine power. For road use, drag has a big effect on fuel economy. A streamlined car has low drag, and will get good gas mileage.

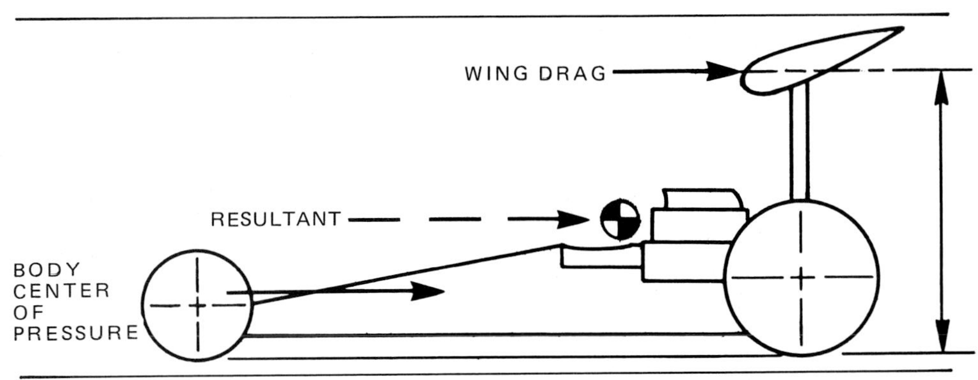

Drag force on the wing changes the resultant total vehicle center of pressure by raising it, thereby unloading the front wheels.

Downforce directly affects the handling. It is a downward force on the body, adding to the vertical force due to gravity. Downforce adds traction in acceleration, braking, and cornering. Because aerodynamic downforce doesn't add centrifugal force in a turn, the effect on the car is higher cornering speeds. Downforce feels the same to the driver as increasing the grip of the tires, but it acts strongly only at high speeds. It increases with the square of speed.

The opposite of downforce is lift. This is an upward force on the body. It tends to make any car handle poorly at high speeds, so lift should be avoided if at all possible. Many road cars develop lift rather than downforce due to the body shape. The VW beetle is a classic example of this, developing severe nose lift at high speeds. The body shape—flat bottom and a rounded top—is largely responsible for the lift on the car and very little can be done to correct it.

Aerodynamics may change the handling characteristics with speed. On many cars the lift gets so bad at high speeds that the steer characteristics change radically. Many cars understeer at low speeds and oversteer at high speeds. This is not a desirable characteristic in any car, and should be changed if racing is considered.

All sorts of devices are used on racing cars to generate downforce and kill lift. Racing cars have low noses to prevent air from passing under the body which tends to create a high pressure between body and road, causing lift. In addition to a low nose, the racing car often has less ground clearance at the front than at the rear. This adds downforce on the entire body, even on a stock sedan body.

Another item used on racing cars to kill lift is the spoiler. They are placed on the front of the body to prevent air from passing under the car. Nose spoilers are common on production bodies which have been modified for racing. At the rear of a racing car a spoiler is mounted on the top of the body. This deflects air upward which adds to the downforce, or reduces lift. These devices are used extensively on cars which cannot use wings in their racing class.

It is common practice to use wings or ground-effect devices on racing cars for increased downforce. The only racing cars that don't benefit from these devices are land-speed record vehicles where minimum drag is the most important consideration. The choice of which aerodynamic device is used for downforce is usually determined by the governing rules—all or none may be used.

Race cars should be designed so their aerodynamic devices can be adjusted when it is desirable to change steer characteristics. Traction can be increased at either end of a car by adding downforce at that end. Adjusting handling with aerodynamic devices is discussed in later chapters.

Funny cars are very prone to aerodynamic problems and spectacular accidents because they are fast, light, and have a large body area. Notice the front and rear spoilers and the extreme body rake. All of these devices usually keep the car glued to the road at 220 mph.

This Chevron B-sports racer has a low nose to prevent air from building up pressure underneath it. The plate below the radiator air intake provides a little extra air flow to the radiator, thus compensating for the air consumed by the frog.

Formula-Ford rules don't permit wings or ground-effect skirts, so an "engine cover" is designed to provide some downforce. This is typical of an aerodynamic design dictated by racing rules. This Crossle, driven by Ed Mertz, also features a streamlined head fairing to provide smooth air flow over the "engine cover." (Tom Monroe photo.)

First step in suspension tuning is to wash the grime off the parts. A trip to the local car wash is a quick way to accomplish this. It is particularly easy if the body comes off the car as on this sports-racer. Don't forget to protect critical parts from water, and lube the car after washing to make sure water isn't trapped inside moving parts.

To check for wear in the steering linkage jack the car up and shake the wheel from side to side as shown. There will be some play in the steering system, but if the tire wiggles more than 1/8" look for worn or misadjusted parts.

Making your car handle is a step-by-step process much like making your engine put out more power. The first step is tuning. In tuning your suspension you set each adjustment to the position you or the manufacturer *think* is best. Then take the car out and test it so you can make final adjustments. Keep in mind that tuning the suspension is a vital part of making your car handle. Take plenty of time to do the job right. With a few tools, suspension tuning can be done in your own garage.

As in working with any part of a machine, first check for broken or worn-out parts. The suspension of a car sometimes gets no care or inspection, and often a handling problem is caused by a part needing repair. To make the job easier, first clean the underside of the car. This can be done in several ways; steam cleaning, elbow grease with solvent, a high pressure solvent gun, or the local do-it-yourself car wash. Wear goggles to protect your eyes. Whatever the method used, a clean underside will pay off later in working on the car.

With the car ready to work on, jack it up and check the suspension for play. Grab a tire with one hand at the front and one at the rear and shake the tire in a steering direction. It should not have slop or play. Then check for play in the other direction by grabbing the tire on top and bottom and shaking it in a cambering direction. Again there should be no play.

If it rattles back and forth a noticeable amount, find out where the movement is coming from. It may be a combination of parts in the suspension, so check all possibilities. Here a friend is of great help in wiggling the wheel while you check the cause of the movement. The play could be in wheel bearings, ball joints, suspension pivots, or steering joints. Once you locate the trouble, fix the parts to as-new condition. Sloppy suspension is impossible to tune properly and a waste of time to work on.

You may find some play in the steering system. You can adjust most steering systems to get rid of play at the steering box with shims or by an external adjustment. Consult your workshop manual. Running with defective or questionable steering is dangerous to you and your fellow drivers. Also, bad steering will make the car handle badly in almost every case. If your steering has worn out beyond repair, replace worn parts with new ones. It pays off in driving pleasure and peace of mind.

When looking under the car, check the springs for breaks. Coil springs sometimes break near the end and wedge in place without obvious indication of the failure. A careful look will locate any breaks. A leaf spring may contain a broken leaf and still hold the car up. Look carefully along the edge of each leaf for cracks. Better yet, disassemble the entire spring. Also check out the rubber bushings in the suspension and replace any that are cracking or are questionable. These are low-cost parts, but they make a big difference.

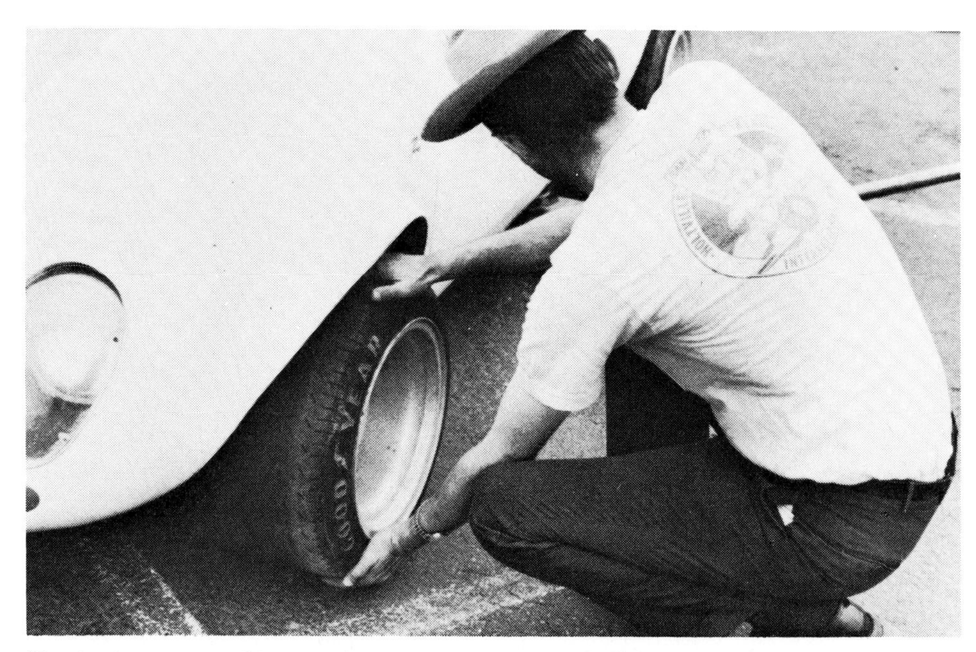

Play in the upper and lower suspension pivots is checked by attempting to move the tire in a camber direction. Play in this direction could be in either the steering pivots or in the suspension linkage pivots on the frame. Loose wheel bearings will also be noticeable with this test.

The adjustment for steering slop is sometimes right on the box. The adjuster on this Datsun steering is the screw with the locknut, sticking out of the top cover plate. Rack and pinion steering often has adjustments by means of internal shims.

The lower rubber washer on this anti-roll bar link is in bad shape—cracks are visible on close inspection. Some rubber bushings in the suspension are not that easy to inspect, and must be removed for close inspection.

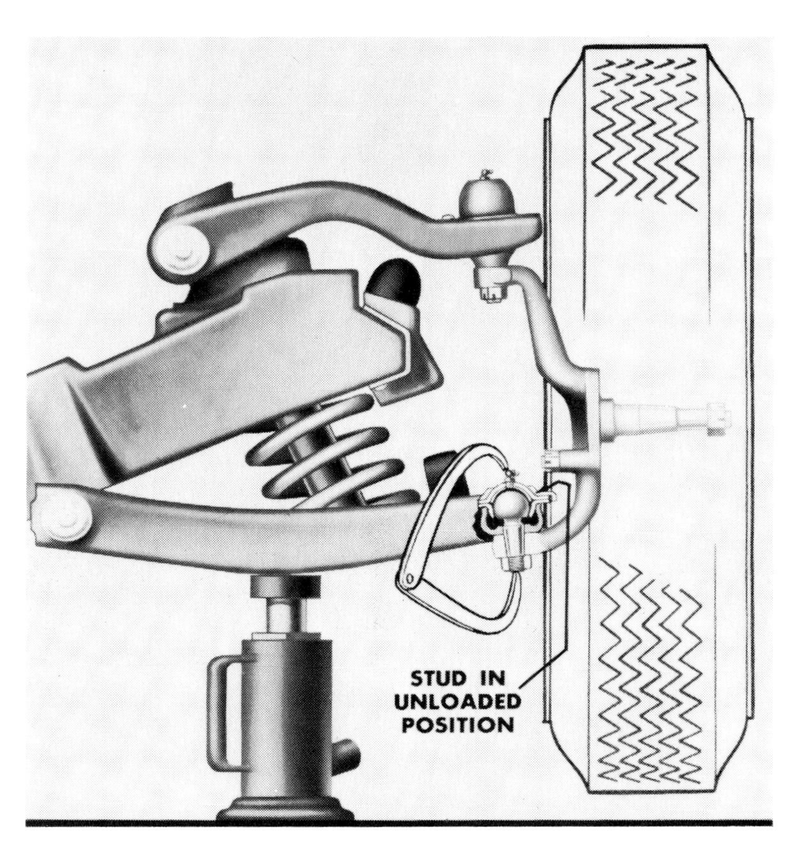

To measure ball-joint wear, support the suspension as shown—depending on whether the coil spring is mounted on the lower or upper control arm. (Drawings courtesy of TRW Inc.)

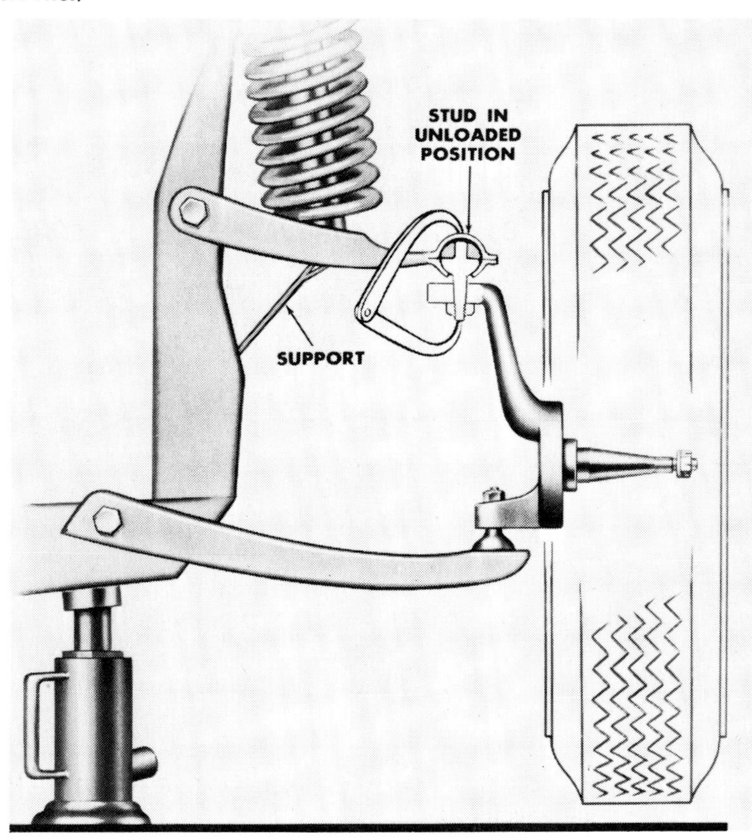

On many cars, just shaking the wheels will not test for worn-out ball joints in the front end. The accompanying drawings and Figure 16 show you the proper way to test ball joints for wear. To do it you will need a jack and some instrument to measure movement—such as a caliper or a dial indicator. Specs in Figure 16 call for an acceptable amount of play. Even new ball joints have play in them. If they didn't they couldn't move freely. For best handling you must be sure the play is *below* the maximum allowed.

The load-carrying ball joint is the one which is loaded by the spring. In the upper drawing that is the lower ball joint. To measure clearance in the ball joint the jack is placed under the lower A-arm. This allows the ball to drop to the lowest point in its housing. The caliper measurement shown is compared to the caliper measurement taken with the weight of the car on the ball joint. The difference between these two measurements is the vertical play in that ball joint.

For cars with springs above the upper A-arm, the jack is placed under the frame as shown in the lower drawing. This again allows the ball joint to drop in its housing. A measurement in this position is compared to one with the car's weight resting on the ball joint. The difference is the vertical play.

Some ball joints have wear indicators which eliminate the need to jack up the car. On these you just take a measurement off the wear indicator as shown in Figure 16. If your car is not listed in Figure 16, consult your shop manual or a book on front end alignment for correct specs on ball joints.

While inspecting the car for worn parts take a good look at the shocks. If you see any sign of fluid leaking replace the shocks. Also check for rust on the shaft where it slides through the seal. This can also cause fluid loss. Bounce the suspension up and down and see how many bounces it takes to come to rest. If more than one full bounce, the shocks are either worn out or they are too soft for high-performance handling. Replace them with a good set. Various types of replacement shocks are discussed in Chapter 4.

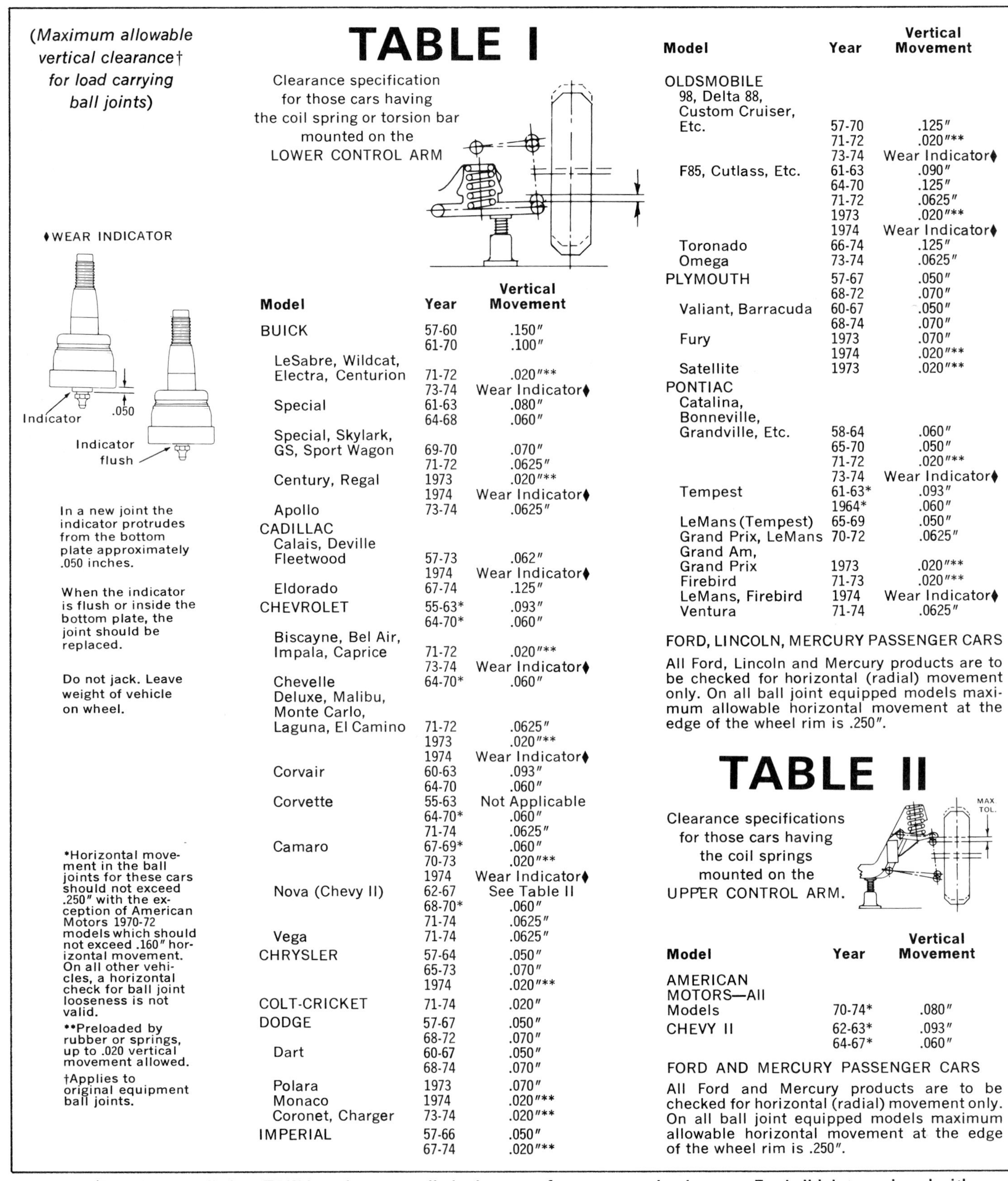

(Maximum allowable vertical clearance† for load carrying ball joints)

♦WEAR INDICATOR

In a new joint the indicator protrudes from the bottom plate approximately .050 inches.

When the indicator is flush or inside the bottom plate, the joint should be replaced.

Do not jack. Leave weight of vehicle on wheel.

TABLE I

Clearance specification for those cars having the coil spring or torsion bar mounted on the LOWER CONTROL ARM

Model	Year	Vertical Movement
BUICK	57-60	.150″
	61-70	.100″
LeSabre, Wildcat, Electra, Centurion	71-72	.020″**
	73-74	Wear Indicator♦
Special	61-63	.080″
	64-68	.060″
Special, Skylark, GS, Sport Wagon	69-70	.070″
	71-72	.0625″
Century, Regal	1973	.020″**
	1974	Wear Indicator♦
Apollo	73-74	.0625″
CADILLAC		
Calais, Deville Fleetwood	57-73	.062″
	1974	Wear Indicator♦
Eldorado	67-74	.125″
CHEVROLET	55-63*	.093″
	64-70*	.060″
Biscayne, Bel Air, Impala, Caprice	71-72	.020″**
	73-74	Wear Indicator♦
Chevelle	64-70*	.060″
Deluxe, Malibu, Monte Carlo, Laguna, El Camino	71-72	.0625″
	1973	.020″**
	1974	Wear Indicator♦
Corvair	60-63	.093″
	64-70	.060″
Corvette	55-63	Not Applicable
	64-70*	.060″
	71-74	.0625″
Camaro	67-69*	.060″
	70-73	.020″**
	1974	Wear Indicator♦
Nova (Chevy II)	62-67	See Table II
	68-70*	.060″
	71-74	.0625″
Vega	71-74	.0625″
CHRYSLER	57-64	.050″
	65-73	.070″
	1974	.020″**
COLT-CRICKET	71-74	.020″
DODGE	57-67	.050″
	68-72	.070″
Dart	60-67	.050″
	68-74	.070″
Polara	1973	.070″
Monaco	1974	.020″**
Coronet, Charger	73-74	.020″**
IMPERIAL	57-66	.050″
	67-74	.020″**
OLDSMOBILE		
98, Delta 88, Custom Cruiser, Etc.	57-70	.125″
	71-72	.020″**
	73-74	Wear Indicator♦
F85, Cutlass, Etc.	61-63	.090″
	64-70	.125″
	71-72	.0625″
	1973	.020″**
	1974	Wear Indicator♦
Toronado	66-74	.125″
Omega	73-74	.0625″
PLYMOUTH	57-67	.050″
	68-72	.070″
Valiant, Barracuda	60-67	.050″
	68-74	.070″
Fury	1973	.070″
	1974	.020″**
Satellite	1973	.020″**
PONTIAC		
Catalina, Bonneville, Grandville, Etc.	58-64	.060″
	65-70	.050″
	71-72	.020″**
	73-74	Wear Indicator♦
Tempest	61-63*	.093″
	1964*	.060″
LeMans (Tempest)	65-69	.050″
Grand Prix, LeMans	70-72	.0625″
Grand Am, Grand Prix	1973	.020″**
Firebird	71-73	.020″**
LeMans, Firebird	1974	Wear Indicator♦
Ventura	71-74	.0625″

FORD, LINCOLN, MERCURY PASSENGER CARS

All Ford, Lincoln and Mercury products are to be checked for horizontal (radial) movement only. On all ball joint equipped models maximum allowable horizontal movement at the edge of the wheel rim is .250″.

TABLE II

Clearance specifications for those cars having the coil springs mounted on the UPPER CONTROL ARM.

Model	Year	Vertical Movement
AMERICAN MOTORS—All Models	70-74*	.080″
CHEVY II	62-63*	.093″
	64-67*	.060″

FORD AND MERCURY PASSENGER CARS

All Ford and Mercury products are to be checked for horizontal (radial) movement only. On all ball joint equipped models maximum allowable horizontal movement at the edge of the wheel rim is .250″.

*Horizontal movement in the ball joints for these cars should not exceed .250″ with the exception of American Motors 1970-72 models which should not exceed .160″ horizontal movement. On all other vehicles, a horizontal check for ball joint looseness is not valid.

**Preloaded by rubber or springs, up to .020 vertical movement allowed.

†Applies to original equipment ball joints.

Figure 16/This data supplied by TRW Inc. shows wear-limit clearances for recent production cars. For ball joints equipped with wear indicators, condition is indicated by protrusion of the indicator shown by drawings at left. If your car is not listed on this chart, get data from manufacturer or a reliable source.

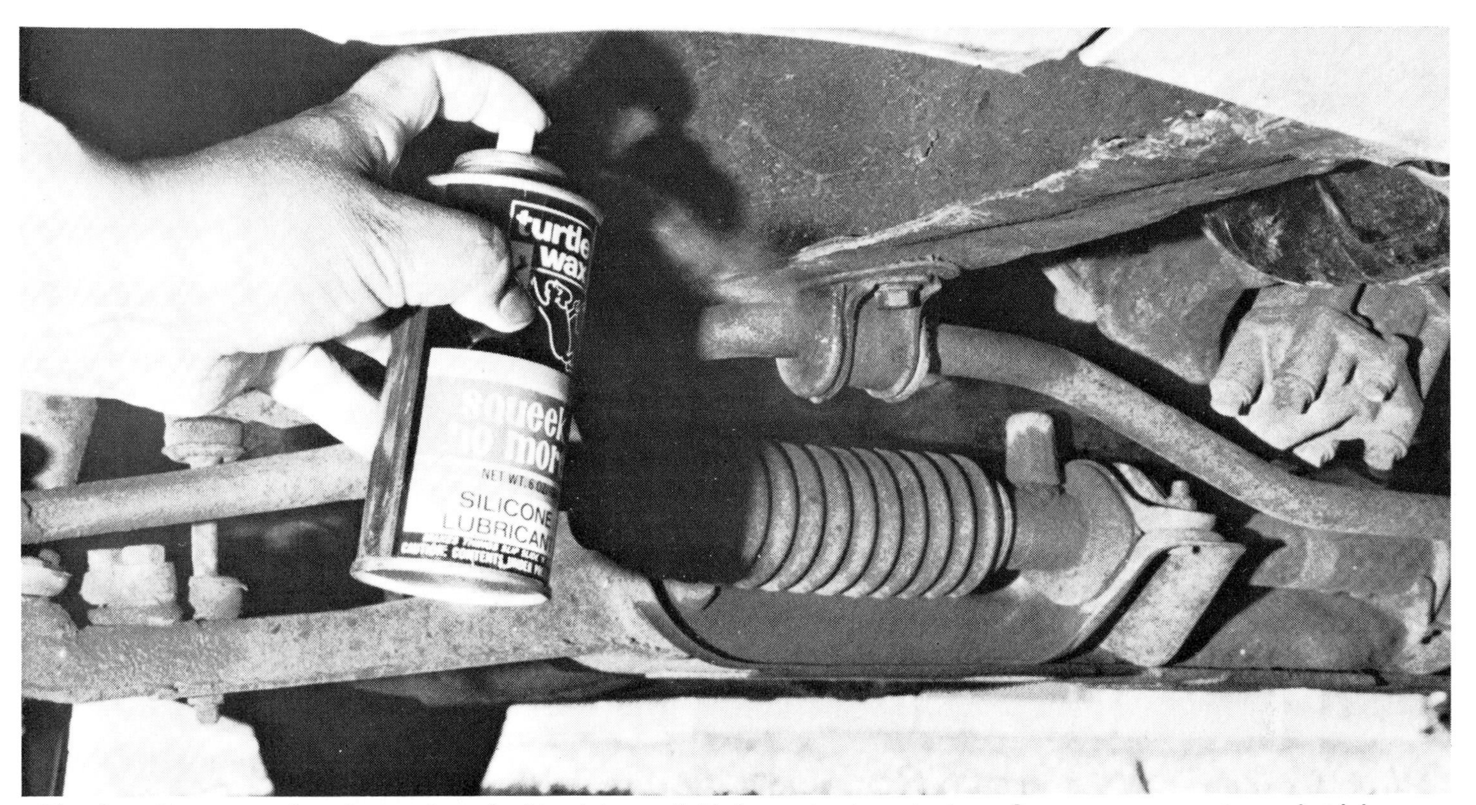

Give the rubber suspension pivots a shot of rubber lube, available in most auto parts stores. Spray cans are easy to use, but lube can also be applied with a pump-type oil can. During the lube job don't forget to inspect rubber parts for cracks.

After checking the suspension, lubricate all moving parts. It's amazing how a good lube job will help the handling of a car by freeing up the suspension. The steering and suspension pivots will usually have grease fittings on them, and these should be wiped free of dirt before using the grease gun. Consult the manufacturer's literature for proper lubricants, or use general purpose chassis grease if information is lacking. Rubber pivots can be lubricated with any of several brands of rubber lube. This will often eliminate a squeaking sound in the suspension. After lubing, move the steering back and forth to work the grease into the parts, and bounce the car on its wheels to do the same for the suspension pivots. Now you are ready to start the tuning job.

A few general observations about suspension tuning are in order. Unless you have a definite reason to do otherwise, always follow the manufacturer's recommendations to the letter. He knows about your car, and won't give you wrong information. You may wish to deviate for some specific reason, but always use the stock settings if in doubt. If you have no stock settings for a special car, start out with settings for a car similar in design.

SUSPENSION INFORMATION SHEET

TIRES						
Size	Front	________	Rear	________		
Manufacturer		________		________		
Compound		________		________		
WHEELS						
Rim & Diameter	Front	________	Rear	________		
SPRINGS						
Rate	Front	________	Rear	________		
TIRE PRESSURE						
Cold	Front	________	Rear	________		
Hot		________		________		
CASTER						
	Front	________	Rear	________		
CAMBER						
	Front	________	Rear	________		
TOE IN						
	Front	________	Rear	________		
RIDE HEIGHT						
Measurement	Front	________	Rear	________		
Reference Points		________		________		
Ground Clearance		________				
Critical Item		________				
CONFIGURATION						
Passengers & Weights		________				
Fuel Load		________				
Other Fluids		________				
WEIGHT DISTRIBUTION						
	LF	________	RF	________		
	LR	________	RR	________		
Total Weight		________				
BRAKES						
Balance Setting		________				
Lining Material	Front	________	Rear	________		
ANTI-ROLL BARS						
Diameter	Front	________	Rear	________		
Settings		________		________		
SHOCKS						
Type	Front	________	Rear	________		
Setting		________		________		
TIRE TEMPERATURES						
Left Front	Inside	________	Center	________	Outside	________
Right Front		________		________		________
Left Rear		________		________		________
Right Rear		________		________		________
TRACK TEMPERATURE						
Air Temperature		________				
Pavement Temperature		________				

COMMENTS:

An organized way to record suspension set-up information is essential. You may reproduce this sheet or make one of your own.

In tuning the suspension it pays to double-check everything. After you have set the various adjustments, go back and check those you worked on first. Sometimes one adjustment affects a previous one, so the final tune is something of a trial-and-error process.

One item you will need is a notebook to record all the settings. Experts use them, so don't feel like you are admitting a lack of talent in writing everything down. If you trust to memory you will certainly have to do extra work later, and you can work more relaxed by not having all those figures in your head. Keep the notebook with you when working on the car. Write down all changes you make and driver comments if any. A form for this purpose is very useful, and a typical one is reproduced here. You can type or write up a similar one or have this one Xeroxed or printed for small cost.

Inflate the tires to the pressures you expect to run. It is not critical that the pressures be exactly the right amount, but it is important that the tires at each end of the car have the same pressure. The front pressure can be different from the rear. If you don't own a good quality tire gage, now is the time to invest in one. The right tire pressure is all-important in the handling of every car.

Because many measurements require the car sitting level, check the garage floor with a level and a long straightedge. If it is tilted or wavy, all is not lost. You can set up four pads to put the tires on for a level reference surface. Squares of masonite or plywood work well for these leveling pads. The straightedge resting on *any* two of the pads should be level. After checking this out, use a paint spray can to mark the pads and the floor so you can quickly set up again. Remember to account for the thickness of the leveling pads when taking measurements from the floor.

RIDE HEIGHT

After all the preliminaries, start tuning the suspension by setting the ride height. Ride height is the distance from the frame to the ground. It can be different in front and rear of the car, and is not the same thing as *minimum ground clearance.* Minimum ground clearance is the distance

Racing mechanic Dave Mundy prepares for suspension tuning by balancing each tire and inflating to the usual operating pressure. Both are critical and should be checked frequently. The balancing tool used here is a simple static balancer. This works, but a more precise job is dynamically balancing by spinning the tires either on the car or on a wheel-balancing machine.

Before starting suspension alignment the floor must be checked for level. This floor had a low spot in it, so a square of masonite is used as a pad to put the wheel on. The pads for the four wheels are checked to each other by placing a long level between them. All four pads should level with each other.

A convenient place to measure the ride height of the 1975 Pinto is under the rocker panel just in front of the rear wheel. There is a convenient flat surface on the bottom of the body at this location. It's easy to see the scale without having to crawl under the car. This car has a ride height of 9 inches. Notice the muffler in the background is closer to the road.

between the *lowest part* on the car and the ground. The minimum ground clearance can be anywhere on the car, and it is often the exhaust system, some other bolted-on item, or the bottom of the differential housing.

It is important to measure the ride height with the car in the same condition as it is driven. This includes the weight of all liquids and the weight of the occupants. Use ballast to simulate the driver's weight unless you have a spare body to sit in the car while you are taking all measurements. Use an average fuel load, say half a tank, and write it in your notebook.

The best spot to measure ride height is the *lower suspension pivot* point on the frame, but any convenient point on the frame can be used. Pick a spot that is easy to measure, preferably near the wheels at the front and rear of the car. On a car with a belly pan, it is usual to measure the ride height from the belly pan to the ground. On cars with a tubular frame, the bottom of the lower frame-member is a convenient spot. Whatever locations you choose to measure ride height from, mark them in the notebook so you can repeat the measurement in the future. Also punch mark the frame if necessary to identify the locations. Double check from the suspension pivot or leaf spring eye to the ground, to make sure the frame is in the same relationship to the ground as the suspension pivot points. Cars are often built with sloppy tolerances and the distance from the suspension pivot to the bottom of the frame may not be the same on both sides. The important reference point for ride height is the suspension pivot.

On many production cars, the place to measure ride height is specified in the factory manual. Try to follow the manufacturer's specifications whenever possible, deviating only for a definite reason.

On many sports cars and racing cars, the driver and passenger are the heaviest items in the car, so their weight greatly affects the ride height. If the car is a sports car or sedan, the driver is off-center and the passenger seat may or may not be occupied. You must decide what condition you want the car set up for. Anything else will compromise performance. For *best* handling the ride height should be the same on both sides of the car.

For racing, set the car up in the as-raced condition. If a passenger is used, ask or estimate his weight and put it in the seat. For a road car a good compromise is half the weight of a passenger. At times, say for rally use where you want the ultimate handling, you may want to set up for the passenger aboard, and let the car handle not quite as well when you are alone driving to work. Whatever compromise you make, be sure to write the weights in your suspension notebook.

Now that you are about to set ride height, what figures should you use? The safest place to start is the factory specs for a car driven on the road. A road car has a ride height designed to give adequate clearance over bumps and dips encountered in average use. If the car has stiffer springs you will probably wish to lower the ride height. Reduced ride height lowers the CG of the car, resulting in less weight transfer during cornering acceleration, and braking. Less weight transfer produces more traction and better handling. You should always use the *lowest possible* ride height for best cornering, with practical limits as described in the following pages.

The obvious limit to a reduced ride height is the minimum ground clearance that results. Insufficient ground clearance will result in the car hitting the ground going over bumps, dips, driveways, or in racing around banked corners. Look under the car and see what limits ground clearance. Check the exhaust system and any other plumbing or systems hanging under the frame. See if any bolts hang down to limit the ground clearance. Think about re-routing lines, raising exhaust pipes, and cutting off bolts to get more ground clearance. If you find fuel lines or electrical wires under the frame, move them so they cannot be scraped on the ground. Such an event can be a disaster.

The part that is most likely to cause a ground-clearance problem is the sprung weight. As the car moves on its suspension the sprung weight gets much closer to the ground than the lowest part of the unsprung weight. Thus the muffler is much more likely to hit the ground than the bottom of the differential.

Also the center of the wheelbase and the extreme ends of the body are more likely to hit the ground than parts near the axles. This will vary with the shape of the bump and the road contour. There is no one part of the car that is always critical, so you should check the following areas:

The ground clearance of this 1975 Pinto is 5 3/4 inches. The lowest point on the car is the muffler. If this car is lowered for better handling, the ground clearance can be restored by modifying the exhaust system. The muffler can be raised closer to the bottom of the body, but heat insulation may be required below the floor. Watch for ground clearance problems with items hanging below the chassis of the car.

The part that is the minimum ground clearance, no matter where it is located on the car.
The lowest part of the sprung weight.
A low part near the center of the wheelbase.
The extreme front and rear of the body, wherever they are closest to the ground.

One of these areas will determine how low the ride height can be, so check them carefully for signs of previous bottoming against the ground. If none has ever hit you might try a little test.

To test which part of the car is critical for ground clearance, tape a small spacer to the bottom of the car at each suspected critical point. A stack of cardboard or rubber pad will work fine. Select a uniform thickness for all the spacers—say one inch. Give the car a severe workout driving over the worst bumps and dips expected. If a one-inch block barely hits in the extreme conditions, you know you could lower the ride height by one inch and have no severe problems—unless the critical part happens to be the oil pan. If this vital part hits the road you are likely to regret it. I suggest at least an extra inch of clearance under the pan.

If any of the spacers are torn off completely, severe bottoming occurred, so reduce the spacer thickness and try again. Don't tape cardboard or other flammable spacers to the muffler. Use something that won't burn, such as a stack of asbestos sheets. Hold it on with wire instead of tape.

One compromise you can make is to allow the car to bottom on the most severe bumps. Some bottoming may not be severe enough to require raising the ride height. One practice has been to install skid pads on all four corners of the "tub" to prevent damage to monocoque-constructed cars.

If you are building a special car or modifying one, you may want to know what ground clearance other people use. The chart in Figure 17 shows ground clearances on a group of production cars. These are stock settings, as reported in *Road and Track* magazine road tests. For road racing you should reduce these figures and for off-road use they must be increased. Notice that most road cars have a ground clearance of from 5 to 7 inches. Road racing cars on smooth tracks can use as little as 2 1/2 inches, but this clearance would cause severe problems on the street and driveways.

On some cars, particularly those using oversized tires, the limit to lowering ride height is clearance between tires and fenders. The problem here is that when the car hits a bump, the tire will strike the underside of the fender before the suspension hits the bump stops. Unless you want to make new fenders—or have the tires do it for you—check this out by measuring the distance between the top of the tire and the underside of the fender at full bump. Do this by removing the springs and letting the car down on the suspension until it bottoms out on the bump stops. There should still be clearance between the tire and the fender, with the steering straight

The fiberglass spoiler on this Datsun has been broken against the pavement. Extra ground clearance may be required if a spoiler is used on rough roads. For racing, a lower ground clearance can be used as long as the tracks are smooth. For a car that is used for both street and racing, the ground clearance can be changed by using lower tires for racing and taller ones for the street.

Ground Clearance (Inches)	Car Make and Model	Ground Clearance (Inches)	Car Make and Model
3.5	Triumph TR7, TR8	5.5	Jaguar E-type V-12
4.0	Triumph Spitfire		Maserati Bora
4.2	Chevrolet Corvette		Renault 12
	Pontiac Grand-Am	5.6	AM Hornet
4.3	Dodge Colt	5.7	Fiat 128
4.4	Peugeot 505S	5.8	Audi Fox
4.6	Chevrolet Camaro		Mercedes 450SE
	Audi 4000	5.9	Datsun 280ZX
4.7	Lancia HPE		Porsche 911
	Ferrari Dino		Volkswagen Beetle
	Fiat 124 Sport	6.0	Capri
	Maserati Ghibli		Jaguar XJ
	Renault 15		Lotus Europa
	Porsche 928		Mazda RX-3
4.8	Ford Pinto		Triumph TR-6
4.9	SAAB Sonnett	6.1	Mazda RX-7
5.0	De Tomaso Pantera		Toyota Corolla
	Jensen Healey	6.2	Audi 100
	Lamborghini Espada	6.3	BMW 2002
	MGB, Midget	6.5	Buick Riviera S-type
	Renault 17		Fiat Strada
	TVR Vixen		Subaru DL-5
5.1	Plymouth Horizon		Honda Civic GL1500
	Ferrari Daytona		Rolls Royce
	Fiat 128SL 1300		Mercedes-Benz 220
	Opet GT		Plymouth Cricket
	SAAB 96	6.6	Honda Prelude
5.2	Chevrolet Vega	6.7	Datsun 200SX
5.3	Fiat 850 Spyder		Toyota Celica GT
	Volkswagen Type 4	6.8	SAAB 99
5.4	Chevrolet Citation	6.9	Toyota Tercel SR5
	Rover 3500	7.0	Mazda RX-2
	AM Matador X	7.1	Datsun 310
	Mercedes 450SL		Toyota Corona
	Oldsmobile Cutlass		Volvo 164
	Open 1900	7.3	Datsun 610
5.5	Volvo 262C	7.5	Avanti
	Aston Martin DBS V-8	7.6	Datsun 210
	BMW Bavaria	11.0	Volkswagen Thing

Figure 17/Ground clearance of late-model production cars.

This car has only 4 inches of clearance between the top of the tire and the underside of the fender. If you wanted to lower this car there wouldn't be enough clearance left. New fenders would have to be made or smaller-diameter tires used.

or turned to the side. Obviously removing the springs can be a lot of work, so take some measurements first.

With the car resting on its springs, measure the distance from the lowest part of the sprung weight to the ground. If the distance between the top of the tire and the underside of the fender is *greater* than the distance between the sprung weight and the ground, you have no problem with the tires hitting; the car will hit the ground first!

On most cars oversized tires will hit the fenders before the body hits the ground, so you must do something. Unless you wish to throw away those tall tires you can do one of the following:

1. Raise the ride height and add enough material to the bump stops to prevent the tires from hitting the fender.

2. Make new fenders which have more clearance.

3. Stiffen the springs to reduce the suspension travel and add to the bump stops to keep the tires from hitting the fender.

4. Drive slower and live with the tires occasionally slamming into the fenders.

Solution **1** raises the CG and reduces the cornering power of the car. A raised ride height may be great for drag racing, but it doesn't help the cornering at all. Solution **2** costs money, but will let you keep a low ride height for good cornering. Solution **3** will reduce the comfort and may hurt the cornering on bumpy corners; **4** is silly.

If you are using lower profile tires than stock, chances are the tires are not the limiting factor to ride height. The bump stops often prevent additional lowering of the car. Before getting out your tool box remember this: Bump stops are *necessary* for proper handling, so don't totally remove them. Without the rubber stops, the suspension will hit metal-to-metal at the end of its travel and something will break!

It is possible to reduce the thickness of the rubber stops without ruining the car. As much as half of the stop thickness can be removed to allow more suspension travel on a lowered car. Just cut the rubber off with a knife or saw. Leave at least 2 inches of suspension travel between static and full bump positions, or you are likely to have frequent bottoming. The amount of cutting of the stops will vary with the use of the car. Smooth-road driving will allow a shorter, stiffer bump-stop rubber than required for off-road use.

If you are planning to drive on really rough roads or off-road *do not cut the bump stops.* By cutting the bump stops you increase the impact loads on the chassis when the suspension bottoms. The shorter stop is stiffer, and cannot snub the suspension motion as easily. These increased impact loads can lead to failure if bottoming is hard and frequent.

When lowering the car always make sure it still has suspension travel before hitting the bump stops. A lack of suspension travel is one of the most frequent causes of horrible handling problems. This is particularly true on racing cars, where hard cornering loads and banked corners can make the suspension rest on the stops. When this happens you are usually in for a wild ride, because the bottoming results in a huge and sudden increase in weight transfer on the tire that bottoms out. The driver usually feels this as a complete loss of traction on that tire.

If you have doubts about the suspension travel, this can be easily tested. Take a small piece of modeling clay and place it on the bump stops. Drive around without hitting hard bumps, but going through hard turns, accelerating, and braking. If the suspension has not hit the clay, the travel is adequate. If you wish to know how much clearance remains, just increase the thickness of the clay and test the car until you see some contact. Many tests are required for an accurate answer, because bumps add to the suspension movement.

Sometimes it is desired to add bump stops to an existing car. The shock absorber can be used as a bump stop if a rubber ring is put over the shaft. Most racing cars use this type of bump stop, and Koni makes some beautiful rubber bump stops specifically designed for shock absorbers.

This Datsun has been lowered so much that there is less than an inch of clearance between the rubber bump stop and the frame. The car is shown resting at the static ride height. To prevent bottoming the rubber stop can be cut down or its mounting bracket can be lowered. The best compromise for street use is to cut the bracket down in height and reweld it. On this car the limit to lowering ride height is where the steering tie rod hits the frame at full bump. The rubber stops are necessary to prevent this from happening.

This Koni racing shock comes with a rubber bump stop on the shaft. The stop (arrow) is almost as large as the body of the shock. It is a hollow tapered part made of special synthetic foam rubber. This snubber gives a long and very soft cushion at the end of suspension travel.

A shock bushing can be made into a bump stop by splitting it on one side as shown. It is then slipped over the shaft of the shock. Various length shock or spring-eye rubber bushings are available to choose from. Get one with a hole size that matches the shaft diameter of your shock. The longer the rubber stop is, the softer the impact will be when the car bottoms.

You can make a bump stop for a shock by using the rubber bushing commonly used in the end of a shock absorber. Slit one side of the bushing, spread it, and slip it over the shock shaft. The bushing is quite stiff and should stay on the shaft by itself. Two bushings can be used to give a softer cushion. If building your own stops be sure they stop the suspension travel before something else hits or binds.

Another limiting factor to ride height is the possible binding of a suspension or steering pivot. This should be checked with the suspension at both extremes of its travel, at full bump and also at full droop. This binding happens when the angular travel of the pivot is all used up and metal-to-metal contact occurs. A broken suspension part is the usual result. On production cars watch out for this condition if you modify the bump stops. The car is designed so that no binding can occur with the suspension on the stock stops, but with the bump stops cut down you are sometimes in for trouble.

With the springs removed and the suspension resting on the stops, check a link with a spherical bearing at each end, such as a steering tie rod. You should be able to rotate it by hand. Also check it at full droop, with the car jacked up. On racing cars using spherical rod ends, binding results in pounding or peening the edge of the bearing race. Check each one very carefully because this is a major cause of broken rod ends.

Besides all the mechanical limitations to the ride height, there is an important geometrical limit. In Figure 9 I showed you how to find the roll center on a car with double A-arm suspension. These cars are designed with the roll center in a certain position with the car at normal ride height. By changing the ride height it is possible to shift the roll center greatly. Sometimes this results in totally different handling. This mistake is common on road racing cars where the owner changes from old-fashioned tall tires to new low ones. Then he raises the ride height to give proper ground clearance and thinks he has the hot set-up. Sometimes this backfires and results in a roll-center position that hurts the car's handling more than the new tires help it. Figure 18 shows this problem in more detail.

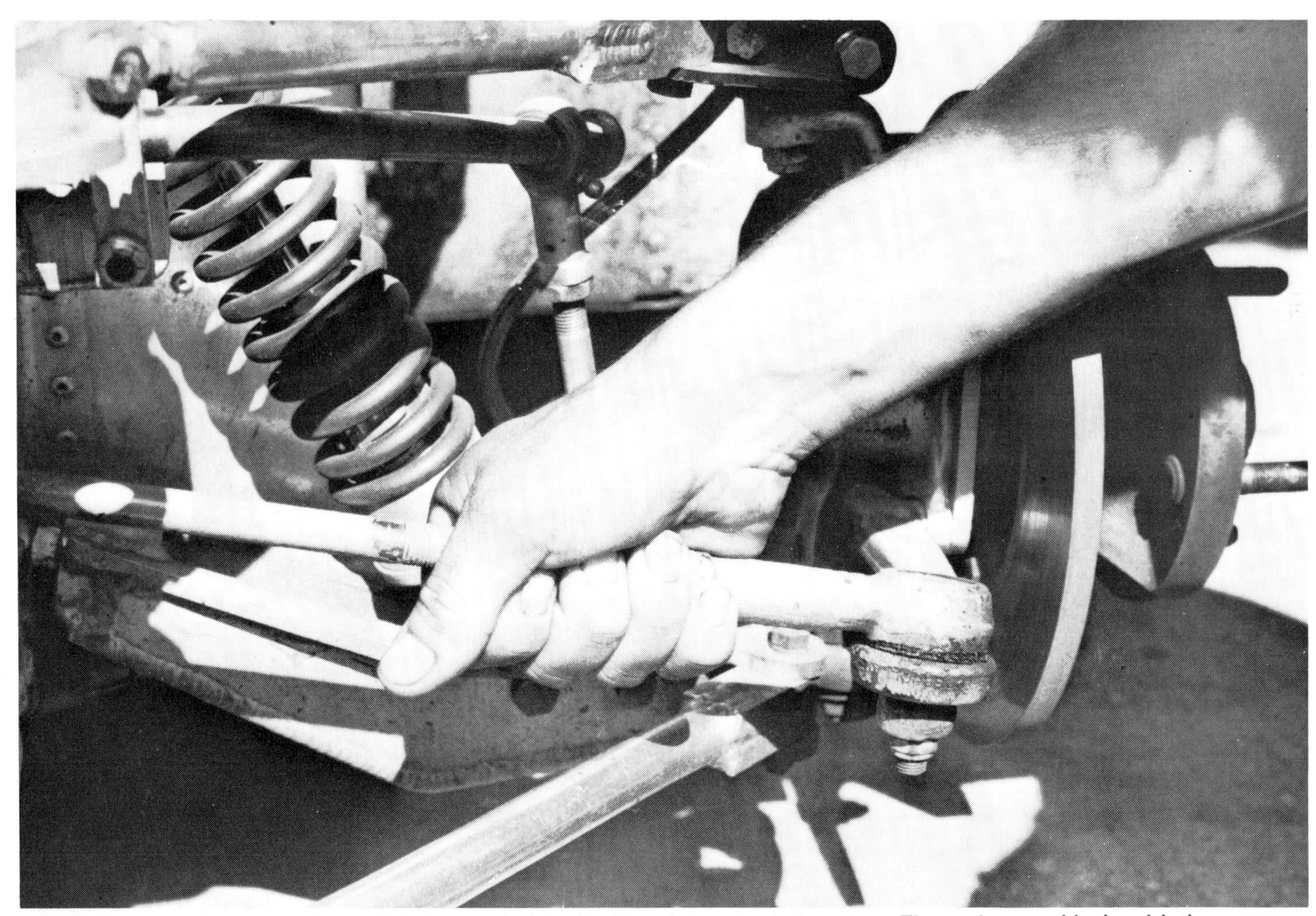

The steering tie rod on this racing car is being checked for binding of the spherical bearings. The car is up on blocks with the suspension at full droop, at the extreme of travel in the droop direction. This should also be checked at full bump and with the wheel turned fully in both directions. Any binding of the pivots could result in a broken steering tie rod.

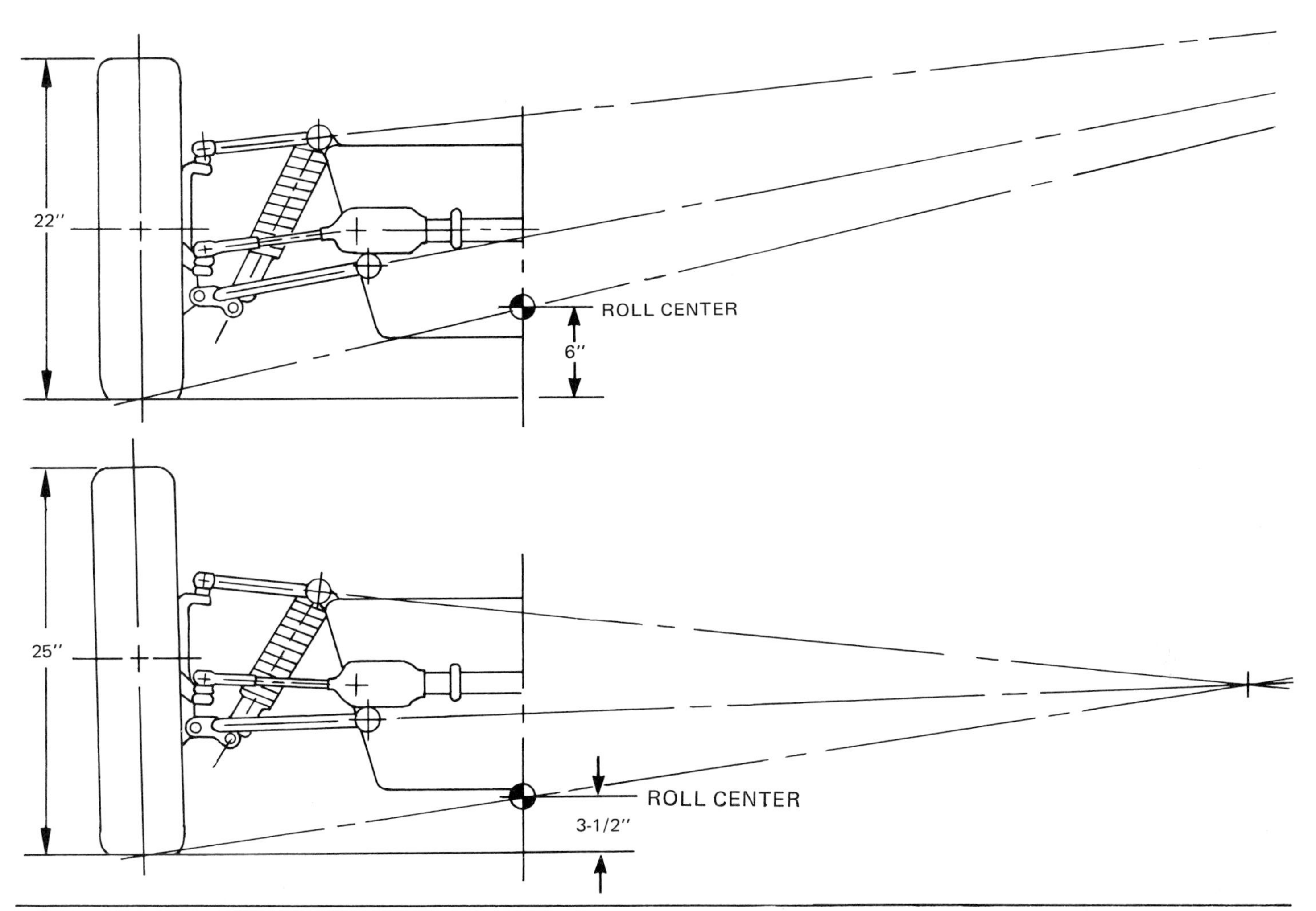

Figure 18/Here's what happens when you install 22" tires on an old car designed for 25" tires. If the car is adjusted to the same ride height with the 22" tires as it had originally, the roll center moves from 3 1/2" to 6" above the ground. In addition there may be problems with the suspension linkage or steering pivots binding at the end of their travel. The car shown here is a 1960 Lotus racing car, which originally came with tall 25" tires. Notice that, with the smaller tires, neither of the suspension A-arms is parallel to the ground.

This geometry problem is not easily spotted, but there are several clues. Look at the suspension links from the front or the rear of the car with the car resting at ride height. If both the upper and lower A-arms are *not* nearly parallel to the ground, you can suspect a possible geometry-related problem. This is not true all the time, but modern cars are usually designed with one of the A-arms parallel or nearly parallel to the ground. If the ride height is changed so it isn't, sometimes a geometry problem is the result. Another tell-tale sign on cars with rack and pinion steering is if the steering tie rod does not line up with the center of the steering rack with the car at ride height. A well-designed car has these links lined up as closely as possible at normal ride height to allow full travel of the suspension and reduce bending of the rack. If the steering tie rods don't line up with the rack on your car, try changing the ride height in a direction to correct it and see if this helps the handling.

Now considering all the limitations, select the proper ride height for your car. As mentioned before, it may be different at front and rear. One reason for this is aerodynamic. For racing it is common to set the ride height so the rear of the car is higher than the front. You have seen this on all types of cars, particularly stock cars and funny cars. This *rake* is to create a downward force on the car at high speed which improves traction. There are limits however, which will have to be learned at the track. Start out with the car set up similar to other racers in your class, and then experiment. *Never set the nose higher than the rear,* as this can cause a dangerous lifting force at high speeds. Normal road cars are set level, as the speeds are not high enough for aerodynamics to have any great effect.

When measuring the ride height on a level surface, bounce the suspension lightly up and down to try to eliminate friction effects. Prior to checking ride height, the car must be rolled backward and forward about a foot to eliminate any scrub that might cause the car to sit high. If your car has adjustable shocks, set them at full soft, or better yet take the shocks off the car. Friction in the shocks may cause a false reading of ride height. Take several measurements, bouncing the car each time to try to get repeatable measurements. If you simply cannot get things to repeat, there is too much friction in the suspension. In this case, take the car out for a short drive, on a flat surface at low speeds. Slowly roll to a stop on the level floor, being careful not to hit the brakes hard causing the car to nosedive. Then measure the ride height with the driver and any ballast still in the car. This will take out the effect of friction in the suspension, since it duplicates the actual riding conditions of the car.

This stock car is set up for high-speed paved tracks and has a noticeable amount of rake. The front ride height is low to prevent air from building up high pressure under the car. The high tail actually helps provide some downforce at high speeds.

HOW TO ADJUST RIDE HEIGHT

After measuring the ride height you may wish to adjust it. This job varies from dead simple to nearly impossible, depending on the design of the car. Let's look at the various types of suspensions and see what can be done with each one.

Before measuring ride height take the friction out of the suspension by bouncing it as shown. On some cars the only way to get accurate ride height measurements is to remove the shocks.

RACE CARS USING COIL SPRINGS OVER TUBULAR SHOCKS

This type of coil shock unit usually has a threaded collar supporting the spring. This is the ride-height adjustment. Moving the collar toward the spring raises the car, and moving the collar away from the spring lowers the car. Turn the collar with a hook wrench or spanner wrench, available from a tool store. In the field you can sometimes use large water pump pliers or a punch. The punch is a last resort, because it tends to mess up the slots in the collar. If the threads are clean and in good condition, the collar can sometimes be rotated by hand with the weight taken off the spring. Lubricate the threads with anti-seize lubricant.

CARS WITH COIL SPRINGS

Most cars use coil springs in the front, and some also in the rear. If there is no adjustment provided, the usual way to raise the ride height is to put a metal shim under the spring. Do not use wedges driven between the coils, as this causes spring failure and drastically alters the spring stiffness. If you wish to lower the

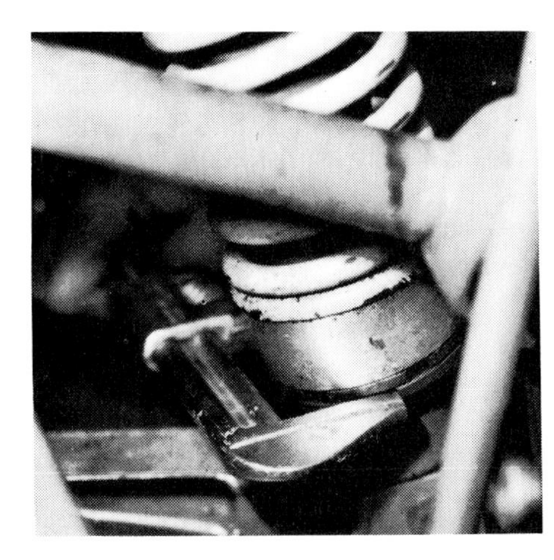
The ride height adjusting nut on this coil-shock unit can be adjusted with a monkey wrench. When doing this it helps a lot to clean and lightly oil the threads. Adjust the ride height with the car jacked up to reduce the friction caused by the coil spring load.

ride height, this can be accomplished by making the spring shorter. There are two ways: by removing metal or by heating and spacing the coils closer together. The first method makes the spring stiffer, so if you want to stiffen *and* lower the car at the same time, cut off part of the spring. I deal with this method in detail in Chapter 4. To lower the car by heating the spring does not change the stiffness, but you must beware of bottoming out the coils before the suspension hits the stops. This method for shortening the spring is also discussed in Chapter 4.

On certain cars there is another method for lowering the car using the stock length springs. On these cars the spring rests on a plate which bolts to the car chassis. Sometimes spacers under this plate will lower the car, or the plate can be made adjustable. See photo on page 71. Study your particular car to see if this will work. It is a better method than shortening the spring, as you have more control over the final ride-height dimension, and it can easily be changed again.

CARS WITH LEAF SPRINGS

Leaf springs are used on the rear axle of most production cars. The easiest and cheapest way to lower ride height is to install lowering blocks. A lowering block is a metal spacer that goes between the axle and the spring, locating the spring closer to the ground. Standard thickness lowering blocks are sold in speed shops in kit form. If special sizes are needed, lowering shims can be made at home from sheet metal. Make sure they are designed like manufactured lowering blocks so they can't fall out.

If a large amount of lowering is required the springs should be re-arched. This should be done by a local spring shop, as it is too difficult a job to try at home. The spring is taken apart, heated to high temperature, and bent into the new shape. Then the metal is heat treated to bring back the high strength required for springs. Re-arching can be done to give almost any ride-height change required, but small adjustments with lowering shims will still be required.

The ride height can be raised with different length shackles. Longer shackles are used to give a car increased rear ride

This racing car has the ride height adjusted by shims between the coil spring and its mount on the shock. These shims increase the ride height. This is not an easy way to adjust ride height, as the coil shock unit must be taken apart to change shims. Small adjustments can take many hours of labor.

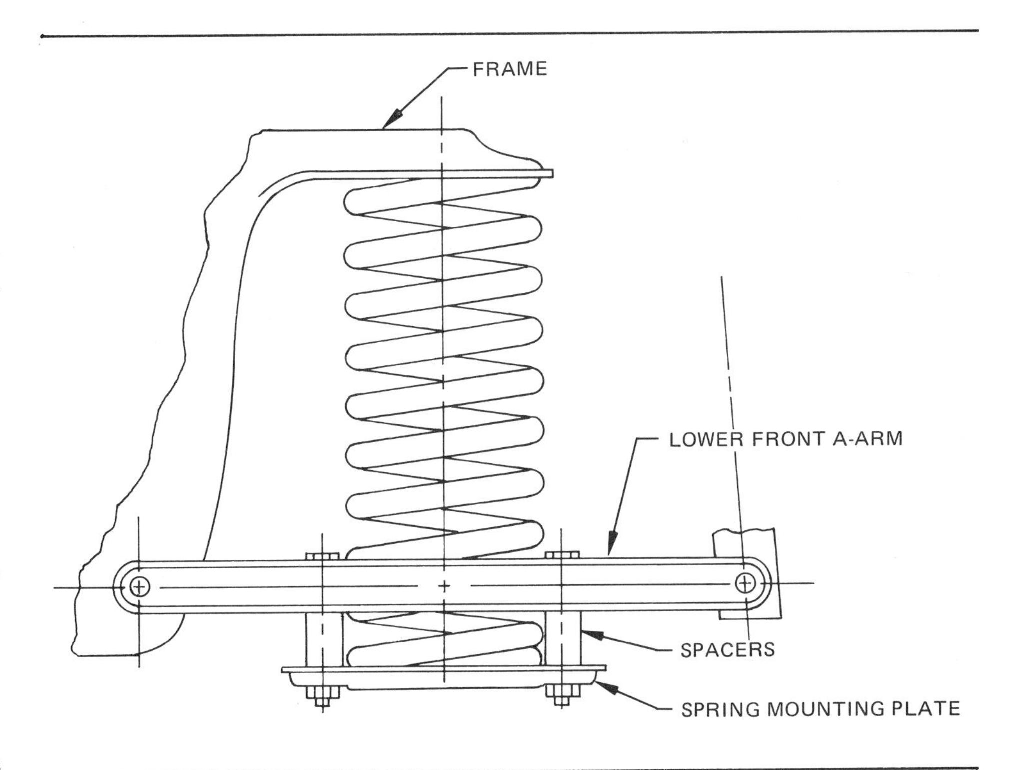

If the spring mounting plate bolts onto either the A-arm or the frame, the car can be lowered by means of spacers. This drawing shows spacers being used between the lower front A-arm and the spring mounting plate. This is an ideal way to lower the car because the ride height can be exactly determined and easily adjusted if necessary.

The old-fashioned lowering block is still a cheap and easy way to lower a car with leaf springs. On this Pinto station wagon, owner Bob Barry wanted to get the car sitting level for daily transportation. This lowers the CG and gives slightly better cornering power. The car comes from the factory with the rear end high to accept heavy loads in the cargo compartment. When Bob goes on his yearly vacation he just removes the lowering blocks.

Long shackles were used to set proper ride height on this Pinto. This car has the spring re-arched too, and the long shackles were used to compensate. The idea here was to end up with the rear spring horizontal rather than tilted down at the front as in the stock set-up. This not only gives adjustable ride height, but it eliminates the roll-steer built into the Pinto rear suspension. The eye-bolt in the bottom hole is the position for maximum ride height and minimum roll understeer.

This sprint car has cross torsion bar suspension. In the foreground you see the moving end of one torsion bar with the driving arm attached with a spline. Behind it is the fixed end of the opposite torsion bar with its ride-height adjuster. To change the ride height, loosen the jam nut and turn the adjusting bolt. Large adjustments can be obtained by moving one end of the bar to the next spline.

height. These are available in kits at the speed shops, or they can be built at home. Long shackles are undesirable because they are weaker than stock ones. Re-arching the spring is a better method to make large ride height changes.

Another change that can alter the ride height is reversing the spring eyes. Again this modification has to be done by a spring shop because the spring must be heat-treated after forming. When the eyes are reversed the spring eye is unwound and bent in the opposite direction. This is a large change in ride height, usually several inches. There is no fine control over the ride height change, so it is usually done to aid re-arching the springs for a large ride height change.

Some people have changed the ride height by adding or removing leaves from the spring. This changes the spring stiffness at the same time, so this should be avoided unless you definitely want two changes at once. Adding leaves stiffens the spring, the longer leaves giving the most effect. A stiffer spring will deflect less and increase the ride height. Removing leaves softens the spring and lowers the ride height. However it also weakens the spring, so this practice is dangerous unless you have also reduced the weight supported by the springs.

CARS WITH TORSION BARS

There are two basic methods of adjusting the ride height of torsion bar suspension. One involves the adjustment of the anchor point at the fixed end of the bar—the mechanical details vary with the car. The other method requires moving the torsion bar on the splines that connect it to the driving and supporting member at each end. The splines are usually one different in number at each end of the bar, so small adjustments can be made by rotation one way at one end and the opposite way at the other end. A small change can result from this method—much less than the distance between two adjacent splines. Consult the factory manual for the exact method used on your car.

OTHER DESIGNS

There are other spring designs used on cars such as rubber springs or air springs. Rubber springs are usually adjustable at their mounts. On the famous BMC Mini, the rubber springs are driven with a metal linkage that can be taken apart and shimmed for length. It is a difficult job, but worth it if you are fine tuning the suspension.

On cars with air springs, the ride height can usually be changed easily by varying the pressure of the air inside the spring. Air shocks and air helper springs that are sold in accessory shops all use this principle for quick and easy ride-height adjustment. Application of these devices as suspension modifications is discussed further in Chapter 4.

After setting the ride height, other suspension settings should be made at that height. It's helpful to make up some wooden blocks to go under the frame to hold the car at the ride height. Weights can be put into the car to push it firmly down on the blocks, or the springs can be disconnected. Save the blocks, label and paint them red to mark them as important tools. Then you can use them next time you set up the car.

On race cars using coil shock units, turnbuckles are sometimes used in place of the shock to hold the car at the desired ride height. Another simple tool that is sometimes used is a metal or wood bar with two holes drilled in it at exactly the right distance. This again is used to replace the shock for suspension tuning. If you want to go all-out, make up one set to simulate the shock length at full bump, and another for full droop. These are helpful in setting up plumbing, mounting the body, etc., and they allow working on the car when it is resting on saw horses.

Air Lift air springs are an accessory that allows easy adjustment of ride height. This one is for cars with coil springs and fits inside the existing spring. It is basically a rubber bag filled with compressed air. The air pressure controls the ride height and is easily changed from outside the car. Very slight pressure changes greatly affect the ride height with this type Air Lift. 4 psi pressure change will raise a typical station wagon 2 inches at the rear.

WEIGHT DISTRIBUTION

The car looks like it is resting on four tires, but it may carry most of its weight on only three of them. If the car does not carry its weight evenly distributed on the four tires it is said to have weight *jacked* into the chassis. The setting of *weight distribution* or *weight jacking* is the adjustment of weight diagonally across the car. This process is like trying to even out the legs of a homemade chair. It requires careful measurements.

If you are driving through both right and left hand corners it is helpful if your car handles the same, no matter which

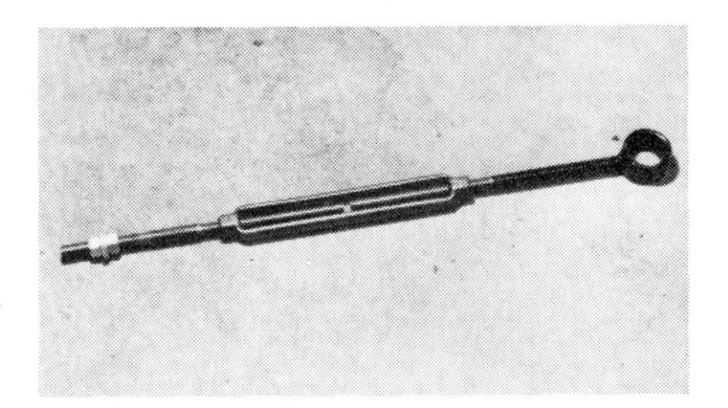

This turnbuckle has been modified to use in place of a shock absorber for adjusting the suspension. The ends off an old shock have been welded to the turnbuckle shaft so it can be bolted on the car's shock brackets. The car can be held at the proper ride height this way. Also turnbuckles are handy for moving the suspension through its range of travel on a car with leaf springs, because the springs usually cannot be removed without the axle falling off.

way you turn. To balance the handling characteristics, it is important that the car's weight be distributed in the same proportion on both the right and left sides of the car.

Imagine a perfectly built car, set up with equal weight on all four tires. Now take the frame and put a lengthwise twist in it so that the front end of the frame is no longer parallel to the rear end. If you put the car down on the ground it rests mostly on two diagonally opposite tires, and carries a smaller amount of weight on the other two. This twisting of the frame has the same affect on the tire vertical forces as altering the weight distribution diagonally across the car.

There are some types of racing cars that benefit from having the weight distribution unbalanced. Drag-racing cars can counteract drive shaft torque reaction on the rear axle by unequally preloading the rear springs, and oval track racers unbalance the weight distribution for improved rear wheel traction in left hand corners. This controlled *weight jacking* by oval-track racers and its effect on the car will be discussed later. First we'll see how to balance the car's weight for both right and left hand corners.

To set the weight distribution you must measure the load or weight at each tire with the car in its normal ride position. There are several ways to do this, depending on the scales available.

The typical platform scales with a sliding weight are great, as are the special low-profile scales made for weighing automobiles. Bathroom scales can be used on very light cars, but these require careful handling to get accurate readings.

Several manufacturers make special weighing devices just for cars. One excellent example is the Wacho Portable Weight Checking Device, manufactured by Wacho Products Co. of Columbus, Ohio. It is a hand operated lever mounted on an adjustable stand. The tip of the lever is placed under the upper edge of the wheel and the wheel is lifted off the ground with hand pressure on the lever. With the wheel barely off the ground, the weight is read on a dial on the top of the instrument. When using this device you must be careful not to lift the wheel any higher than necessary, because errors are introduced as the car rolls to the side.

If you have to turn *both* right and left, weight jacking cannot be used to an advantage. What you gain in one direction you lose in the other and the result is reduced overall performance. For road racing or street use weight jacking should be zero.

The Wacho Portable Race Car Weighing Device. This is a very quick and convenient way of making weight checks, but a set of four platform scales would be more accurate. However, platform scales cost 10 times as much as this device and the scales are too big to carry to the races. (Photo courtesy Wacho Products Co.)

Bathroom scales are not the best, but they will work on a very light car. Be sure to allow the scale to roll sideways as shown so as not to jam the scale mechanism. Bounce the scale lightly with your hand and make sure it repeats the reading. If the readings are inconsistent the scale is jamming or is overloaded.

If you are using conventional scales under the tires, the best way is to use four scales of equal height. The scales give immediate readings and you do not have to move the car around. However, if your budget can't stand the price of four scales, two will do. It just takes a little more work. To get an accurate reading with two scales, place the scales under the two front tires and put blocks equal to the scale height under the rear tires. After taking the readings, switch the scales to the rear tires and put the two blocks under the front.

The blocks are used to keep the car level while weighing. There is a small error if the car is tilted. This error is not noticeable on a very low car with a long wheelbase, but on a production sedan the error would be noticeable. Make a test with and without the blocks to measure the error on your car. Maybe you won't need blocks.

There are some bathroom scales that

go up to 300 pounds, and these can be used to weigh very light cars. If you use bathroom scales you will have to prevent the tires from scrubbing sideways when they are placed on the scales. With independent suspension there is usually some sideways motion of the tires as the suspension moves, so if you jack up the car and let it down on the scales the tires try to slide the scales sideways on the floor. This motion will jam the scale mechanism and give you false readings.

To avoid sideways jamming the scales, rest them on some boards with pipe or tubing rollers underneath. The tubing should be lined up with the direction of the tires so the scales can move sideways freely. Always bounce the car gently to be sure the scale mechanism is not jammed.

Platform scales have wheels on them. The wheels should be pointed across the car so the scale can move sideways freely. This prevents jamming the scale mechanism as described above. Platform scales are designed for more rugged use, and they are usually not troubled by side loads on the platform.

Another way to get good scale readings is to set up ramps at the same height as the scales. The car is rolled onto the scales rather than being jacked up and lowered onto the scales. The ramp method eliminates the sideways movement of the suspension, since it remains at the static position at all times.

Remember, the car should be weighed as driven, with the driver or equal weight aboard. An average fuel load should be included. If you jack up the car, bounce the suspension to remove the effects of friction. Repeat the measurements several times to be sure you are getting consistent results. Record the weight on each tire.

After weighing each tire calculate the front-to-rear weight distribution for the car. To find the total weight of the car, add all four tire weights together. The percentages at each end of the car are:

$$\text{Percent on front tires} = \frac{\text{Total weight on both front tires}}{\text{Total weight of car}} \times 100$$

$$\text{Percent on rear tires} = \frac{\text{Total weight on both rear tires}}{\text{Total weight of car}} \times 100$$

As a check, the percent on front plus the percent on rear must equal 100%.

For ideal or balanced weight distribution, the left-side tires must have the same front-to-rear weight distribution as the whole car. Likewise, the right-side tires must also have this same weight distribution. Note that the left and right sides of the car may not necessarily have equal weights, because the CG may be offset to the side. It is only the front-to-rear *ratios* that should be identical. Compute the ideal weight distribution as follows:

$$\text{Right front ideal weight} = \frac{\text{\% on front tires} \times \text{total weight on both right tires}}{100}$$

$$\text{Right rear ideal weight} = \frac{\text{\% on rear tires} \times \text{total weight on both right tires}}{100}$$

$$\text{Left front ideal weight} = \frac{\text{\% on front tires} \times \text{total weight on both left tires}}{100}$$

$$\text{Left rear ideal weight} = \frac{\text{\% on rear tires} \times \text{total weight on both left tires}}{100}$$

To say it another way, if the front-end of the car carries 40% of the total car weight, then the left front tire should carry 40% of the left-side weight.

You can easily check the ideal weight calculations. The sum of all four ideal weights must equal the total weight of the car. The total never changes no matter how the weight is jacked in the chassis.

Compare the ideal weights to your measured weights. If they are different by more than 10 pounds per wheel you may wish to adjust the weight distribution. If they are off by more than 50 pounds you certainly should adjust the weight distribution.

When there is a difference between actual and ideal weights, the amount of error will be the same for each tire. It will be high on two tires and low on the other two. The two high tires will be diagonally opposite each other on the chassis, and the two low tires will also be diagonally opposite.

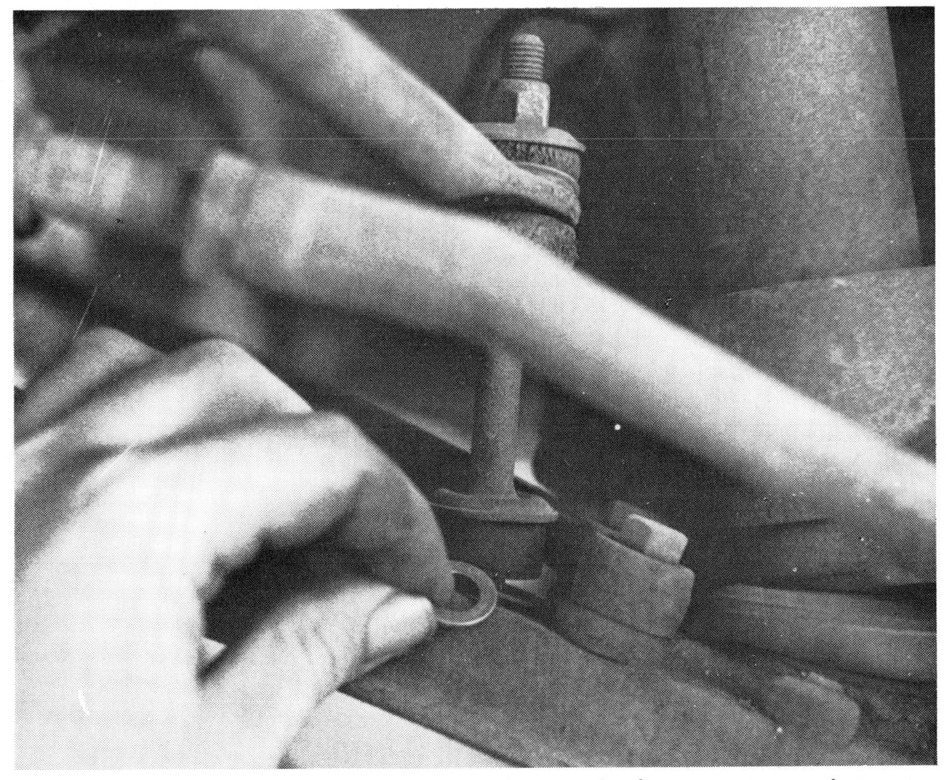

If you determine that the anti-roll bar has a twist or preload, you can correct it on some cars by changing the position of one end of the bar. In this design, the end of the anti-roll bar is located by a short vertical driving link. Remove the nut on the bottom side, install spacer washers between the rubber pad and the mounting bracket as shown. This effectively lengthens the driving link. If it's not clear which driving link should be lengthened to take the preload out of the anti-roll bar, do it by trial-and-error.

Before adjusting weight distribution, disconnect anti-roll bars. Now weigh the car again and see if the weight distribution changed any. Sometimes a preload in the anti-roll bars causes the unbalanced weight distribution and disconnecting the bars restores the balance. If this turns out to be so, adjust or shim the anti-roll bar driving links so there is zero preload when installed on the car. That will correct the unequal weight distribution when the anti-roll bar is in place.

Some anti-roll bar driving links are threaded rods. On these the length can be adjusted by tightening or loosening the nut on the end. If you fail to make the proper adjustment within the limit of the nut position, use washers or shims at the ends of the link. In drastic cases the rod can be cut or lengthened.

If the anti-roll bar links are non-adjustable, the entire bar can be tilted by shimming under one of the mounting bearings for the bar. Tilting the bar is a trial-and-error process much like changing the link length. Make a new weight check after making an adjustment to the anti-roll bar. If you are using only two scales, you do not need to weigh both ends of the car. When one tire reaches the ideal weight the other three will be there too. You should, however, make one last check of all four tires just to make sure of your measurements and adjustments.

If the car does not have ideal weight distribution with the anti-roll bars disconnected, you must adjust it at the springs. This is done by changing the ride height on a pair of wheels at either end of the car, say by raising the left ride height and lowering the right. Then at the opposite end of the car do the reverse—raising the right side and lowering the left side. The reason all four corners are adjusted is that you have already set the car at exactly the proper ride height, and working on only one end or one corner will rake or tilt the chassis. However, for small changes in weight distribution you can work on only one end of the car without noticeably changing the overall ride height.

To increase the weight on one tire, *raise* the ride height on that corner. To reduce the weight, lower the ride height on that corner. A change on any one corner will change the weight on all four corners an equal amount. So make the changes in smaller increments than required to entirely correct the weight distribution, and make the remaining changes at the other corners. If all four corners are to be adjusted, then 1/4 of the total change is made at each corner. If you are going to adjust only two corners, then half the total change can be made at each corner.

As an example, assume the car has the following tire weights when originally weighed:

Right front = 400 lb.
Left front = 400 lb.
Right rear = 500 lb.
Left rear = 700 lb.

Total weight on front tires = 400 + 400 = 800 lb.
Total weight on rear tires = 500 + 700 = 1200 lb.
Total car weight = 400 + 400 + 500 + 700 = 2000 lb.
Total weight on right tires = 400 + 500 = 900 lb.
Total weight on left tires = 400 + 700 = 1100 lb.

$$\text{Percent weight on front} = \frac{800}{2000} \times 100 = 40\%$$

$$\text{Percent weight on rear} = \frac{1200}{2000} \times 100 = 60\%$$

Now we can compute the ideal weights on each tire to get the front/rear ratio the same on each side:

$$\text{Right front ideal weight} = \frac{40 \times 900}{100} = 360 \text{ lb.}$$

$$\text{Right rear ideal weight} = \frac{60 \times 900}{100} = 540 \text{ lb.}$$

$$\text{Left front ideal weight} = \frac{40 \times 1100}{100} = 440 \text{ lb.}$$

$$\text{Left rear ideal weight} = \frac{60 \times 1100}{100} = 660 \text{ lb.}$$

In this example, the car is heavy on the left side, probably because of the driver's weight on that side of the car. The ideal weights compared to the measured weights show each tire to be off by 40 pounds. The right front and left rear are heavy, and the left front and right rear are too light. Notice the error is reversed at each end of the car, and the error is the same number of pounds for each tire. It will always work out this way if your measurements are accurate and your calculations correct.

To correct this car we start with the right front tire. Because it is 40 pounds too heavy, we will lower the ride height enough on the right front to change the tire weight by 10 pounds. This is 1/4 of the total error of 40 pounds. Then we raise the left front ride height enough to change the weight another 10 pounds. Then raise the right rear 10 pounds worth and lower the left rear 10 pounds worth. This should bring all tire weights to the ideal values.

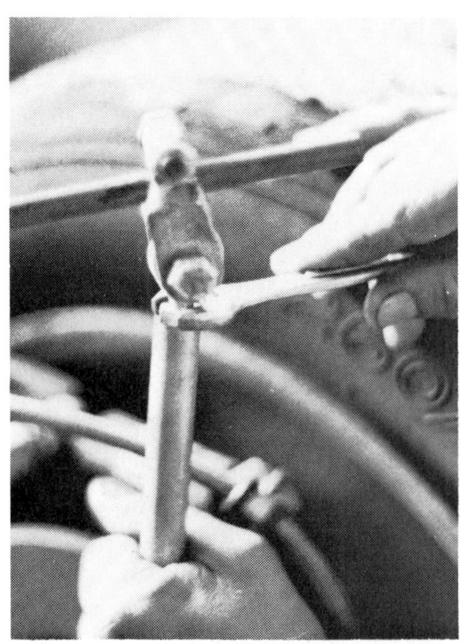

This is a tubular driving link for the anti-roll bar. It has a spherical rod-end threaded into each end of the link and held by a jam nut. Loosen the jam nuts at each end and rotate the tubular link to adjust its length. You can make one link longer and the other link shorter to take out preload of the anti-roll bar, or to put some in for a minor bit of weight jacking.

After setting the weight distribution, install the anti-roll bars, and adjust the connection links so as to put zero twist in the bar when it is bolted up. If the bar is twisted, it will change the weight distribution. Check this with the scales after installing the bars, and make sure they are still at the ideal weights. If the anti-roll bar links are not adjustable, then the ride height will have to be changed again to counteract the weight jacking put in by the anti-roll bars. This is a tedious job; I hope your bars can be adjusted.

That big threaded rod with the jam nut on it is the weight jacker on this stock car. The rod is threaded through the frame above the coil spring. Ride height or weight jacking can be quickly adjusted by turning the rod.

Here is what happens when you try accelerating hard with a stock car using a solid rear axle. The right tire breaks loose before the left one, reducing the maximum possible traction. Jacking the weight to add vertical load to the right rear tire will give better performance at the drag strip.

When the anti-roll bars can be adjusted easily at the driving links, you can change weight distribution by putting a twist in the bars without touching the ride height adjustment at the springs. This is not as satisfactory, but it will work for small weight changes and for quick changes at the track. The usual way to adjust anti-roll bar preload is to shorten or lengthen the driving links as described previously.

Oval-track race cars have weight jacking screws easily accessible for fast adjustments at the track. For best performance, start out with zero weight jacking in the shop, and make final adjustments at the track. The typical setting is to reduce the weight on the left front wheel for left turns. This makes the rear wheels more nearly equally loaded in a turn, allowing more throttle to be applied. This weight jacking will also change the car's steer characteristics. There is too much weight jacking if the left front tire lifts off the track in a turn. It should barely remain in contact with the pavement.

If you have drag raced a car with a solid rear axle you notice that it tends to spin the right tire easier than the left tire. This is because the torque on the drive shaft produces an additional vertical force on the left tire and reduces the force on the right tire. This has the same effect as unbalancing the weight distribution when you put your foot down hard. To counteract this, you can deliberately jack the weight so that the load on the right rear tire is increased. Then when you leave the line at the strip, the rear tires will be more equally loaded and that will improve the total traction. Just raise the ride height on the right rear of the car, and adjust the amount at the drag strip for best performance. On a typical sedan, about 50 pounds of weight added to the right rear tire will put you in the ballpark.

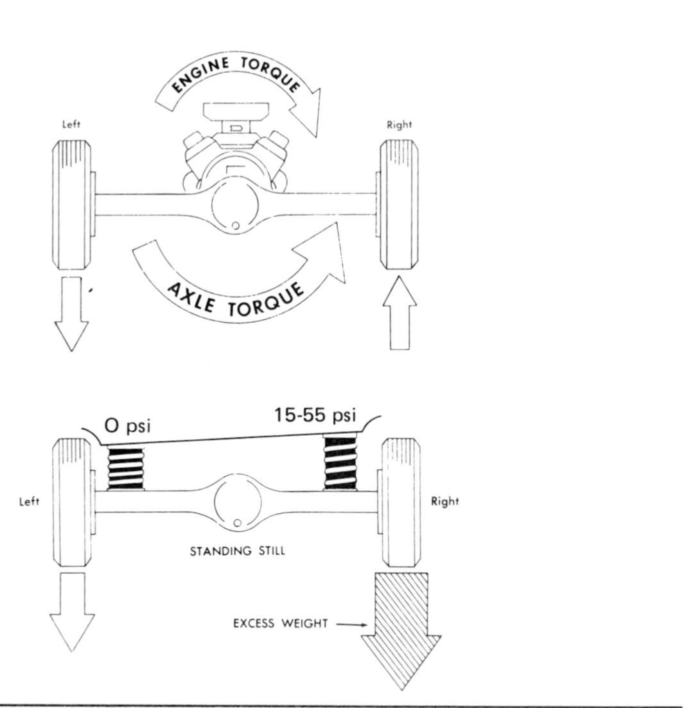

The upper drawing shows how drive-shaft torque tends to lift the right rear tire on a car with a solid rear axle. This reduces traction during hard acceleration. The cure is to increase the ride height on the right rear spring. The lower drawing shows how this is done using Air Lifts inside coil springs. The air pressure in the Air Lifts is shown, with final adjustment done at the drag strip. Air Lifts are available for leaf springs too, and are a very quick and easy way to tune your suspension for drag racing. Drawings courtesy of Air Lift Company.

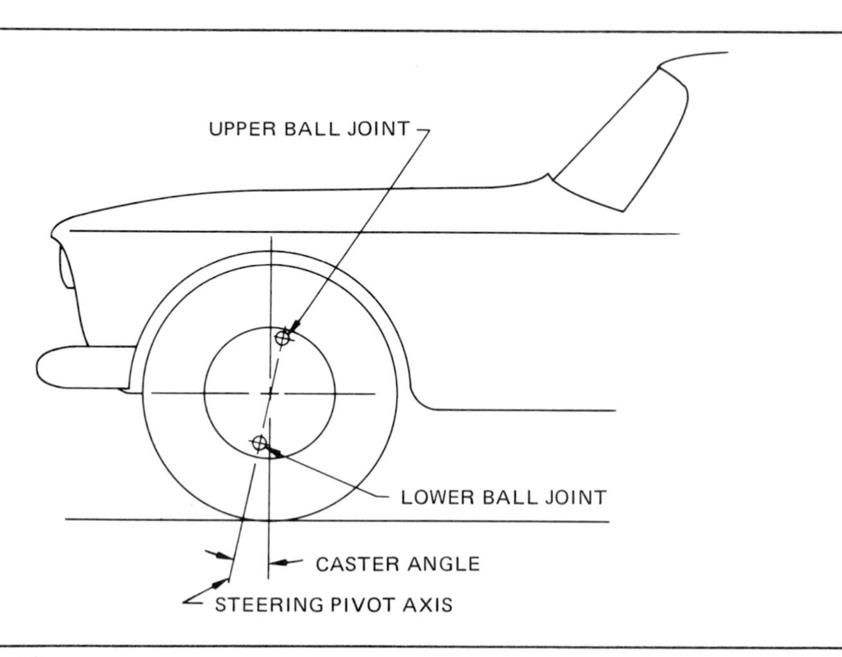

Figure 19/The caster angle is the angle between the steering pivot axis and a vertical line as shown above. Positive caster is illustrated, where the upper end of the steering pivot axis tilts toward the rear of the car. Most cars have positive caster. On cars with kingpins, the center of the kingpin is the steering pivot axis. MacPherson strut front suspensions have the steering axis on the center of the strut.

CASTER

The word *caster* as commonly used refers to a setting of the front wheels. It is the angle made by a line between the upper and lower steering pivots and a vertical reference line, viewing the car from the side. This is illustrated in Figure 19, which shows a *caster angle* measured in degrees.

In the figure, notice that the tire contact point on the ground is *behind* the steering axis. When the wheel is turned, the tire contact point is moved to the left or right of the direction of travel of the car. This generates a force which tends to return the wheel so it points in the direction the car is traveling, called a *self-centering* force.

Self-centering force is the reason the front wheels return to straight-ahead after or during a turn if you release the steering wheel.

The caster angle shown in Figure 19 is defined as a positive angle and all caster angles which are positive will cause the tread contact point to fall behind the steering axis.

If the tire contact point is in front of the steering axis, the tendency is reversed. When you turn the wheel a small amount it would tend to go all the way to full lock on account of the leverage of the forward contact point.

All cars are designed to have a self-centering steering action so it returns to straight-ahead automatically. Not all of the self-centering effect comes from caster alone.

When a wheel is turned, the car will also turn, but at a smaller angle. This was discussed earlier as the slip angle of the tire. Due to slip angle, some of the rubber of the tire is distorted and pulled away from the centerline of the tread. Rubber is elastic and the rubber which is distorted due to slip angle "wants" to return to its normal location on the tire. The result is another self-centering force due entirely to the rubber of the tire.

The total self-centering action is the sum of that caused by caster and that caused by the tire. If the tire exerts a very high self-centering force, some of it can be canceled or counterbalanced by a small amount of negative caster in the suspension. The net result is always self-centering action rather than the reverse, however, the *mechanical* caster setting at the wheel can be slightly negative in some cases.

In general, a lot of caster causes steering to feel heavy at low speeds but makes the car stable at high speeds in a straight line because the front wheels tend always to point in the direction of travel even if momentarily deflected by a rock or road surface irregularity.

A small amount of caster or self-centering action makes a car steer easily at slow speeds which is handy when parking or maneuvering. The car will feel less stable at high speeds and may tend to wander a bit on the straights.

It is important that caster be set at the manufacturers' specs, and be equal on the right and left side of the car. High speed instability can result from the wrong caster setting. If the steering feels too heavy in a corner, the caster can be reduced a small amount, but be careful not to go too far in one step. For cars that are unduly sensitive to steering, an increase in caster may help. Again it is a matter of experimentation out on the track.

The best caster angle has a lot to do with the weight of the car, as well as the tire diameter. Light cars and small-diameter tires use larger caster angles. The table of Figure 20 shows some typical values. If you don't have any specs for your car, copy a car of similar weight and usage.

Caster is measured with a special tool that can be purchased at a good auto parts store. Its use is explained in the instructions, and involves rotating the steering 20 degrees one way and then back the other way. The caster gage converts the rise and fall of the spindle into a caster angle measurement. It is difficult to measure caster directly on the car due to parts interfering with the measuring tools, but you can get a rough figure that way in a pinch. Just use a protractor with a level in it, and eyeball it parallel to an imaginary line drawn between upper and lower steering pivots. This sort of measurement should only be used in an emergency as it is very difficult to get an accurate reading.

TYPICAL CASTER SETTINGS

Caster	Car Make and Model
−2°	68-75 Oldsmobile Toronado
−1-1/2°	68-72 Pontiac Intermediates
−1°	75-79 Buick Skylark with manual steering
	75-80 Nova with manual steering
	68-74 Cadillac
	71-72 Cadillac Eldorado
	73-74 Camaro Z-28
	68-74 Chevelle
−3/4°	75-80 Buick Skyhawk
	71-80 Chevrolet Vega
	75-80 Chevrolet Monza
	75-80 Olds Starfire
	75-79 Pontiac Sunbird
−1/2°	74-77 Valiant, Dart Barracuda, Challanger with manual steering
	75-80 Maverick, Comet, Monarch, Granada, Versailles
	75-78 Fury, Charger with manual steering
	80 Pontiac Sunbird
−1/3°	70-73 Toyota Pickup
0°	73-77 Hornet, Gremlin
	73-80 Cadillac Eldorado
	71-75 Chevrolet Camaro
	75-78 Oldsmobile Toronado
	70-77 Pontiac Firebird
	70-73 Mustang, Cougar
+1/2°	79-80 Cadillac
	73-74 Buick Apollo
	68-70 Chevrolet Camaro
	68-74 Chevy II, Nova
	73-74 Oldsmobile Omega
	71-72 Peugeot 304
	70-71 Toyota Crown
	73 Volvo 140 Series
	71-75 Pontiac Ventura
+3/4°	80 Fairmont, Zephyr
	74-77 Valient, Dart, Barracuda, Challanger, with power steering
	75-70 Cordoba
	75-76 Chrysler, Imperial
	78-80 Chrysler
	75-78 Charger, Fury
	79-80 St. Regis, Magnum XE
+7/8°	74-78 Ford Mustang II
	78-79 Fairmont, Zephyr, Mustang, Capri
+1°	70-74 AMC Javelin
	75-77 Matador, Pacer
	78-80 Concord
	79-80 AMX
	77-80 Pinto, Bobcat sedan
	80 Mustang, Capri, Thunderbird, Cougar, Fairmont and Zephyr sedan
	75-79 Buick Skylark with power steering
	76-80 Chevrolet Camaro
	75-80 Nova with power steering
	79-80, Monte Carlo with manual steering
	68-76 Corvette with manual steering
	75-79 Olds Omega with power steering
	78-80 Cutlass with manual steering
	75-79 Pontiac LeMans with manual steering
	78-80 Pontiac Firebird, Grand Prix with manual steering
	71-73 VW 1700, Type 4
+1-1/4°	74-76 Pinto, Bobcat sedan
	70-71 Simca
+1-1/2°	74-76 Pinto, Bobcat station wagon
	75-76 Pontiac, Olds 88 & 98
	75-76 Chevrolet Bel Air, Caprice, Impala
	75-76 Buick Electra, LeSabre, Riviera
	67-70 Datsun SRL 311, SPL 311
	73 Volvo 164
+1-3/4°	73 Capri
	77 Pontiac LeMans with power steering
+2°	78-80 AMC Pacer
	73-78 Ford, Mercury
	74-80 Continental Mk IV
	75-77 Buick Century, Regal
	75-76 Chevrolet Chevelle
	76-78 Cadillac Seville
	79-80 Eldorado
	71 Datsun 510
	68-70 Jaguar XK-E
	70-73 VW 1600, Type 1
+2-1/4°	68-75 Corvette with power steering
	76-80 Corvette
	71-72 Jaguar XJ6
	73 Volvo 1800ES
+2-1/2°	76-80 Aspen, Volare, Diplomat, LeBaron
	79-80 Olds Toronado, Buick Riviera
+2-3/4°	72 Triumph TR6
+3°	79-80 Ford, Mercury
	80 Continental
	78, 80 Buck Century, Regal, with power steering
	77-80 Electra, LeSabre
	77-78 Cadillac
	79-80 Chevrolet Malibu, Monte Carlo, with power steering
	77-80 BelAir, Caprice, Impala
	77-80 Olds 88, 98
	78-80 Cutlass with power steering
	78-79 Pontiac LeMans with power steering
	78-80 Pontiac, Grand Prix with power steering
	67-69 Sprite
	71-72 MG Midget
	73 Opel GT
+3-1/2°	71-72 Fiat 124 Spyder
	72 Pantera
	72 Triumph GT6 Mk3
+4°	75-79 Ford Thunderbird
	78-79 Continental MkV
	75-79 Torino, Montego, Cougar, Elite
	72-73 Renault 12, 15, 17
	72 Triumph Spitfire Mk IV
	70-73 VW 1600, Type 3
	72 Mercedes 220, 280SE, 280EL, 4.5, 300SEL 4.5
+4-1/2°	76-80 Chevrolet Chevette
	73 Opel 1900, Manta
+5°	75-77 Chevrolet Monte Carlo
	77 Pontiac Grand Prix
	69 MGC GT
+5-1/2°	69-71 Austin America
+6°	67-68 MG 1100
	70-73 Porsche 914
+6-1/2°	73 Porsche 911
+7°	71-72 MGB, MGB GT
+9°	73 Fiat 850 Spyder

Figure 20/Caster settings for a number of late model cars. Notice that heavy cars have smallest angles and light cars have largest. The negative caster angles are possible because the tires have a self-aligning torque. Also cars with negative caster tend to be heavy and strongly understeering, and thus do not need positive caster for high-speed stability.

Using a Craftsman magnetic level you can get a quick indication of caster. This is not an accurate method, but it's better than nothing if you can't use a caster gage. Try to get the level lined up with the axis between the upper and lower steering pivots.

Here is a caster gage in place ready for a mreasurement. This is a magnetic-base gage that attaches to the iron hub. The car is resting on a turntable that allows accurate measuring of the steering angle. It is necessary to turn the steering 20 degrees each way to measure the caster. If the turntable is not available, the steering angle can be measured relative to a parallel reference line or computed with trigonometry.

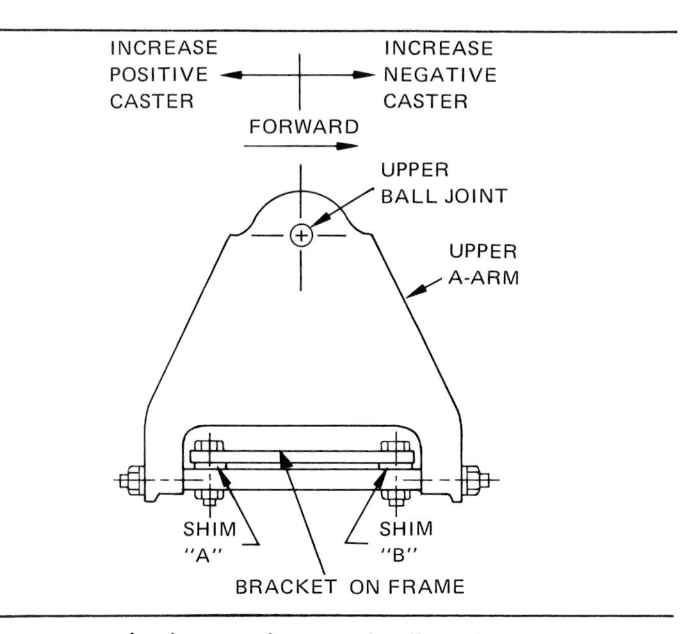

On many production cars the caster is adjusted by changing shims on the upper front A-arm. To increase positive caster the upper ball joints must be moved to the rear. Thus you use a thinner shim at *"B"* and a thicker shim at *"A"*. On other cars caster is adjusted at the lower A-arm by changing the length of the drag strut.

If you don't wish to invest in a caster gage, a front-end shop can measure caster for you. If you get a fair price, it's a lot cheaper than buying a precision tool. Caster is seldom changed once a car is set up, so you don't have to run back to the front end shop all the time. Also, caster is not the most critical of the settings, compared to camber and toe-in as discussed later.

Caster is adjusted by shims, eccentrics, or other mechanical means on production cars. Consult your manufacturer's manual for instructions. On racing cars, caster is usually set by changing the length of an adjustable link in the suspension. On some racing cars with rigid A-arms, caster is changed by screwing one rod end in and the other one out on the inboard end of the A-arm. The exact method will vary with the car.

On some cars with independent rear suspension there is a caster angle specified for the rear suspension. The *rear caster* is the angle of tilt of the rear upright off the vertical, viewing the car along the centerline of the rear axle. It is similar to front caster, but obviously the rear caster has nothing to do with front-end steering. It may affect rear-end behavior.

Usually the rear caster is set at zero—with the rear upright vertical. If there is no specification for rear caster on your car set it up at zero caster angle. Later when you bump steer the rear suspension you may wish to change the rear caster. Its function is to aid in correcting bump steer problems. On racing cars rear caster is adjusted by changing length of rear radius rods. Most production cars have the rear caster set at zero and there is no way to change it.

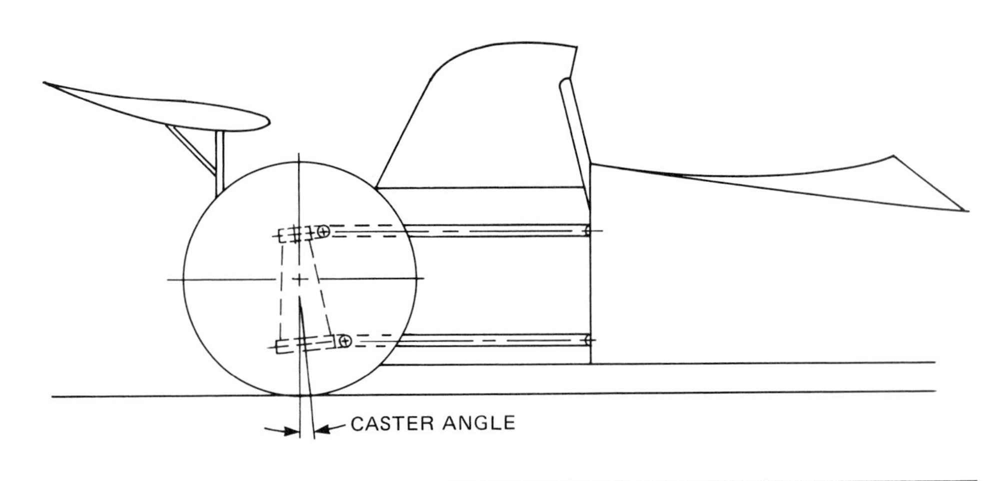

Racing cars with independent rear suspension have adjustable rear caster. The caster angle is the angle between the centerline of the rear upright and a vertical line, as shown above. Positive caster is shown in the sketch. Rear caster is used to adjust bump steer of the rear suspension.

TOE

The toe setting is one of the most important in determining how your car will handle. A very small change in toe can have a large effect in the way a car feels. The adjustment is simple and should *always* be checked when tuning the suspension.

Toe-in and toe-out are shown in Figure 6, Chapter 2. The majority of cars are set up with toe-in on the front wheels and zero toe at the rear. The basic reason for this is that the forces which act on the front tires tend to cause toe-out, and the initial toe-in setting is to counteract this tendency.

Most production cars have the suspension mounted in rubber bushings. This isolates road noise and vibration, but it allows some flexing at the bushings. On the front suspension some of this flexing is in a fore-and-aft direction at the tire. When the car is in motion the road drag on the tire pushes it towards the rear of the car. This force is resisted by the rubber suspension pivots, but some flexing occurs. This tends to make the front tires toe out, and it increases at higher speeds. When you hit the brakes the flexing becomes much greater because the braking force on the tire is much greater than road drag.

If the car is to handle properly the front tires must not toe out in normal driving.

When the tires are set at toe-out, there is a small slip angle on both front tires. The forces due to these slip angles try to steer the car, but the two tires each steer in an opposite direction. As long as the car has exactly equal slip angles everything is balanced and the car goes straight ahead.

The problem starts if the car hits a small bump, a gust of wind, or a change in road surface—anything to unbalance the slip angles on the two front tires. As soon as one slip angle is reduced slightly the opposite slip angle is increased and the car steers itself off to one side. The result is *wandering,* which the driver must correct by steering. It is also known as *high-speed instability.*

The change in slip angle can also be caused by the driver turning the wheel to enter a corner. The effect of toe-out is oversteer during transitional cornering. The handling may be different during steady-state cornering conditions.

Toe-in does the same thing as toe-out except in reverse. The result is that the car tends to keep going straight, improving *high-speed stability.*

Toe-in causes stability and initial understeer in a turn. Toe-out causes instability and initial oversteer. Usually you want just enough toe-in to keep the wheels from going into toe-out at speed. On most cars this is about 1/8 inch measured at the tire tread. See your car's factory specs for the recommended setting on your own machine. You can do fine tuning of the toe setting later at the test track.

TYPICAL FRONT AND REAR TOE SETTINGS

Front Toe (Inches)	Car Make and Model
1/16 toe-out	79-80 Pontiac Sunbird
	80 Firebird
	75-80 Chevrolet Monza, Vega
	69-71 Austin America
	67-68 MG 1100
	72-73 Renault 12, 15, 17
1/32 toe-out	70-71 Simca
	79-80 Chevrolet Monza, Vega
0 toe	79-80 Buick Riviera
	77-78 Cadillac Seville
	80 Seville
	74-80 Eldorado
	75-80 Olds Toronado
	79-80 Horizon, Omni
	69 MGC, MGC GT
	73 Porsche 911
	All Corvairs
1/32 toe-in	79-80 Cadillac
	79 Seville
	68-75 Olds Toronado
	72 Triumph GTG Mk III
1/16 toe-in	75-80 Buick Skyhawk
	75-77 Apollo, Skylark, Century, Regal
	75-76 Electra, Le Sabre, Riviera
	76-78 Cadillac Seville
	73-76 Chevrolet Chevelle
	75-77 Nova, Monte Carlo, Camaro
	76-80 Chevette
	75-76 Corvette
	75-80 Olds Starfire
	75-77 Omega, Cutlass
	75-76 Olds 88, 98, Pontiac Firebird
	73-77 Pontiac Grand Prix
	75-77 LeMans, Astre, Sunbird, Ventura
	80 Phoenix
	67-69 Austin Sprite
	71 Jaguar XJ6, XKE
	73 Opel GT
3/32 toe-in	80 Olds Omega, Buick Skylark
	67 Healey 3000
	71 Datsun 510
	68-70 Jaguar XKE
	71-72 MGB, MGB GT
1/8 toe-in	68-80 AMC, All Models
	80 Continental
	74-78 Ford Mustang II
	77-80 Pinto, Bobcat
	77-79 Thunderbird
	75-79 Torino, Montego, Cougar, Elite, LTD II
	75-80 Maverick, Comet, Monarch, Granada, Versailles
	78-80 Buick Century, Regal
	78-79 Skylark
	77 Electra, LeSabre, Riviera
	78 Chevrolet Nova, Camaro, Malibu, Monte Carlo
	75-78 Bel Air, Caprice, Impala
	77 Olds 88, 98
	78-79 Omega
	78-80 Cutlass, 88, 98
	78-80 Pontiac, Grand Prix
	79 Firebird, Phoenix
	80 LeMans
	67-70 Datsun SRL 311, SPL 311
	73 Capri
	72 Pantera
	71-72 MG Midget
	71-73 VW 1600, Type 1 & 3, 1700 Type 4
	73 Volvo 140 Series, 164
5/32 toe-in	74-77 Valient, Dart, Barracuda, Challanger
	76-80 Aspen, Volare, Diplomat, LeBaron
	75-79 Cordoba
	75-76 Chrysler, Imperial
	78-80 Chrysler
	77-78 Monaco, Fury, Charger, Magnum
	79-80 St. Regis, Magnum XE
	78-80 Buick Electra, LeSabre
	79-80 Chevrolet Camaro, Malibu, Nova, Monte Carlo
	78-79 Pontiac LeMans
	78 Firebird, Phoenix
	73 Opel 1900, Manta
3/16 toe-in	75 Dodge Coronet, Charger, Fury, Monaco
	75-80 Ford, Mercury
	75-79 Continental Mk V
	80 Fairmont, Zephyr
	77 Pontiac Firebird
	70-74 Camaro Z-28
	71-75 Pontiac Firebird
	75-76 Ford Thunderbird
7/32 toe-in	70-71 Toyota Crown, Pick-up
	72 Fiat 124 Spider
1/4 toe-in	80 Ford Thunderbird, Cougar
	74-76 Pinto, Bobcat
	77-78 Corvette
	68-74 Corvette
	71-75 Chevrolet Vega
	75 Monza 2+2, Astre
9/32 toe-in	71 Fiat 124 Spider, Sport Coupe
	73 Fiat 850 Spyder
	70-71 Toyota Corona Mk II
5/16 toe-in	80 Mustang, Capri

Rear Toe (Inches)	Car Make and Model
1/16 toe-out	78-80 Dodge Omni, Horizon
0 toe	75-78 Corvette
	70-75 Datsun 240Z, 260Z, 280Z
	63-65 Porsche 356B
	68-72 Porsche 911, 912, 914
	69-72 Triumph GT6
	73 Fiat 850 Spider
	All cars with solid rear axle
1/32 toe-in	62-72 Fiat 850 Coupe, Spider
	67-72 Triumph TR6
1/16 toe-in	66-72 BMW 1600, 2002, Sport Coupe
	68-74 Corvette
3/32 toe-in	71-72 Triumph Spitfire
3/16 toe-in	72 Fiat 128

Toe settings for a number of late-model cars. The amount of toe depends on the fore-and-aft stiffness of the suspension system, the weight of the car and the steer characteristics desired by the manufacturer. Notice that most front-wheel-drive cars use toe-out.

On some cars, direct measurements to set steering straight ahead are difficult to make. On this Chevy van it's easy. Measure from the center of the ball-joint bolt at the wheel to the center of the lower A-arm pivot at the frame. Check both sides—the distance should be the same when steering is centered.
If you can't find a good way to make a direct measurement, measure from the wheels to a set of parallel reference lines along each side of the car.

Many cars with front wheel drive have toe-out specified on the front wheels. This is because the driving forces on the front wheels are greater than the road drag and cause toe-in at speed. Thus a toe-out setting is used to prevent excessive toe-in due to suspension deflection. A front wheel drive car is inherently stable anyway, so having extra toe-in is not necessary or desirable.

Cars with independent rear suspension also have a toe setting at the rear. Because the driving forces tend to cause toe-in, the rear wheels are usually set at zero toe. The effect of toe-in or toe-out on the rear is much like the front, toe-out causing instability and oversteer. You should set the rear toe at the car manuracturer's specifications and do final adjustments at the test track. Usually zero or a slight toe-in will be the final setting, with high powered cars having the greatest need for a toe-in setting.

The toe setting at the front is controlled by a link in the steering system.

This car has rack and pinion steering with a toe adjuster at each end of the rack. Here the lock nut is being loosened to allow the adjuster to be moved. Each car differs in design, so consult your shop manual for method of setting toe.

There are various types of steering linkage, but they all control the toe by means of some adjustment, usually in the tie rods. Check your car's shop manual for the means of toe adjustment.

Most cars have a steering system that uses an equal length tie rod on each side of the car. Toe is adjusted by changing the length of the tie rods by means of a threaded adjustment. A clamp bolt is loosened and the tie rod is rotated in the threaded fittings to change its length between pivots.

If your car has equal length tie rods, before starting the toe adjustment make sure the tie rods are exactly the same length. Measure the distance between pivots with a steel tape and change each tie rod the same amount in opposite directions to make them equal length. Doing this makes the steering geometry the same on both sides of the car.

Now center the steering. It is necessary for the front wheels to be straight ahead so they will be in the normal straight line driving position. Due to the geometry of the steering linkage the toe setting changes as the steering is turned. A slightly turned steering will give false toe measurements.

Check the steering for centering by measuring under the car. Using a steel tape, measure between the outboard tie rod pivot and a suspension pivot on the frame. This distance should be equal on both sides of the car with the steering straight ahead. If it is off, turn the steering wheel slightly and measure again.

If the steering wheel is turned when the steering is centered, it can be taken off the shaft and put on a different spline. Do this after the suspension tuning is done and you have driven the car on a test run.

After centering the steering, toe is measured in one of three ways:

Direct measurements between the tires.
Angular measurement between a pair of tires.
Measurements between the tires and a pair of parallel reference lines.

Direct Measurements Between The Tires—Measuring between the tires is the simple way to get toe setting. It requires accuracy, so don't just rush out and measure between the tire treads with a steel tape. First of all, the tire treads are not very accurate. Second, the steel tape must be stretched straight across the car at axle height, and this is seldom possible without drilling a hole through the car.

To get an accurate place on the tire to measure from you must scribe a line on the tire. Jack up the car, rotate the tire, and scribe a thin line along the center of the tread, all the way around the tire with an awl or a sharp pencil. Take a good look at the scribe line with the tire rotating. It must not wiggle sideways. If it does, do the job again. Also check for slop in the wheel bearings. If the bearings have play the scribe line won't be accurate. Adjust the wheel bearing play to the factory recommended setting before scribing the tires.

To measure the toe setting you will probably need a gage.

A simple toe gage that can be purchased or even built in your own shop is a trammel bar. This is a length of metal, usually steel tubing, with two pointers mounted on it. The trammel bar is slipped under the car and rotated so the pointers are at the height of the center of the tire. The pointers are then lined up with the scribe lines, clamped in place on the bar, and then taken out to measure the other side of the tire. The difference between the two measurements is the toe figure.

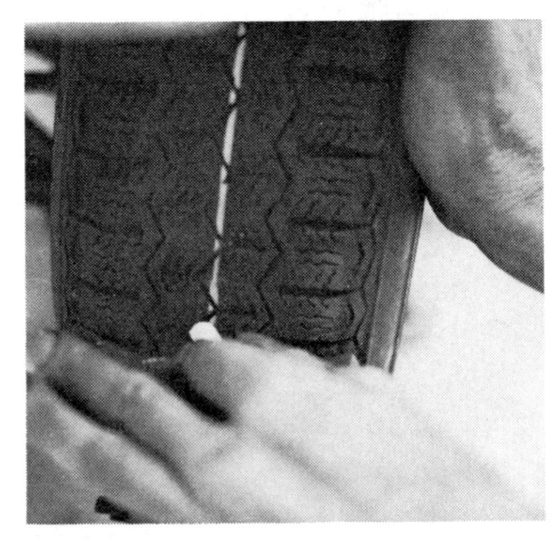

Before measuring toe, rotate the tire with one hand and mark it with the other. Here white chalk is being used for clarity, but a sharp pencil mark or scribe line is more accurate. After scribing the line, check it by spinning the tire to make sure the line runs true.

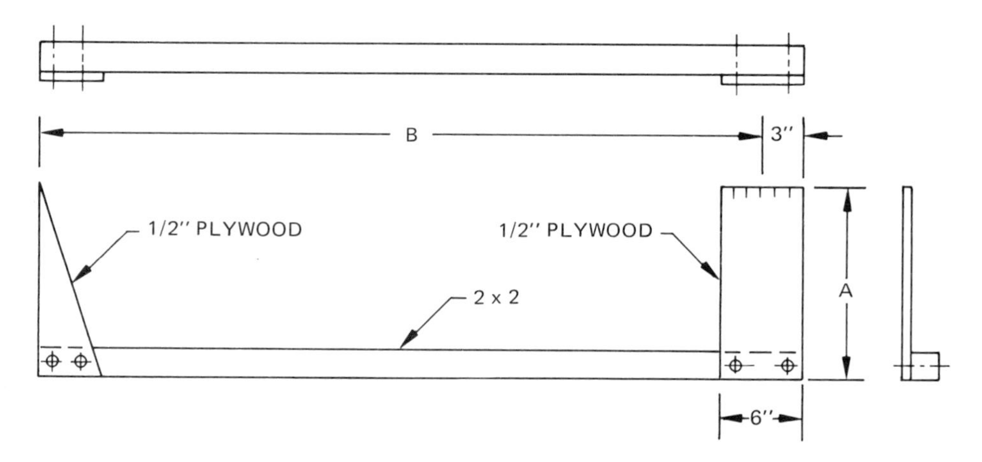

Plans for a simple trammel bar made of wood. The dimensions are determined from the car you are measuring. It is not adjustable, but it will work for cars of similar dimensions. The pointer on the left is lined up with the scribe line on one tire, and the scribe line on the opposite tire is compared to marks on the right hand plywood. A lighter, easier to use trammel bar can be fabricated from steel tubing.

A home-built trammel bar in position to take a toe measurement at the front of the tires. The pointer is lined up with the scribe line on the right front tire, and the left tire scribe line is compared to a pencil mark on the top surface of the trammel bar. A strip of masking tape on the trammel bar makes it easy to place reference marks. Not very elegant, but entirely practical.

The car has toe-in when the rear of the tires are spaced farther apart than the front. Be careful not to get this relationship mixed up! A car that is set with toe-out instead of toe-in will give the driver some wild surprises in the first corner. It's easy to mix up the two, so be careful.

The trammel-bar gage measures toe at the outside diameter of the tire, but there are other gages that take the measurement at the wheel rim. These gages are like a huge C-clamp that touches the outboard edge of each wheel rim. The measurement is taken at the front and at the rear of the wheel rim and the difference of the two measurements is the toe setting.

There is one problem using this type gage. The toe setting measured at the wheel rim is not the same as the toe setting measured at the tire tread. The setting at the rim will be a smaller number because the two reference points on the rim are closer to each other than the two reference points on the outside of the tire. To solve this problem the toe setting should be converted to an angular dimension.

The graph shown in Figure 21 will help. Start with the toe setting recommended by the car manufacturer. Entering the graph with this toe setting and the diameter it is measured at—usually the outside of the tire—convert this to an angle. Then this angle can be converted back to a toe setting at any other diameter.

Let's say for example you have a car that comes stock with 26-inch outside diameter tires and a factory toe-in setting of 1/8". In Figure 21 the factory setting corresponds to an angle of 0.275 degrees. If you wish to measure the toe setting at the rim of a 14-inch wheel, the toe setting at a 14-inch diameter is about 9/128 inch, or 0.07" as closely as I can read between the curves.

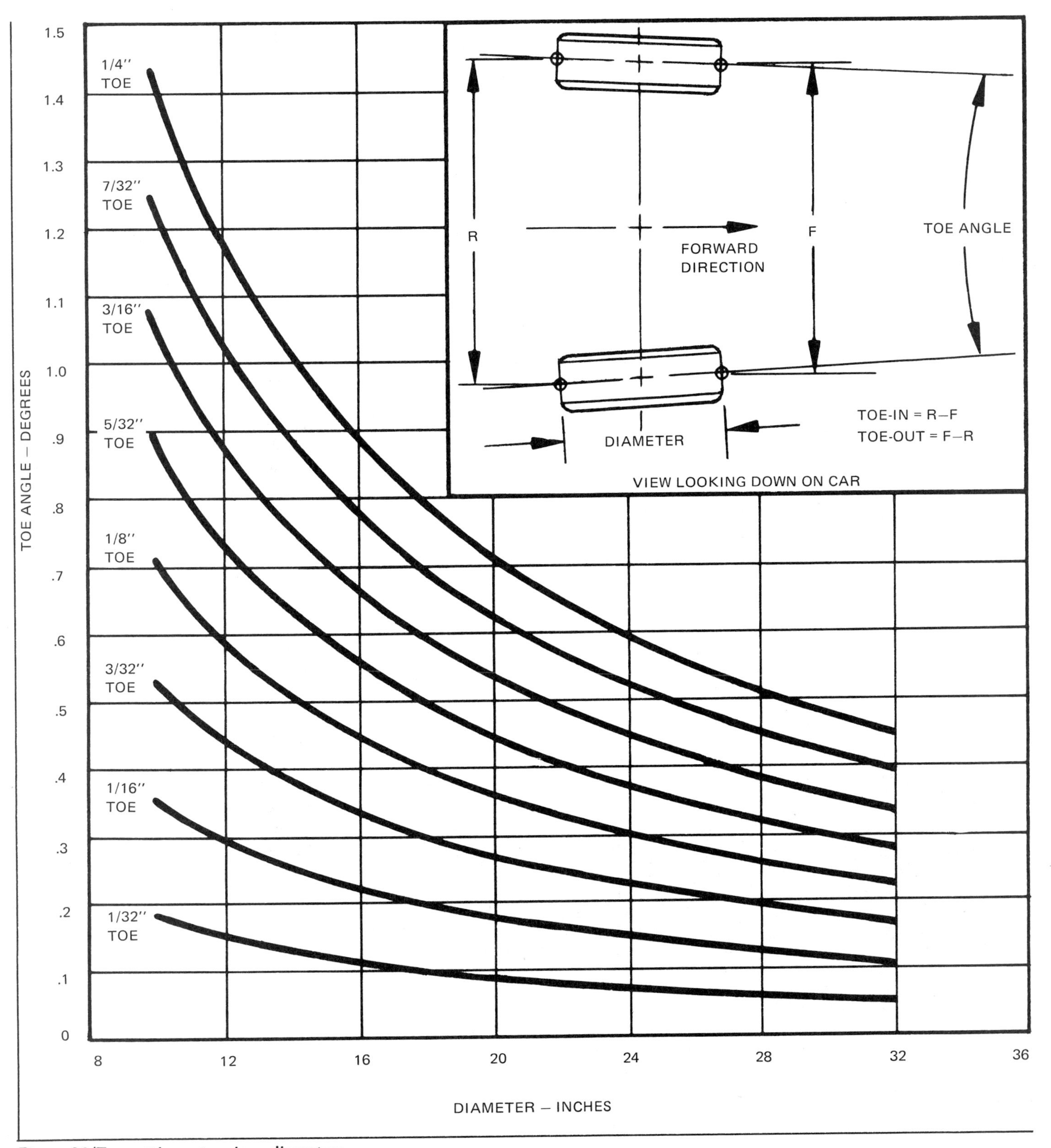

Figure 21/Toe settings at various diameters.

If you don't like to interpolate between curves, you can do it by proportion. You know what the toe is supposed to be, call it T_a, at some measuring diameter we can call D_a. For the same toe angle, you want to know what the toe will measure at some other diameter, call these T_b and D_b.

It's true that: $\frac{T_a}{D_a} = \frac{T_b}{D_b}$

or, $T_b = T_a \times \frac{D_b}{D_a}$

In the example used above,

$T_b = 0.125'' \times \frac{14}{26} = 0.067''$

which is pretty close to the 0.07" I was guessing from the curves of Figure 21. You can see that you don't measure toe with a piece of chalk and an old cloth tape from your mother's sewing basket.

ANGULAR MEASUREMENT BETWEEN A PAIR OF TIRES

The direct measurement toe setting can be converted to an angular measurement as shown in Figure 21 or by simple trigonometry. Toe can be measured directly as an angle by a gage. A very precise and useful portable gage is made by Dunlop. This optical instrument uses mirrors to make the measurement.

The Dunlop gages have two parts. One rests against each wheel, with pointers against the outboard surface of the wheel rim. Obviously this is only as accurate as the wheel, so rotate the wheel to be sure it runs true. If it does not, take several toe readings with the wheel rotated to various positions and average the readings.

Dunlop gages are sold through a number of dealers such as LRE of Millerton, New York. These gages are portable, but they are large precision instruments that require special handling. They can be taken to the test track if you have plenty of room and use care to pack them well. Dunlop gages are commonly used at the track for quick toe changes.

Front-end shops usually have large non-portable gages for measuring toe. These are mechanical or optical devices, and they offer very quick measurements. This type of gage is great for turning out work in a hurry, but they are too big and expensive for home use.

MEASUREMENTS BETWEEN THE TIRES AND A PAIR OF PARALLEL REFERENCE LINES

Measurements to parallel reference lines allow not only the toe to be determined, but also the squareness of the chassis and wheels. It is possible for the rear wheels to be steered to the side. It is also possible to have the rear wheels displaced sideways from the front wheels. The car is not square in such cases, and it will travel down the road in a sideways attitude. Perhaps you have followed a car with this condition on the freeway. A car with this condition is subject to instability and has different handling characteristics in right and left turns.

Using parallel reference lines does not require a toe gage. This method is handy in your shop, and a toe gage is then used at the track for quick adjustments. Once the car is checked for squareness with the parallel reference lines, there should be no need to check it every time you change the toe setting. The toe gage does not allow you to check squareness, but that shouldn't change unless the car is wrecked or a great amount of suspension adjusting done.

When using parallel reference lines, all measurements start with the centerline of the chassis. The centerline is measured midway between the suspension pivots on the frame. Don't assume that the outside of the frame or the body is true, because it generally is not built that closely. Only the suspension pivots matter, and all measurements should be taken from them.

The toe-in on this car is being checked with a Dunlop optical gage. The two parts are set on each side of the car against the wheel rim, and the mechanic looks through the telescope in front of him. Opposite is a mirror facing the telescope, and the toe-in is observed from a scale reflected in the mirror. This is a very accurate and expensive instrument.

On a car with a solid axle and leaf springs, the front spring eyes represent the suspension pivots.

Use the upper or lower pivots on the frame, whichever are the easiest to measure from. Take careful measurements and mark a point on the frame exactly halfway between a pair of pivots. Do this at the other end of the car and then make a reference line between the two midpoints. This can be done with a string stretched tightly between two points and just touching the marks on the frame. This is the centerline of the chassis; mark it near the ends of the car in a spot convenient to measure from.

Now set up two straight reference lines, parallel to this centerline, on each side of the car. Have the car resting on its tires at ride height, and place the reference lines outside the tires a few inches away from the wheels. You measure toe from the reference lines to the wheel rim, so elevate the reference lines to a height above the floor even with the center of the wheels. Use a straight piece of metal for the reference lines, or string stretched tightly between two points. Adjust them so they are exactly parallel to the centerline of the chassis. After you get them set up, kill anyone who stumbles over them.

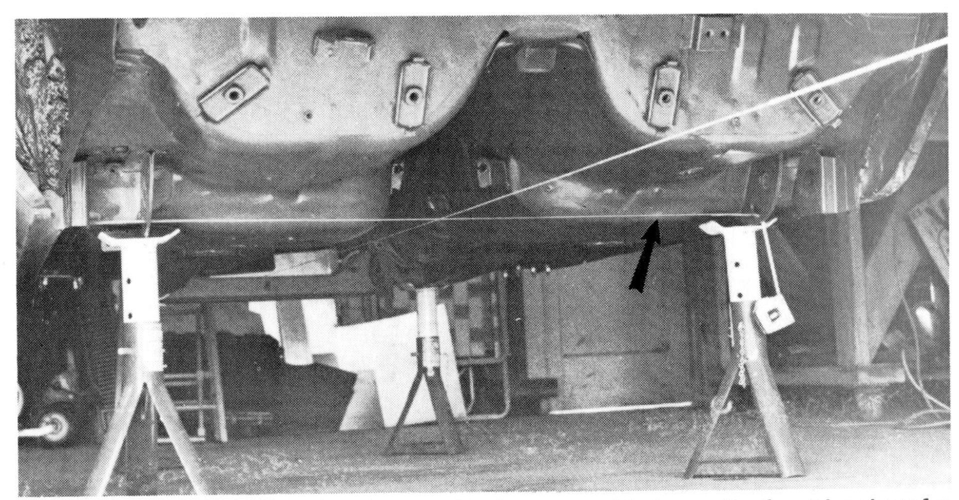

At the rear of this Pinto the frame centerline is measured between the front brackets for the leaf springs. The steel tape (arrow) is stretched between the two brackets and the string is set up at the center just touching the tape. The centerline is marked on the car at the rear of the body for future use.

On this Pinto the centerline is found by measuring between the lower front suspension pivots. Then it is marked on the front frame crossmember. The car shown here has the engine and suspension removed, which makes the job really easy. A bolt is slipped into the suspension pivot hole in the frame. A plumb bob string is wrapped around the bolt on each side of the car, and the tape measures between the two strings. Halfway between the strings is the centerline.

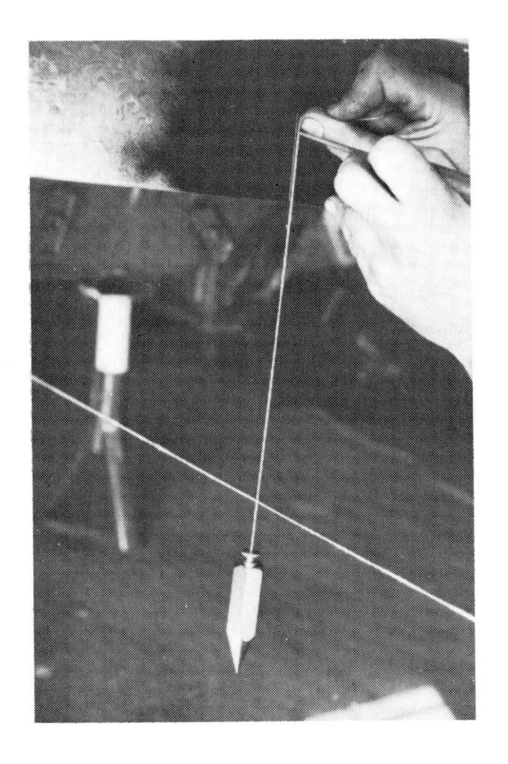

To scribe the centerline on the body or chassis drop a plumb bob down to the string running down the centerline of the car. Here the spot was first covered with black spray paint to make the scribe line easier to see.

After you have a centerline marked on the car, setting up parallel reference lines is easy. Measure from the centerline to the reference line at front and rear. Having two people to do this job is very helpful.

Toe measurements are made with a scale from the reference line to the outboard surface of the wheel rim. Pick a spot on the wheel that is clean and true. Before measuring check the wheel for trueness by rotating it. If the wheel runs out laterally you should take an average of several toe measurements. This is a long and time-consuming process, so hopefully the wheels are true.

Remember when setting the toe to parallel reference lines you are only measuring one wheel at a time. Thus each wheel has a toe setting *one half* of the total toe setting between a pair of wheels. For example if the car manufacturer specifies 1/8″ toe-in at the front, each front wheel would have 1/16″ toe-in when measuring to its reference line. Also remember that you may not be measuring toe at the diameter implied in the specs. You know how to convert the numbers by the proportion given earlier.

Check the toe setting on the rear suspension by measuring from the front and the rear of each wheel rim to the reference line. The distance to the front of the rim minus the distance to the rear of the rim is the toe-in for that wheel. If the toe setting is zero, then the two distances must be equal. With independent rear suspension it is possible to have one wheel with toe-in and the other with toe-out. This

Setting rear toe is easy on this racing car. First the jam nuts locking the adjustable lower radius rod are loosened as shown. Shortening this rod moves the upright forward, increasing the toe-in. This adjustment changes the caster of the rear upright unless the upper radius rod is adjusted a like amount.

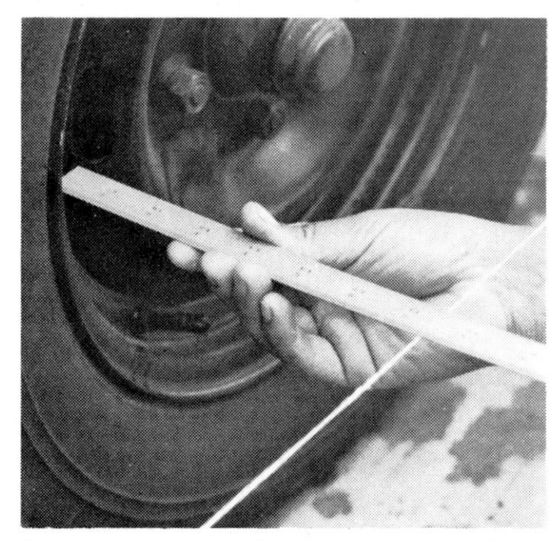

Here the toe-in is checked by measuring to the reference lines set parallel to the center of the chassis. The reference lines are light string stretched between two jack stands at the height of the wheel centers. This toe-in measurement is taken at the wheel rim on an accurate machined surface.

situation can be corrected by this parallel reference line method, because each rear wheel is checked individually.

Now set the front toe, first making sure the steering is centered. Checking this is easy with parallel reference lines, because if the steering isn't centered, one wheel will have toe-in and the other toe-out. If this condition exists, just turn the steering wheel slightly and measure again.

SQUARING THE CHASSIS

Perhaps this section could also be titled "frame alignment." You check that the four corners of the car are relatively square and not warped to one side or the other. You have previously set the toe on all four wheels, but the rear wheels must also be centered directly behind the front wheels. Also the wheelbase should be the same on both sides of the car, and the wheels on the left side of the car should be lined up fore-and-aft with those on the right side of the car. All these measurements come under the general topic of squaring the chassis.

A car can be out of square because of three things—sloppy manufacturing tolerances, wrong suspension adjustments, or a bent frame. It is often hard to tell what the cure is, and sometimes there is no cure. However, it is good to know what the situation is, so let's check the chassis.

To begin you must establish four points to measure from, one at each corner of the car. You are most concerned with the tires, so select points as close as possible to them. From the center of each wheel spindle drop a plumb bob down to the floor. Try to do an extremely accurate job here, making sure the plumb bob is lined up with the center of the spindle exactly. You may have to remove the wheels. If so, support the car under the suspension so it is sitting as close as possible to the normal riding position.

With the four points established on the floor you are ready to take measurements between them. Measure all distances shown in Figure 22. For a perfectly squared chassis, **C** = **D** and **E** = **F**. The distances **A** and **B** do not necessarily have to be equal.

The problem is making the necessary corrections. Some cars can be adjusted and some cannot, depending on the suspension design. The error may be in the suspension adjustment or on the frame itself.

To set up for testing chassis squareness, center the steering, plumb bob from the center of each spindle or axle. Masking tape is on the concrete floor for easy marking of the spot where the plumb bob touches. After all four points are marked on the floor the measurements shown in Figure 22 are made.

The way to check the frame is to drop your plumb bob from the suspension pivots on the four corners of the frame. Make measurements on the floor for the frame as shown in Figure 22. If the error at the frame is the same as at the wheels, then the frame is bent. If the suspension pivots on the frame are square, but the wheels are not, then the error is in the suspension.

If the frame is in error, no adjustment will work properly. The car should be taken to a frame-alignment shop and straightened if possible. On some cars this is relatively simple and others not. A frame-repair expert can tell you.

If the error is in the suspension, then adjustment is possible. If distances **C** and **D** are equal but distances **E** and **F** are not, then the rear suspension is displaced sideways relative to the front suspension. On a solid rear axle, this sort of problem is corrected by moving the entire axle sideways on its mounts. If the lateral location of the axle is with a linkage, this linkage can be shimmed or adjusted. If leaf springs locate the axle, then they can be moved sideways slightly in their mounts. A large movement would require new brackets, but even this is not impossible.

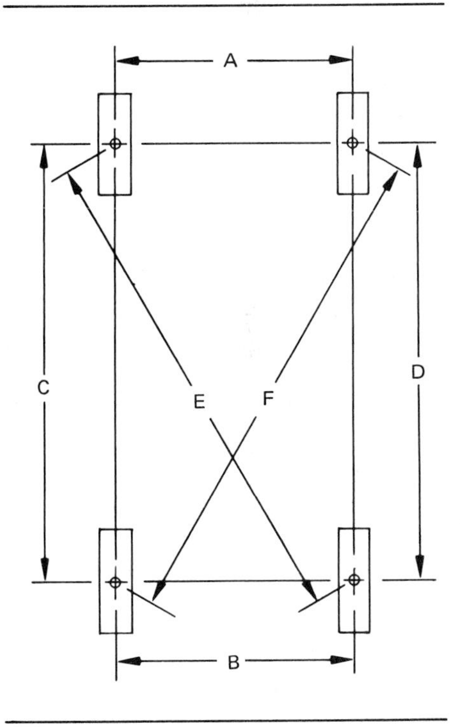

Figure 22/This shows a view looking down on the car with the small circles representing the reference points for checking chassis squareness. Locate the reference points with a plumb bob as close to the tire as possible. It is not necessary to use the same width between points at the front and the rear of the car. Just use the same points on the chassis at the right and left side of the car.

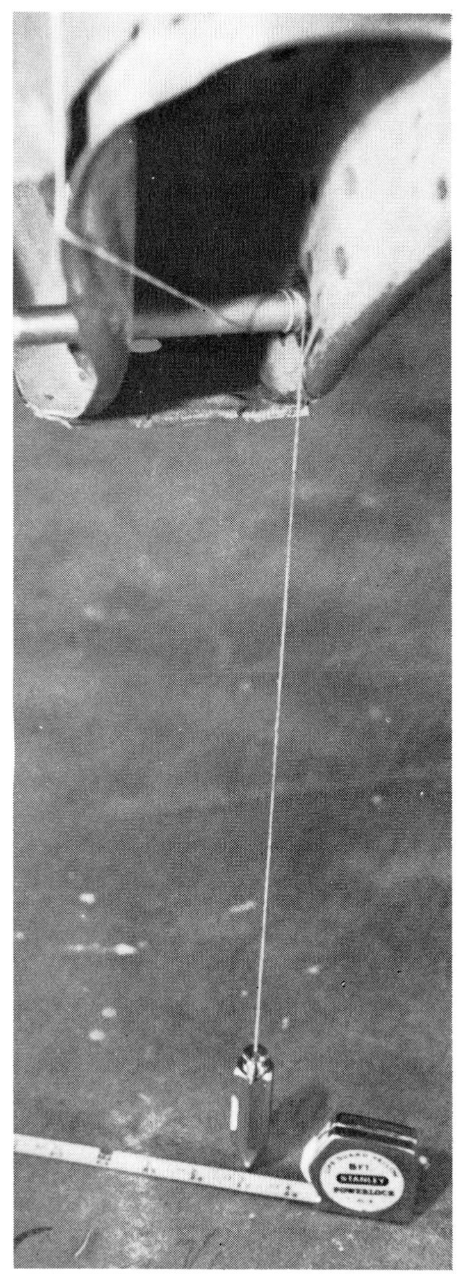

Before mounting the suspension on his racing Pinto, Guy Ober is checking the frame for squareness. The plumb bob string is wrapped around the lower A-arm mounting bolt with the string against the forward side of the bracket. A similar measurement is made on the other side of the frame. Measurements are taken between points marked on the floor.

It is also possible for one wheel to be ahead of where it belongs. If for example the left front wheel is too far forward, then both **C** and **E** would be too long relative to **D** and **F**. In such a case it is usually possible to move the front wheel back by shortening the strut locating the lower suspension arm. On many cars this strut is adjustable. However, watch out for caster changes when making such an adjustment. If the caster cannot be corrected another way, then you will have to leave the chassis out of square.

Locating the exact cause of the error can be a matter of trial-and-error. If there is an easy adjustment to make, try it and see how this affects the various dimensions. If the ones in error get closer together, then you did the right thing. However, always be aware that an adjustment in wheelbase or in lateral location can change the caster, camber, or toe setting of that wheel. The correct toe, camber, and caster are *more critical* than a perfectly square chassis.

As a final check for squareness, check that any adjustable suspension links on both sides of the car are the same length between pivots. At times unequal adjustments can cause errors in squareness.

Some errors may be unavoidable. If the distances **C** and **D** are within about 1/2", then the car will most likely handle reasonably well. Obviously you would like to be closer, but don't throw the car away until you test drive it with all other settings correct.

CAMBER

Camber is the angle the plane of the tire makes with the vertical. This is illustrated back in Figure 5. Camber is a very critical setting, particularly if your car has wide-tread tires. Again, start with the manufacturer's settings if available. If not, try zero camber for general road use, and about 1/2-degree negative for a racing car. The final camber setting will be determined out on the test track using tire temperature as a guide.

Camber is measured with a specialized level called a *camber gage.* These are available at auto supply stores. A really good camber gage suitable for all cars is manufactured by Wayne Mitchell Engineering. This precision gage features correction for non-level floors and fits all types of wheels. It offers more precise measurements than the common gage used by front-end shops.

You can set some race cars with a very simple tool, providing the body is removable. You need a straightedge long enough to lay across the tires, and a carpenter's framing square. Rest the straightedge across the tops of the tires and put the long side of the square against the outer face of the wheel or tire. Slide the square downward until the short leg just touches the straightedge and measure the gap as shown in Figure 23.

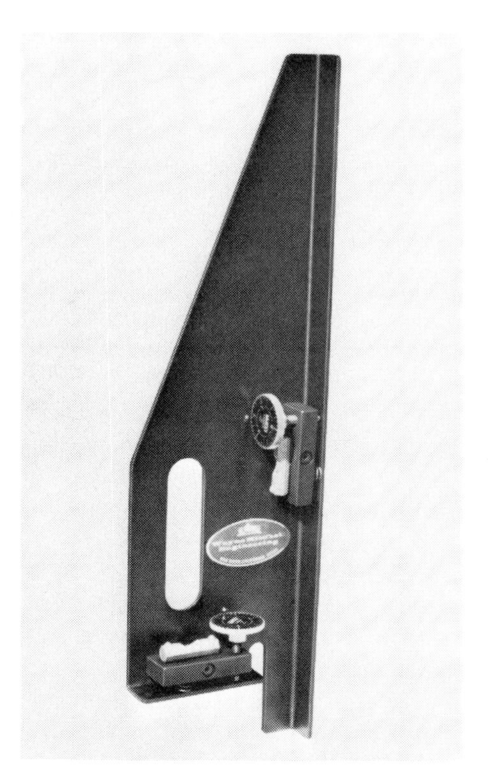

The Wayne Mitchell Engineering camber gage was developed for use on racing cars, but it works great for any vehicle. The vernier adjuster on each level allows very precise camber measurements to be taken. The gage is held against the outer surface of the wheel to measure camber.

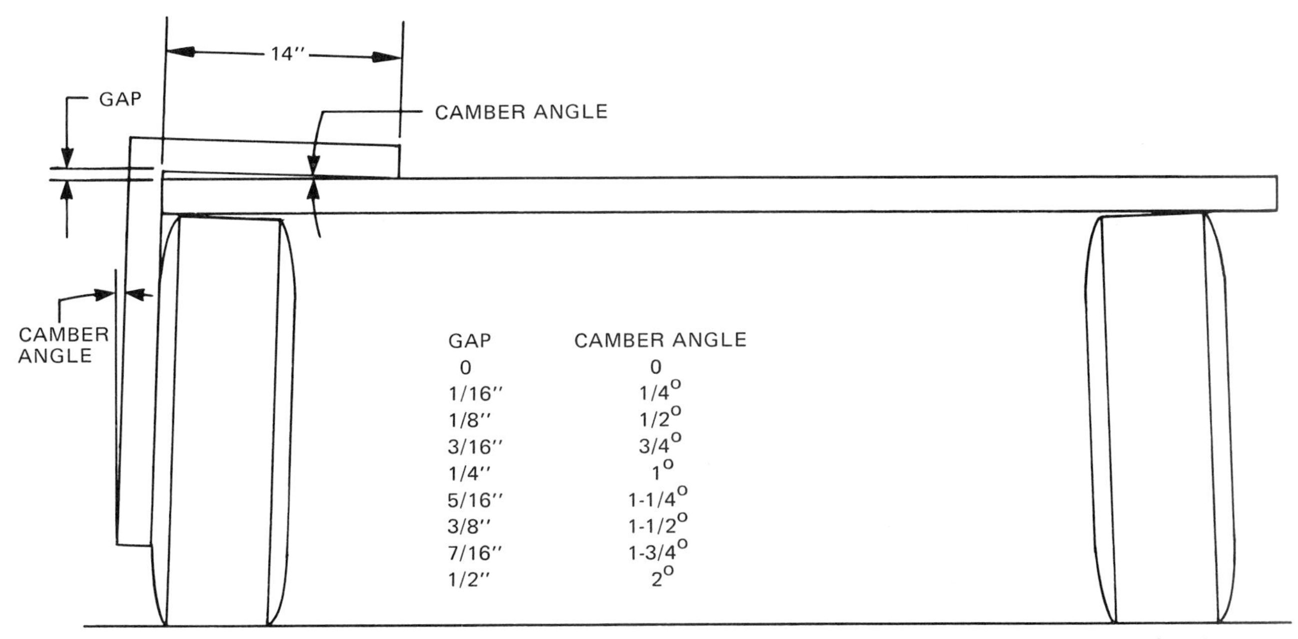

GAP	CAMBER ANGLE
0	0
1/16"	1/4°
1/8"	1/2°
3/16"	3/4°
1/4"	1°
5/16"	1-1/4°
3/8"	1-1/2°
7/16"	1-3/4°
1/2"	2°

Figure 23/A standard framing square can be used to measure camber on a low car. The above drawing shows the framing square in position. The gap between one end of the square and the straightedge can be converted directly to a camber angle. This method works on non-level ground. Make sure the framing square measures 14 inches long as shown.

The camber is measured on this open wheel car very simply with a straight piece of steel tubing and a framing square. This only works on cars with a removable body and a very low chassis between the wheels. The conversion of gap between the square and the straightedge into degrees of camber is shown in Figure 23.

The adjustment for camber takes various forms. On most race cars it involves changing the length of one of the A-arms. On a typical production car, camber can be adjusted with shims, with an eccentric bushing, or some other means. Consult your shop manual for the method of adjustment. On cars with solid axles, bending the axle is the only way to set camber, and this should be done at a qualified front end shop with all the big tools necessary.

When adjusting camber, you should write in your notebook how much the camber is affected by a given amount of mechanical change. Record the camber change for a certain shim thickness, or the camber change for one turn on a rod end. This is a big help when trying to make a fast adjustment at the track.

Ideally the car should have zero camber under all conditions of straight-line driving and cornering. However, most suspension systems cause the camber to change with travel, so a compromise is required. To help make an intelligent choice for static camber setting you should test the car for camber change during vertical suspension movement and body roll.

This Datsun has MacPherson-strut suspension with a modification. The camber is adjustable by means of sliding the top end of the shock strut in the slots shown. The mounting bolts are being loosened prior to making the camber adjustment. Normally these cars do not have a camber adjustment.

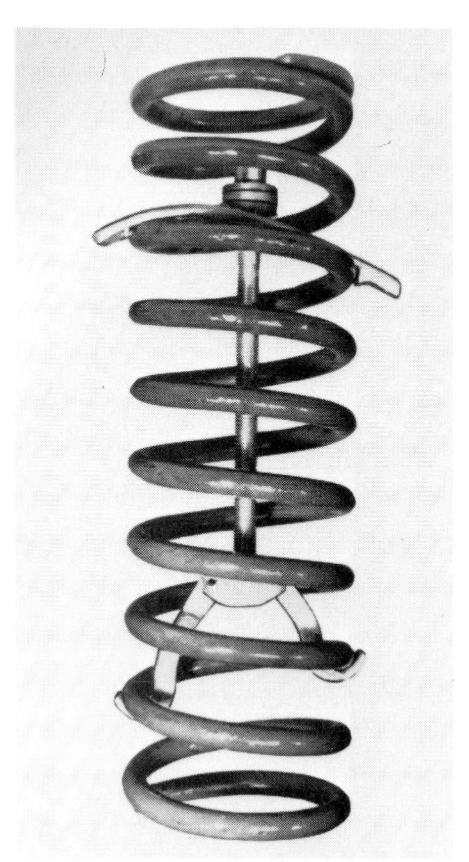

This TRW spring compressor is a real help in installation and removal of coil springs. It can be used on most applications including the front springs on intermediate Ford products. Using a ratchet wrench on the compressor bolt compresses springs rapidly and the holding hooks are adjustable to the spring diameters.

This table shows average camber settings for a number of late-model cars. Average figures are used, because many cars specify a different figure for the right and left side of the car. Putting more positive camber on the left front wheel compensates for a crowned road surface and makes the car run straighter. For high-performance driving the camber should be the same on each side, and the driver must live with a small amount of pull to the side on a crowned road.

The rear camber settings are sometimes different than the front on cars with independent rear suspension. Some of these are also shown.

To test the camber change you must first remove the springs. This is also required when bump steering the car as described in the next section. A spring compressor may be required. A good unit is the Coil Spring Compressor Tool made by TRW Replacement Division, Cleveland, Ohio. This is inserted into the front coil spring and tightened down to compress the coil. Once the preload is taken off the spring it can be removed from the car easily. Just be careful of a compressed coil spring—it can kill you if it gets loose.

To test for camber change with vertical suspension travel use a jack to move the chassis to various ride heights and measure the camber at each position. Plot camber versus ride height on a graph. This will be useful for future reference because it tells you how much to change the camber if you adjust the ride height. The tires are resting on the floor during the test,

TYPICAL CAMBER SETTINGS

Front Average Camber	Car Make and Model
−3/4°	71-72 Camaro Z-28
−1/4°	80 Pontiac Sunbird
	74-76 Cadillac Eldorado
−1/8°	71-76 Cadillac
	74 Cadillac Eldorado
0°	79-80 Buick Riviera
	77-78 Cadillac Seville
	77-80 Eldorado
	78-80 Olds Toronado
	73 Ford, Mercury
	72 Pantera
	72 Jaguar Series 3 V-12
	72 Mercedes 350SL, 600
	69 MGC, MGC GT
	73 Porsche 911
	70-73 Porsche 914
	All Corvairs
+3/16°	74-75 Ford, Mercury
+1/5°	77-80 Chevrolet Chevette
	78-80 Monza, Vega, Olds Starfire
+1/4°	74-80 AMC, All Models
	75-80 Buick Skyhawk
	75-77 Chevrolet Monza
	71-77 Vega
	70-80 Ford Maverick
	75-80 Granada, Versailles, Comet, Monarch
	80 Mustang, Capri
	75-77 Olds Starfire
	77 Olds Toronado
	68-71 Jaguar XKE
	75-79 Pontiac Sunbird
	75-77 Astre
	70-71 Simca
	72 Triumph TR6
	73 Volvo, All Models
+5/16°	78 Omni, Horizon
+1/3°	80 Pontiac Poenix
	79-80 Chrysler Omni, Horizon
+3/8°	68-80 Dodge, Plymouth, All Models
	75-79 Chrysler, Cordoba
	76 Cadillac Seville
	74-78 Ford, Mercury
	74-79 Torino, Montego
	78-79 Fairmont, Zephry, Mustang, Capri
+2/5°	80 Ford Thunderbird
	79-80 Cadillac
+7/16°	80 Fairmont, Zephyr, Sedan
+1/2°	75-76 Chevrolet Chevelle, Monte Carlo
	79-80 Citation, Malibu, Monte Carlo
	78-80 Olds Cutlass
	80 Omega
	78-80 Pontiac LeMans, Grand Prix
	78-80 Buick Century, Regal
	80 Skylark
	77-78 Cadillac
	79-80 Ford, Mercury
	80 Continental
	80 Fairmont, Zephyr, Wagon
	74-78 Mustang II
	77-80 Pinto, Bobcat
	71 Fiat 124 Spyder, Sport Coupe
	71-72 Jaguar XJ6
	72 Mercedes 220, 250, 280SE/SEL 4.5
	71-72 Peugeot 304
	70-73 VW 1600, Type 1
+5/8°	Peugeot 504
+3/4°	75-77 Buick Apollo, Skylark, Century, Regal
	75-77 Electra, LeSabre, Riviera
	77 Pontiac Ventura, LeMans
	79 Phoenix
	75-77 Grand Prix, Pontiac
	75-77 Olds Omega, Cutlass, 88, 98
	75-76 Ford Thunderbird
	74-76 Pinto, Bobcat
	75-76 Chevrolet Nova
	77 Monte Carlo
	68-80 Corvette
	67-69 Austin Sprite
	69-71 Austin America
	71-72 MG Midget
	73-74 Camaro Z-28
+4/5°	77-80 Chevrolet Nova, Bel Air, Caprice, Impala
	78-79 Olds Omega
	78-80 Olds 88, 98
	78 Pontiac Phoenix
	78-80 Pontiac
	78-79 Buick Skylark
	78-80 Electra, LeSabre
+1°	70-80 Pontiac Firebird
	70-80 Chevrolet Camaro
	75-76 BelAir, Caprice, Impala
	71 Datsun 510
	71-72 MGB, MGB GT
	73 Opel GT
	70-73 Toyota Pick-up
+1-1/4°	70-71 Toyota Corona MkII
+1-1/3°	67-70 Datsun SRL 311, SPL 311
	70-73 VW 1600, Type 3
+1-1/2°	73 Opel 1900, Manta
	72-73 Renault
	72-73 Toyota Corona MkII
+1-3/4°	73 Fiat 850 Spider
	67-70 Ford Cortina 1600
+2-3/4°	72 Triumph GT6 Mk3

Rear Average Camber	Car Make and Model
−3-1/4°	71-72 Triumph Spitfire
−2°	66-72 BMW 1600, 2002
	75 Fiat X-19
−1-1/2°	66-72 BMW Sport Coupe
−1-1/4°	71-73 VW, Type 3
	PRE-71 VW 1500 & 1600, Type 1 With Double U-Joints
−1°	78-80 Dodge Omni, Plymouth Horizon
	73 Fiat 850 Spider
	69-72 Triumph TR6
	71-73 VW, Type 4
−7/8°	69-74 Corvette
	68-73 VW, Type 2
−3/4°	68 Corvette
	70-72 Datsun 240Z
	62-72 Jaguar XKE
	63-72 Porsche 911, 912
−11/16°	75 Corvette
−1/2°	70-73 Porsche 914
	79-80 Corvette
−1/4°	72 Fiat 128
0°	69-72 Triumph GT6
	All cars with solid axle rear suspension
+3/4°	63-65 Porsche 356B
+1°	PRE-71 VW 1500 Sedan with swing axles
	71-73 VW Beetle, Type 1
+3°	PRE-68 VW, Type 1 40HP, Type 2 with swing axles

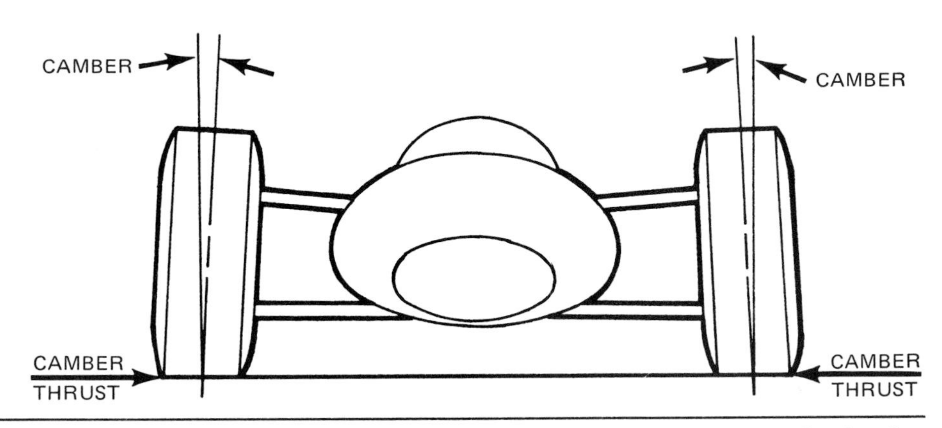

Camber thrust acts inward on a tire with negative camber. This is shown in the sketch above. If the car had positive camber the camber thrust would act outward on each tire.

and the chassis is moved with the jack. Make sure there is no body roll. Measure both front and rear, unless the car uses a solid rear axle.

Now test for camber change with body roll by tilting the chassis at various angles. Measure the angle of roll with a level and protractor. Plot a graph of camber on the outside tires versus body roll. Don't worry about the camber on the inside tire, as it does very little work in a corner. When making this test put a block under the chassis in the center of the car to keep the average ride height constant. If you change ride height as well as body roll you won't get accurate results. The block under the frame should be as thick as the static ground clearance at that point on the frame.

Once you know the relationship of camber and body roll at each end of the car you can make an intelligent choice on static camber and how much to limit the roll angle. For best cornering power, the camber should be zero on the outside tire in a corner. Let's assume that your car has a camber change of 2 degrees positive at at body roll of 3 degrees. If you select 3 degrees as the maximum body roll, the static camber would have to be set at 2 degrees negative to make it zero in a corner. Perhaps a better compromise might be to limit the body roll to 2 degrees by increasing the anti-roll bar diameters. At this body roll let's say the camber change is 1 1/2 degrees positive. A reasonable compromise would be 1 degree negative static camber setting, which would make the camber 1/2 degree positive in a corner.

With some suspensions it is impossible to get zero camber in a corner without using a very large static negative camber. A large amount of negative camber has disadvantages. It adds drag in a straight line and it wears out the tires faster. Also it reduces braking traction, because camber affects traction in that direction too.

For street driving a car with negative camber is apt to be unstable. The camber causes a side load on the tire when driving in a straight line. This force is called *camber thrust,* and it acts inwards on a tire with negative camber. The camber thrust is equal on both tires when driving on a perfectly straight smooth surface, but if one side of the car is disturbed by a bump there is an unbalanced side force on the car. This will cause wandering on bumpy surfaces. In addition the camber thrust causes additional deflection in the suspension linkage, which changes the ideal toe setting. The toe may have to be re-adjusted after changing the camber. The amount of toe change will be very small, and it will vary with the design of the car.

There have been rule-of-thumb statements that a certain amount of camber is equivalent to so much toe-in, but these rules hold true only for a specific car. It is best to set the toe on the test track or on the road according to driver feel.

When you tested the camber change with body roll you made sure the ride height did not change during the test. However, in actual driving the ride height may be affected by the cornering loads on the suspension. This is known as *jacking effect,* and it is serious on any car with a high roll center and independent suspension. Jacking effect causes the car to increase its ride height in a corner. The greater the cornering force the higher the ride height.

The reason for this is the side load acting through the roll center causes forces on the suspension tending to raise the car on its springs. This happens at the same time as body roll, so the two effects are combined. The resulting camber is a combination of the two effects, usually positive camber on the outside tire.

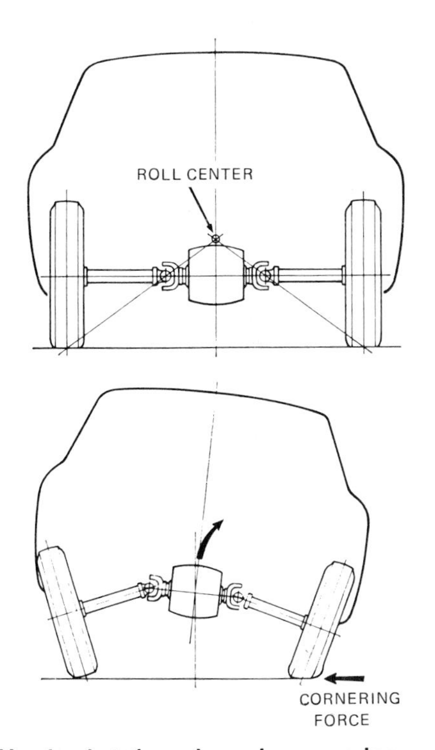

Here's what the swing-axle suspension does in a turn. The jacking effect raises the ride height which causes positive camber on both tires. This reduces the cornering power and causes oversteer. Old VWs, Renaults, Fiats, and early Corvairs all suffer from this problem. All these cars have changed to a more modern rear suspension, having a lower roll center and less jacking effect.

This effect is particularly bad on cars with swing-axle suspension, as they have the highest roll center of all.

To combat the jacking effect it is necessary to make the springs very stiff, use extra devices that add vertical stiffness without adding roll stiffness, and set the car up with a lot of negative camber. Modifications to the suspension of swing-axle cars are discussed in detail in Chapter 4. The negative camber should be determined by test, but it will be a poor compromise for any car used for both street driving and racing. The proper camber for racing is totally unacceptable for the street, usually over 2 degrees negative. A swing-axle car set up properly for the street will usually not be ideal for racing unless it is so stiff it jars your teeth. All facts considered, swing axle cars are not ideal for high-performance driving.

The old-type VW rear suspension is a swing axle. The camber changes a great deal as the chassis moves on its suspension. On this racing car the suspension is set at negative camber so the jacking effect in a turn causes a minimum of positive camber. The results are a compromise at best. That's not an anti-roll bar, it's a Z-bar—described later in the text.

In the tests for camber change with suspension movement and body roll it is necessary to determine the characteristics of the front and rear suspensions separately. If you have an independent rear suspension the camber may act exactly the same as the independent front suspension. This is an ideal situation for predictable handling. If the camber change is different between the front and rear, the car will be harder to drive. A single wheel bump or a rapid change in direction may require some really expert reactions on the part of the driver. Whatever the camber characteristics are, you will be forced to live with them unless you re-design the car. The only hope is to understand what is happening and make changes that are good compromises.

An example of a car with widely different camber characteristics at each end is the old VW. This chassis is commonly used for home-built cars, dune buggies, and racing cars. These VWs use trailing-arm front suspension giving one degree of positive camber on the outside tire for each degree of body roll.

The rear suspension is quite different. It has a large amount of camber change with vertical suspension movement, going into positive camber with increasing ride height. With body roll the outside tire has small negative camber slightly increasing with body roll. In a corner the very large jacking effect more than cancels out the negative camber due to body roll, resulting in considerable positive camber on both rear tires.

If this car is set up with typical camber angles for racing it has zero on the front and say two-degrees negative at the rear. When the driver hits the brakes, the rear rises causing less negative camber in the rear. This is pretty good for braking. Now the car is turned into a corner. If the driver has let off the brakes the car settles down into the negative camber position and the car is initially neutral. However, if the car hasn't settled yet, the lack of negative camber results in initial oversteer.

Then, in the turn, an application of power changes the camber and upsets the handling even more than the power should alone. Any bumps in the turn compound the problem. If the corner is banked the rear camber goes more negative and the front does not change. What is neutral handling on a flat corner may become oversteer on a banked corner.

You can see how the variables can upset the handling of the car in so many ways. The suspension system takes away some of the driver's control with all this variation in camber between the front and the rear. Balance is the key to control, and this type of suspension is far from balanced in its properties.

BUMP STEER

Adjusting bump steer is an advanced suspension-tuning technique and often makes the difference between average and outstanding handling. Many elusive twitches and wiggles can be blamed on bump steer, and often it is the cause of high-speed stability problems. Bump steer is the change in toe setting as the wheels move up and down. It can happen at front or rear. Ideally you want zero bump steer—the toe setting should remain constant no matter where the suspension moves. Here's how to check bump steer and adjust it.

First you need a tool to check bump steer. Of course it could be checked with a toe gage, but a bump-steer gage is easy to build and often a lot handier to use. A drawing of a home-made gage is shown in Figure 24. This is made from two pieces of plywood and a piano hinge. The gage is placed on the floor next to the wheel, and one side is held down with a heavy item such as a battery or a toolbox. The other side of the gage rests against the wheel in a nearly vertical position. The two bolts through the gage are adjusted to just touch the wheel rim. A dial indicator can be substituted for one of the

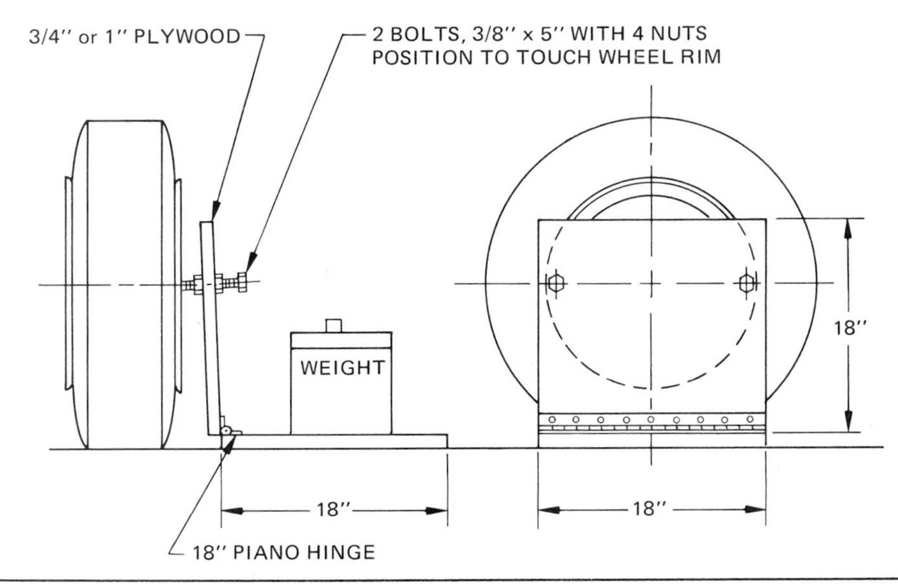

Figure 24/You can build this handy bump steer gage out of two pieces of plywood, a piano hinge, and two bolts. The gage is positioned on the floor, held by a heavy item such as a battery. The gage can have several hole patterns for various size wheels. If more accuracy is required, substitute a dial indicator for one of the bolts.

bolts if a more precise measurement is desired. Remember this gage only measures the *change* of toe, not the actual toe setting.

To check bump steer remove the springs so the suspension can be moved through its full range of travel.

With the tires resting on the garage floor, the car chassis is moved through its range of travel with a jack under the frame. On some cars it may be necessary to place blocks under the tires so the frame can be lowered to its full bump position with the jack. Do not attempt to go beyond the normal range of suspension travel.

Start with the car at full bump position. Make sure the bump steer gage has both bolts against the wheel rim. Then raise the jack about 1/2-inch and look at the bump-steer gage. If both bolts are still touching, there has been no toe change. A change will show up as a gap between one bolt and the wheel. If the gap is on the front bolt, the wheel has increased toe-in. A gap at the rear bolt indicates toe-out. Record the gap and the ride height above the floor.

The bump steer gage shown in Figure 24, ready for use. It is held in place with the concrete block and set up with the bolts just touching the wheel rim.

Move the jack again and take another reading. Record the toe change if any. Continue this process until you have measured the toe change all through the suspension travel to full droop. The results can be plotted on a graph as shown in Figure 25.

If you are using a dial indicator the gap shows up as a reading on the indicator. Start out with the dial set at zero, so the reading is the change from the initial position. This method is more accurate than measuring the gap between the bolt and the wheel.

If there is a significant toe change you will wish to make an adjustment. This will vary with the design of the car, and is different on the front and rear suspensions. First we will look at the front.

On the front suspension, the bump steer can be adjusted by changing the height of one end of the tie rod. This assumes that your car has a conventional steering system using equal-length tie rods. This type of steering system is almost universal with independent front suspension of the A-arm or MacPherson-strut type. Other types of front suspension such as sliding pillar, trailing arm, or swing axle generally cannot be bump steered. This is one reason these suspension systems are not widely used.

The method of changing the height of the tie rod will depend on the design of the car. If your car has rack and pinion steering, perhaps the rack can be raised or lowered on its mounts. Some cars lend themselves to this if they have rack mounting brackets mounted on a horizontal frame rail. Shims under the rack brackets or machining off the bottom of the brackets will work for the adjustments. Make sure you don't bend or bind the steering shaft or any of the mechanism with these adjustments.

On cars with a steering box the usual method of adjustment is to modify the steering arm. On many cars this is a removable part, and it can be bent to move the tie-rod end up or down. To do this you should heat the steering arm with a torch and carefully bend it to the proper shape. The best place to make the bend is toward the tie-rod mounting end of the arm. After checking the bump steer with the modified steering arm on the car, it must be heat treated to its original strength. Heating softens and weakens the steel. Also the part should be Magnafluxed to check for cracks caused by bending. Steering arm modification is a critical job, and is not the place to "Mickey Mouse."

To avoid steering-arm modifications, the steering system can be converted to spherical rod ends in place of tie-rod ends. This allows shims to be used between the steering arm and the rod end for small bump-steer adjustments. Large changes cannot be made this way, but it is a good way of getting an extremely precise small adjustment. To do this, the steering-arm

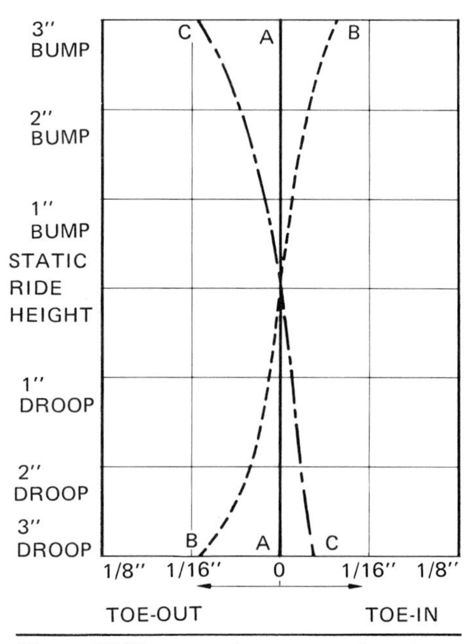

Figure 25/Here are three examples of bump steer curves. Curve A is ideal, with zero change all through the travel. Curve B has toe-in with increasing bump travel, with nearly the same error at both bump and droop positions. Curve C has toe-out in bump and toe-in in droop, with the error greatest at full bump. Many other curves are possible, but only Curve A is ideal.

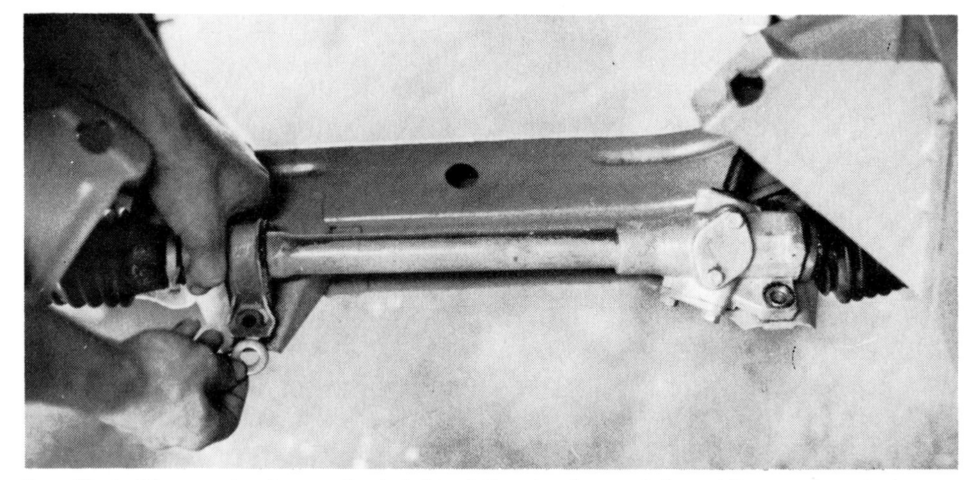

On a Pinto it's easy to change the height of the steering rack by adding washers between the rack brackets and the frame. If the rack has to be lowered the rack brackets or frame brackets have to be modified. Some cars do not allow such easy bump-steer adjustments.

To correct bump steer on his Datsun 510, owner Dave Redding is modifying the steering arms. Here he heats the arm to yellow color. The heat is applied to the compression side of the arm to prevent cracking—arm is pulled toward him to bend it. Careful measurements are taken every step of the way. The arm is Magnafluxed and heat treated back to its original strength after forming. The steering arm is a critical part, and this is a job best done by experts.

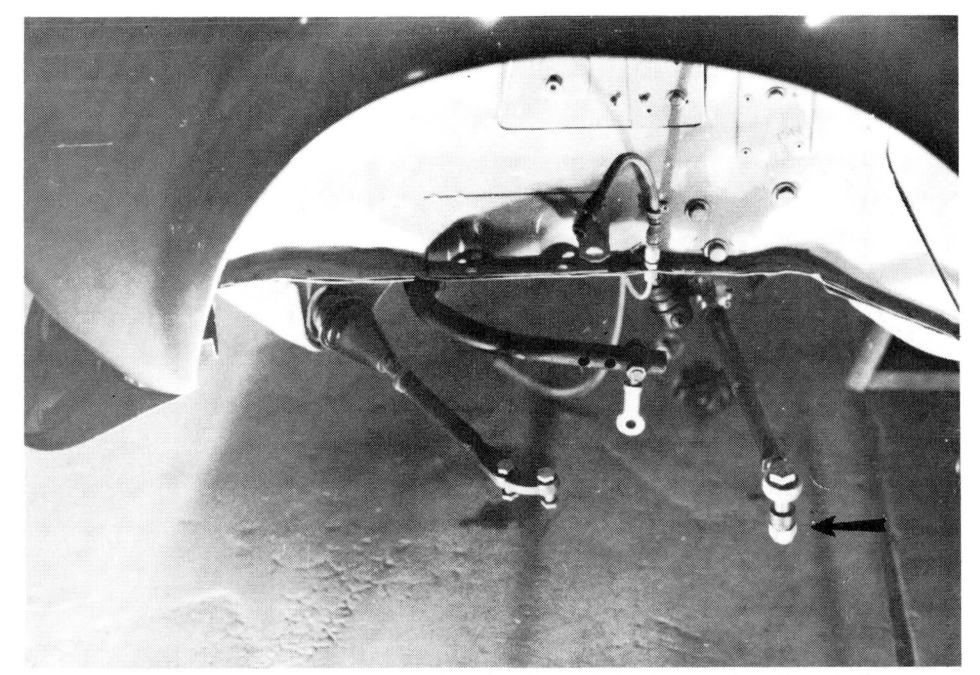

This Datsun front suspension uses a rod end and a bolt in the steering tie rod, shown in the lower right hand corner of the photo. You can also see the spacer under the rod end that is used to adjust bump steer. In this car, the steering arm is also bent to prevent the spacer from being too long. A long spacer causes twisting of the steering arm, resulting in excess deflection or even structural failure.

tapered hole for the tie rod is drilled out to accept a bolt. The spherical rod end is bolted to the steering arm using a lock nut for safety. Small OD flat washers between the rod end and the steering arm allow free movement of the spherical bearing and also allow bump steer adjustments. After installing the spherical rod ends be sure to check for free movement in all position of suspension motion and steering travel.

The curves in Figure 25 show the ideal bump-steer characteristic and two other ones with some error. All sorts of curves are possible, most of them terrible for good handling. You should try like crazy to get a curve with zero bump steer both at the front and the rear. The time spent will be worth it.

The amount of change and directions of the change is not an exact science, because so many suspension dimensions can cause bump steer. The best way is to use a trial-and-error approach. Make a small change to the car, in the direction that you guess will help the bump steer toward zero. Then make another test and plot the new bump-steer curve. Write on the curve how much you changed the car and in what direction. The curve will tell you the next step. If things are improving make another change in the same direction. If things got worse, make a change in the opposite direction from the first change. Don't forget to write everything in your notebook, so you won't get confused and have to repeat a test.

If you get a nearly ideal curve except for the last bit of suspension travel, don't worry too much about it if it won't improve. You seldom get into the last inch of suspension travel anyway.

Try really hard for zero bump steer at the front. A car with bump steer in the front suspension will be unpredictable in a turn and unstable during braking. The car is very sensitive to toe-in changes. With bump steer, toe changes can happen with every dip in the road or when you hit the brakes. Both toe-in and toe-out errors give terrible handling and should be avoided in the front suspension.

Cars with independent suspension may also have bump steer at the rear. The adjustment is usually difficult or impossible on road cars. If you have one of those unfortunate cars with designed-in bump steer, the only satisfactory way to combat it is to stiffen up the suspension so it moves very little. Many cars with semi-trailing-arm rear suspension fall into this category.

Racing cars with fully adjustable rear suspension have the advantage of being adjustable for rear bump steer. Here you change the caster of the rear upright to adjust bump steer. This is usually done by changing the length of the radius rods to tilt the upright off vertical. This also is a trial-and-error procedure. Try a different setting and record the results in your notebook. Again, zero toe-in change is the best, but if this is impossible shoot for toe-in which increases as the wheel rises. Avoid the opposite characteristic like the evil spirit!

ROLL STEER

A characteristic that is linked with bump steer is roll steer. This is steering of the rear wheels as the car leans. Even cars with a solid rear axle can have roll steer, and many indeed have this characteristic built into the car. Production car manufacturers use roll steer to give the car a high degree of understeer in a corner. Sometimes it is used to counteract the oversteering characteristics of a rear engine car. This is called *roll understeer* where the axle causes an understeering tendency. Roll understeer is the case where the rear wheels steer in the same direction that the front wheels are turning into a corner. The opposite, *roll oversteer,* would be where the rear wheels steer away from a corner. Roll oversteer is seldom deliberately designed into a car because it causes stability problems.

On a racing car, rear caster is adjusted to change rear bump steer. Here a gage is resting on top of the rear upright to measure the caster angle. The adjustment is usually made by changing the length of the radius rods that locate the rear upright in a fore-and-aft direction.

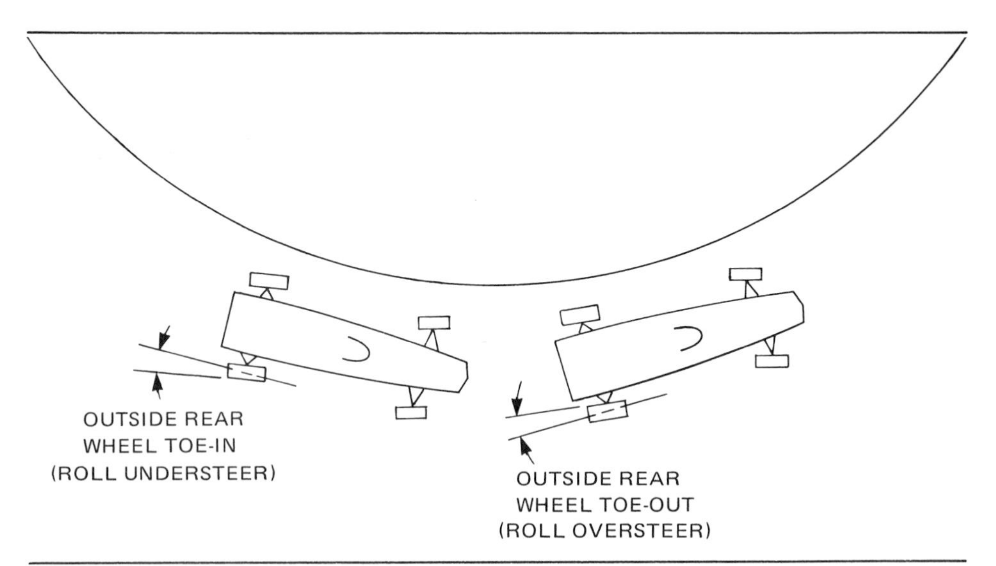

The car in front has roll oversteer and the car behind has roll understeer. This can occur with independent rear suspension or with solid-axle rear suspension. On independent suspension the toe change of the outside tire determines the roll steer, as the inside tire has a reduced amount of vertical force on it in a corner. Notice how the racing drivers are steering the front wheels to compensate for all that help from the rear wheels.

Rear spring of a stock Pinto. The front of the car is toward the right, outside of the picture. Notice that the front of the rear spring is lower than the rear. This causes roll understeer when the car leans in a turn. Lowering the rear end of the spring by increasing rear shackle length is a cure for this and similar designs.

The roll steer of this Pinto has been reduced by using longer rear shackles. The spring is thus horizontal in the normal riding position, giving the axle a straight up-and-down motion. The longer shackles raised the car, so the spring was re-arched to reduce the ride height.

On cars with a solid rear axle, roll steer is created by having the leaf-spring mounts at different heights. This causes one side of the axle to move forward and the other side rearward as the car leans. A similar roll steer can be built into a car with independent rear suspension by using the proper geometry.

Roll steer should be zero for good handling. It only affects the transient handling anyway, and there are better methods for changing the steer characteristics of a car. If possible try to eliminate roll steer on your car, and you will find it more predictable in fast corners.

To change the roll-steer characteristics of a car requires building some new parts, and this can be a very complex task. On certain cars, it is almost impossible to change, such as on cars with semi-trailing-arm rear suspension. A new suspension would have to be built to help these cars.

On a car with a solid rear axle, changing the shackle length changes the roll-steer characteristics. Because this modification changes the ride height also, you should be aware of this double change. Sometimes it helps both ways.

Roll steer is almost never used on racing cars because it can make the car less predictable. When one wheel hits a bump it acts the same as body roll, causing rear-wheel steering. When driving on the limit of adhesion this sort of thing reduces driver control.

PREPARING FOR TESTING

After the suspension is set up in the shop, it's time to drive the car. Testing can vary from a simple drive in the country all the way to a full-scale test program. The results you get are worth whatever effort you can put into testing. If you are serious about getting the best possible handling from your car, track testing is the only way. In all types of motor racing, you will find the guys who win the races are the ones who do the most testing.

The first problem that always arises is finding a place to test. If you have a racing car, obviously the best place to test is the track where you will be doing most of your racing. If you don't have a race track for your type racing available easily, don't forget the other tracks in the area. An oval-track car can be tested on a road course with a suitable left corner, or a road-racing car can be driven on a paved oval track if a road course is not available. Any straight piece of pavement will work for a drag racing machine, at least to test performance off the line. Use your imagination, and don't overlook possible test sites.

If there are no tracks at all, try the small airports in your area. Noise is no problem, and they usually have lots of room to run. Some airports are not too busy and will give you permission to run on a secondary runway or taxiway. It helps to be friendly with the airport manager.

Parking lots are sometimes available from a business on a weekend. Because of insurance and liability problems, many businesses are reluctant to let a car run, so you will have to convince the owner that you will be responsible for any damage to the property and that you will cover your own losses. Usually you will be able to get permission if the man you are talking to likes racing and understands the problems and hazards involved. Beware of creating too much noise if you are using a parking lot, and muffle the exhaust if there are any houses around.

This racing car is ready for testing on an old airport runway. This is an ideal place to test, but hard to find. Sometimes the use of a private airport can be obtained by asking permission from the appropriate person.

In a Solo I event each car runs against the clock for the fastest lap times. The real value of this type of event is the practice sessions, where you have ample opportunity to try different suspension settings. Solo I events usually have much more practice time than races, and the entry fee is less too.

This is the sort of equipment you need to thoroughly test a race car. The rack in the foreground holds anti-roll bars of various stiffness, the stack of tires in the rear is a variety of sizes and compounds. This equipment belongs to Balboa Datsun, which fields a first-class racing team.

Although it usually doesn't work as well as a private session, some testing can be done at an actual race. Pick an event that gives you the most running time, so you will have time to make changes and evaluate the car's performance. Don't try to win, just test. If you get serious about winning, you forget about testing and sorting out the car, so pick an event that you don't care if you win or not.

For drag racing, testing at a strip is best because you get measured times to compare different settings. Best is a small local event, where very few cars are likely to show up. Run "time only" so you don't get hung up with beating the other guy, and make all the changes you can. It helps to redline at a conservative figure, so the car won't be worn out in the test session. Make as many runs as possible, and compare one with the next to see if your suspension changes do any good. Ignore the track record or what everybody else is turning, and save that for your first all-out race. Surprise the competition by turning up your redline to its real maximum, and blowing their doors off!

For road-racing cars, a super event for testing is a Solo I event. This is a time-trial competition where one car at a time runs against the clock. Usually they have long practice sessions to allow drivers to learn the track, and that is when you can do some serious testing. They have full safety crews out on the corners, and you can drive hard without worrying about getting in trouble on a lonely part of the course. Remember this is only a test session, so stop at the pits for whatever changes you have to make, and don't worry about what the other cars are turning. Set a low redline until you are ready to see exactly how the car can do against the competition. Testing can wear out a racing engine if it is pushed to its maximum all the time.

Testing a car to be driven on the road is something of a problem, unless you wish to enter it in some sort of competition. You can, of course use a parking lot or airport for testing, but be sure you get permission first. Some drag strips are OK for testing if there are also corners you can use. The return road often has corners, or perhaps you can set up artificial turns on the strip using rubber cones or cardboard boxes. Some people resort to a public road for testing, and this is OK if you stay within the law and do it in a safe manner. The problems and dangers in using public roads are obvious, and you should be aware of them before trying it. Whenever you test a road car, have some people along in another vehicle to assist you and to reduce any dangers that might be associated with the test. Remember the police may not agree with your ideas or mine as to what constitutes safe driving while testing.

No matter what type of testing you are doing, be aware that it can be as dangerous as racing. Always wear a crash helmet and the type of driving suit required for your type of competition. This applies to testing road cars too. An airport manager is much more likely to give you permission if you look like you will take every possible safety precaution. If you are using a test track, get as many friends as possible to assist you, and station them at various points around the track. If you go off course on the far end of the track, it is comforting to know there is somebody close by to help you. Take as many fire extinguishers as possible and have them at all possible parts of the track. Above all, if something doesn't seem right in the car, stop and check it out. There is no reason to keep on driving if you *suspect* a problem.

Plan your test session in advance and bring all the tools and equipment you will need. Start out by writing a test plan. This is a written list of what things you are trying to find out, and a step-by-step plan of how you will do the testing. As you are writing the test plan, think of all the test instruments, tools, and parts you might need to do the job, and write them down too.

Always assume that you will have to make changes, and bring all the spare parts and tools required for the change. If there is any doubt, bring it, as you can sometimes lose a full day of testing for lack of a small part. Put the list of parts and supplies in your notebook, and it will help you when you go to the next race.

Start your test day early, so you can have the maximum amount of time. Lack of time is usually what limits a test, so take full advantage of every opportunity you have. You can always leave the track early if you feel you have nothing else to learn.

Change only one thing at a time! If you make more than one change, you will never know what change caused the effect, and also one change may work against the other to give you a false result. Be methodical and repeat any test that you are not sure of. It pays to go back to a known setting from time to time, just to be sure. Make certain that you always write down the change and what the result was. Too much written information never hurts, but too little often does. This effort will pay off later when you are pushed for time and need accurate information fast.

TROUBLESHOOTING HANDLING AT THE TEST TRACK

Once you are out and running on the test track you will want to make adjustments to improve the handling. Change one item at a time, and if it doesn't seem

If you can get the use of a parking lot you can set up a small test corner using rubber cones or cardboard boxes for markers. On this lot the circular marks can also be used as a skid pad as described later.

If your car lifts a front tire off the road in a turn it has too much front roll stiffness, too little rear roll stiffness, or a combination. A car on three wheels generally cannot corner as fast as on four wheels due to the excess roll after the tire lifts off. Change the roll stiffness so that the tires all remain on the road in a steady-state corner. Adjusting roll stiffness to tune the suspension has no effect on a three wheeled car. (Don Larsen photo.)

to work, go back to what you started with. It pays to be methodical here and repeat tests if you don't fully understand the results.

The two items of most interest are steer characteristics and cornering speed. Usually steer characteristics can only be measured by driver evaluation, while cornering speed can be measured with a stop watch. On a racing car, lap times are the best indication of proper handling. Always be aware of the lap times, but try to eliminate the variables such as the driver pushing the car harder. The results of a test session will depend both on measurements and on driver reaction to the changes.

If the car has already been to a skid pad, the camber and tire pressure for fastest cornering have already been determined. Start with these settings if available. If not, you can determine the best settings with a tire-temperature check. Here you should have a pyrometer, because holding your hand against the tread just isn't accurate enough. More details are in the later section on using a skid pad.

The car should be run around the test track for a few hot laps and then come in for an immediate tire temperature check. Don't delay, as the temperature will change with time once the car is stopped. Check the temperature at three places on the tread, the inner side, the center, and the outer side. For best camber, the inner and outer sides should be nearly the same temperature. If the outside is hotter, there is too much positive camber in a turn. This can be corrected by either increasing the negative camber setting or by reducing the body roll. Body roll is reduced with stiffer anti-roll bars, or an adjustment. Keep checking the camber, and get it at the best setting for cornering as indicated by temperature. If camber is more than about one-degree negative, body roll should probably be reduced. This is a compromise because high roll stiffness will reduce adhesion on bumpy corners.

It is often useful to know what the roll angle of the body is. It can be measured by running at the limit of adhesion in a curve without banking. A quick and handy way to get a pretty close measurement is to take a head-on picture of the car with a Polaroid camera.

Draw a line parallel to the road surface and another line across two points on the body which should be level except for body roll. Measure the angle between the two lines with a protractor.

The best tire pressure is also determined with the three temperature readings taken across the tire tread. If the center of the tire is the same temperature as the average of the two edges, then the pressure is about perfect for best handling. If the temperature is higher in the center of the tire, the pressure is too high. If the sides of the tread are hotter than the center, the tire pressure is too low. Play with this until you get it right.

After arriving at the best camber and tire pressure, the steer characteristics should be adjusted to give the best driver control. Drivers vary in what they like, so leave this up to your driver. The steer characteristics can be altered by changing either the toe-in or the relative roll stiffness between front and rear. Each adjustment should be done separately so as not to make two changes at once. Generally steer characteristics will be adjusted with the suspension for low-speed corners, and any aerodynamic devices will be set for high-speed turns. Aerodynamic devices

Wrench used to loosen the pinch bolt on the anti-roll bar fitting so the fitting can be moved along the bar. Moving the fitting toward the end of the bar reduces stiffness; moving the fitting away from the end stiffens the bar. Both sides of the bar must be adjusted at the same time. If you run out of adjustment a different diameter must be installed.

are not effective at low speeds.

Steer characteristics are usually varied by adjusting the anti-roll bars. Most racing cars are equipped with adjustable anti-roll bars at both ends of the car, and the secret here is to vary only the ratio between the front and rear and thus not change the angle of roll. To do this, one bar is made stiffer and the other softer. Changing the angle of roll changes the camber in the corner, and this is two changes at once.

To make the car understeer less, reduce roll stiffness on the front and increase it on the rear. To make the car oversteer less do just the opposite. We are assuming of course, that no other changes are made at the same time.

Perhaps the car runs out of adjustment on the anti-roll bar. Then you must change to a heavier or lighter bar, depending on which direction you need to go. The maximum limit to roll stiffness is when a tire lifts off the road in the turn. This will happen on the front of a conventional racing car. If a tire lifts, you have gone too far, and other methods have to be used to adjust the handling. Back off on the front roll stiffness until the tire just stays on the road.

Toe-in adjustments are another way of changing the handling. This usually has the most effect on transient handling, just as the car is turned into the corner. Increasing toe-in on the front will make the car understeer more, and reducing toe-in will make it understeer less. The limit to reducing toe-in is when you have high-speed stability problems. You should try to run close to zero toe-in for minimum rolling drag, but sometimes changing it is the only way to get the handling you need.

Rear toe setting should be close to zero, with perhaps a tiny bit of toe-in to minimize oversteer on very powerful cars. If you ever have toe-out on the rear you can expect all sorts of wild things to happen, mostly instability and oversteer. Check for rear toe-out if scary things are happening out there.

Notice in the previous discussion there was no recommendation to use either camber or tire pressures to adjust steer characteristics. While these devices do work for this purpose, they tend to reduce the total cornering power of the car, and thus should only be used in extreme cases where nothing else works.

In the old days, tire pressures were often the only means used for adjusting steer characteristics. The old rule of thumb was to increase the front pressure to reduce understeer and increase the rear pressure to reduce oversteer. These tips basically only applied to road tires, where the normal setting is underinflated. Thus increasing the tire pressure at either end increases the adhesion. However, for best cornering you should find out from skid-pad testing what tire pressure gives maximum cornering power, and stick to it. The same goes for camber settings.

If your car has adjustable shocks, try the various settings. Shocks settings are a matter of trial-and-error to find the best compromise. Shocks are most important on bumps, particularly in a corner. Find the bumpiest part of the course and see if you can improve the traction there. Also check the transient handling after making a shock adjustment. Shock stiffness can affect weight transfer just as the car is turned into a corner. Once the car reaches its steady-state roll angle, the shocks do nothing unless the wheels hit bumps.

The method for arriving at the best shock setting is a matter of trial-and-error. However, it is good to use a systematic series of adjustments so you spend a minimum amount of test time arriving at the answer. Koni recommends using the following procedure for setting up their double-adjustable shocks.

To start testing, the shock should be set at the softest settings on both bump and rebound adjustments. Drive the car and note the amount of wheel hop over the bumps. Adjust the bump setting stiffer and test the car again. Keep increasing the bump setting until wheel hop is reduced to a minimum.

These photos and instructions courtesy Kensington Products Corporation, United States Representative for Koni Shock Absorbers.

1/Rebound or extension adjustment with eye-mount shock: The small disc under the eye has eight holes of 3mm diameter. One "sweep" from left to right or right to left covers two holes and is one quarter turn. Moving the adjuster from left to right increases rebound damping, from right to left decreases it. There are 12 "sweeps" from minimum to maximum. Count the number of "sweeps" from minimum and write them down for reference. When adjusting off the car it is necessary to hold the piston rod with your other hand so that it does not turn as you adjust. The adjustment becomes a little heavier as you increase the damping but never use more than gentle pressure (5 ft. lbs.).

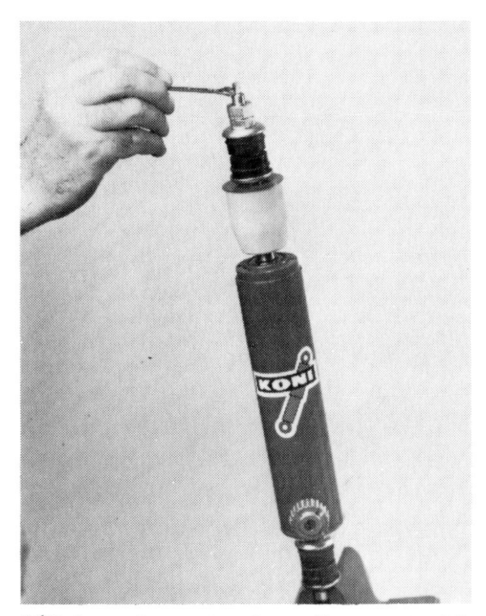

2/Rebound or extension adjustment with pin-mount shock: For the square adjuster use a 3/16" or 5mm wrench. For the round adjuster a 2.5mm Allen wrench is a good tool. When adjusting off the car hold the rod still by grasping the bumper and the mounting rubbers with your other hand.

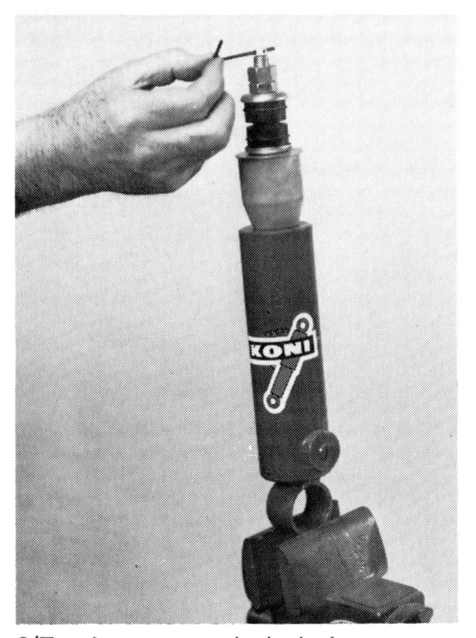

3/Turning *counter-clockwise* increases rebound damping. Each quarter turn is an adjustment. There are twelve quarter turns from minimum to maximum. Count the number of quarter turns from minimum as described under 1. *Never* hold the rod by the square when tightening or unscrewing the nuts.

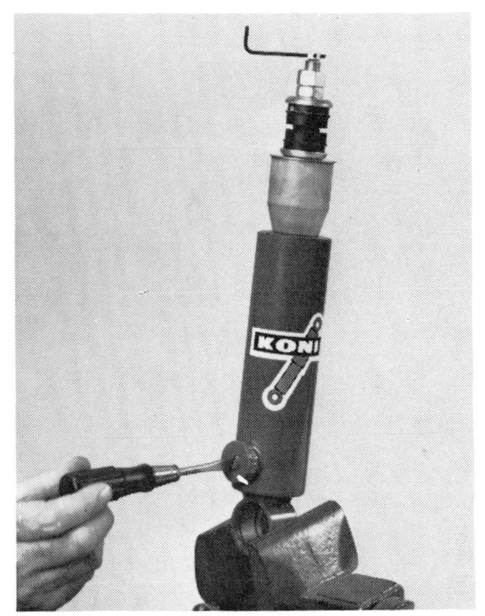

4/Bump or compression adjustment: The knob adjuster which can be turned by hand or with a screwdriver is most commonly used for bump adjustment. The screwdriver adjuster without knob is used when there is insufficient room for the knob.

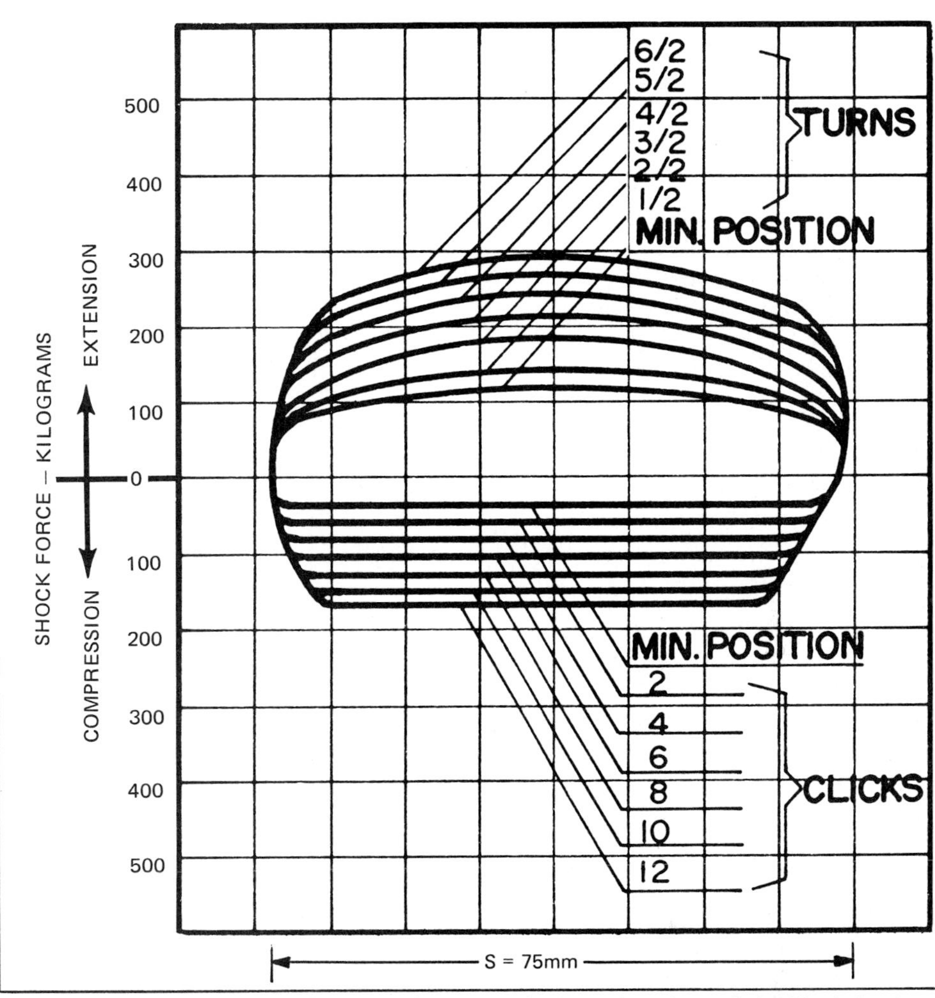

This diagram shows the wide range of bump and rebound settings of the Koni shock absorber. This unique design of double adjustments permits control of bump and rebound separately.

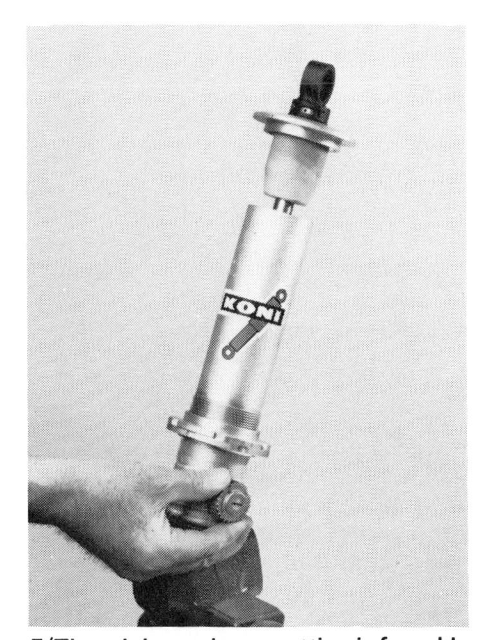

5/The minimum bump setting is found by turning *counter-clockwise* to the stop. There are 12 click positions from minimum to maximum. As you adjust *clockwise* for heavier damping a little more effort is required as described for the rebound under 1.

Shock fading and foaming are discussed on pages 145, 146. The advantages of gas pressure shocks are detailed on pages 146, 190.

The angle of attack of this wing is adjusted by the two small turnbuckles. The wing pivots about its mounting points on the large struts. This wing is set at a high angle of attack for maximum downforce.

Then increase the rebound setting until floating is reduced to a minimum. *Floating* is the car trying to rise up on its suspension, caused by not enough rebound stiffness in the shocks. The basic idea is that the shocks can expand or lengthen readily but bump stiffness opposes compression. Over a rapid series of road undulations, the shocks will gradually lengthen and the car will seem to float. After the rebound is at the best setting, minor adjustments can be made in both settings to arrive at the best overall compromise. Koni recommends that you use the minimum amount of shock stiffness consistent with good handling.

Although other adjustable shocks may not be adjustable in bump and rebound separately, the same sort of test plan can be used. Start with a soft setting that you are sure is not stiff enough and gradually increase stiffness for improved handling. When you cannot feel any difference between settings, you have gone too far. Go back to the last setting that caused a noticeable change.

AERODYNAMIC TESTING

Aerodynamic downforce is used on many modern racing cars to obtain different steer characteristics. This adjustment only works at higher speeds, so keep that in mind. Most cars understeer on slow turns and tend to lose this characteristic on fast turns. Having aerodynamic devices that promote understeer at high speeds will tend to make the car handle more nearly the same under all conditions. Many times suspension adjustments are made that promote neutral steer on low-speed turns and result in horrible oversteer on fast turns. Aerodynamic downforce can be used to cancel out the high-speed oversteer. Many racing cars are set up this way.

The devices used for this are wings, tabs, and spoilers. They are a powerful tuning aid in track testing.

Many modern cars use aerodynamics to influence handling. Even road cars use aerodynamics with spoilers, wings, and other add-on devices. Many of these

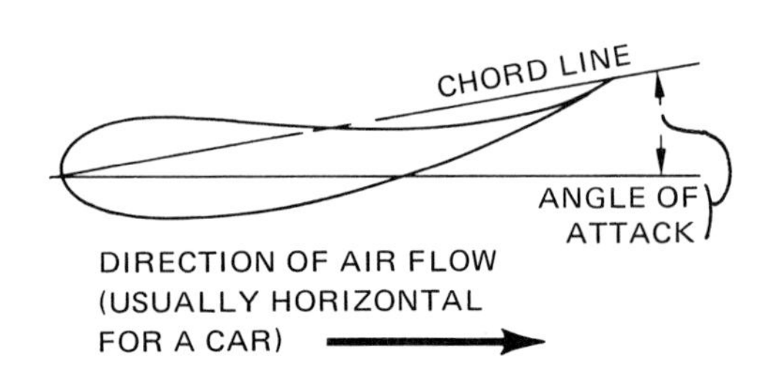

The angle of attack is the angle between the chord line of the wing and the direction of air flow. The *chord line* is a line joining the front of the wing leading edge with the tip of the trailing edge as shown. This chord line is not necessarily parallel to the top surface of the wing, and is difficult to measure from. Angle of attack is usually adjusted by trial-and-error out on the test track.

devices can be used to adjust the handling qualities of the car, and at times aerodynamics are the only way to obtain proper handling at high speeds.

Because normal road speeds are relatively low, aerodynamics have little effect. The most useful effect of aerodynamic devices for road cars is improved stability in a cross wind. This can be tested on a gusty day at road speeds. If you have adjustable devices, their effect can quickly be determined by seat-of-the-pants feel.

Our major concern with aerodynamic devices will be in speed ranges well above legal highway speeds. This pertains mainly to racing machines. The most common device used on a racing car for downforce is a wing. Wings take various forms, but are usually mounted near the tail of the car with provision for adjusting the angle of attack. This is aircraft language for the angle the plane of the wing makes with the direction of air flow.

This D-Sports racer has only 50 cubic inches pushing it, so the rear wing is relatively small—a compromise between cornering and straightaway speeds.

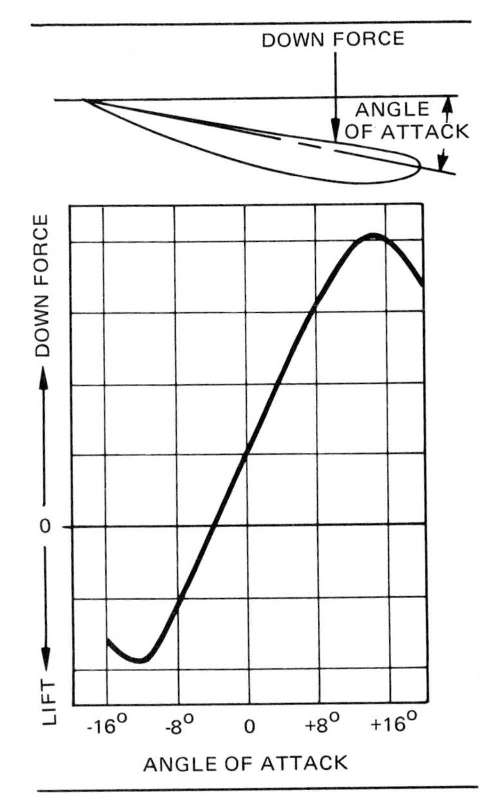

Figure 26/This curve shows how downforce varies with angle of attack. For the wing shown the maximum downforce occurs at an angle of attack of about 15 degrees. Any angle greater than this results in less downforce. Also notice that at zero angle of attack there is still some downforce. This is due to the shape of the wing cross section. The amount of downforce depends on the size of the wing.

Wings on cars all use negative angles of attack, with the front edge lower than the rear edge. More downforce is put on the car if this angle of attack is increased.

A typical relationship between angle of attack and downforce is shown in Figure 26. Notice that there is a limit to how much downforce the wing can develop, and exceeding this maximum angle of attack results in a sharp drop in downforce. When this happens, the wing is stalled. It is a common mistake to

The front wings on this formula car are adjustable for angle of attack. An increase in angle of attack increases the front downforce and increases the cornering power of the front tires. This is a way powerful cars can be tuned to handle properly at high speeds.

ignore the possibility of stall when using a wing to get the maximum possible downforce. A stalled wing merely adds a lot of drag, which slows the top speed of the car.

For effective use of aerodynamics to adjust steer characteristics there must be two separate sources of downforce—one at the front and another at the rear. The more downforce on one end of the car the better those two tires will stick to the road. The balance between front and rear downforce results in a change in steer characteristics. For less understeer put more downforce on the front tires. For less oversteer put more downforce on the rear tires.

Keep in mind that high downforce on the chassis will compress the suspension and reduce the ride height. With independent suspension this increases negative camber at high speeds. Stiffer springs are used with high downforce, so this additional modification may be required for best handling. The camber will usually have to be reset after making large downforce adjustments. Final setting will be a compromise made on the basis of best overall handling and lap times.

Determining the combination of downforce and drag that will result in the best racing performance is always of interest. This is a difficult trade-off to make, because adding more downforce always adds more drag, and these two things work against each other in racing performance.

A few general rules can be stated. If the car has limited horsepower, it will be more sensitive to an increase in drag. Thus small displacement cars usually run much smaller wings than their bigger-engined brothers. If you are racing on a fast track, where the car is at or near its top speed, much of the time, the extra drag caused by a large downforce usually results in slower lap times. High downforce is most beneficial on slow tight courses. In these conditions, cornering power is relatively more important than top speed in determining a fast lap time. The final trade-off will be determined in testing.

The test is very simple. Just measure the lap time of the race car with various amounts of downforce. Try to keep the steer characteristics of the car constant during the tests by adding or subtracting downforce proportionately at both the front and rear of the car. Changing the steer characteristics will introduce another change in the car's performance, and it will be difficult to tell what caused a change in the lap time. The downforce of spoilers or tabs sometimes can be varied by adjusting the amount of spoiler sticking up into the airstream. On non-adjustable devices it is necessary to vary the actual size of the tab or spoiler by adding or cutting off material.

TUFT TESTING

Another useful test is to determine the air-flow pattern on the surface of the car body. This can then be used to decide where to place air dams, duct openings, air exits, or fairings. This type of testing is easily done with bits of string or yarn, and it is called *tuft testing.* This test shows you the direction the air is moving over the body surface, and it also shows if the air stream is smooth, turbulent, or totally separated from the surface.

To make a tuft test, use bits of light string or knitting yarn. Cut off pieces about 5-inches long and tape them to the body with transparent tape. It works best to place the tufts in a uniform pattern in straight parallel lines. Place them far enough apart so they don't touch each other. After taping on the tufts, drive the car at speed, making runs in opposite directions if possible. Observe the tufts. They will point in the direction of air flow if the air is flowing along the surface of the body. If the tufts are straight and stable, this indicates the air is closely attached and relatively undisturbed. If the tufts are straight but whip back and forth in a fluttering motion, this indicates that the flow is turbulent—still acceptable as far as drag is concerned. However, if the

This early Can-Am McLaren has over 400 cubic-inches of brute power, so it sports a huge wing. Top speed is not attainable on a road course, so maximum downforce is the primary design goal. Powerful cars usually use the largest wings permissible to obtain maximum traction and cornering speeds because these have more effect on lap times than top-end speed.

This custom-built aluminum spoiler is adjustable by means of turnbuckles. Tilting the spoiler will have small effect on the aerodynamics of the car. Nose spoilers are usually adjusted by the actual size of the spoiler rather than its angle.

Wool tufts have been installed on this car and it is ready to test. The bits of yarn are taped to the body in rows even though they do look all mixed up with the car at rest Normally, tufts on only one half of the car are sufficient—assuming the body is symmetrical.

tufts fly about in random directions, this indicates flow separation. Separation always occurs behind any blunt object, such as the tail of the car. The effect of separation is a low-pressure area and a large amount of drag. Separation should be avoided as much as possible with gently flowing contours towards the rear of the car. A stalled wing has separated flow on the underside of the wing, and this can be checked with tufts.

If the driver cannot easily observe the tufts, perhaps a fast chase car can be used. Also movies are handy, or still photos with a long lens. Just make sure you have some way of observing the tuft motion and direction.

Tuft testing is useful for checking flow into and out of duct openings. If a duct is placed wrong, it may be getting little flow or even reversed flow, and tufts can tell you this. Tufts can tell a story about your windscreen's efficiency. Many windscreens can have the effect of causing separated flow on the tail of the car. Check this carefully, because separated flow can cause lifting of the tail and high-speed oversteer. Check for separation behind the driver. There may be a way to incorporate a head fairing to eliminate this separation and thus reduce drag. If you are installing a wing on a car, use tufts to check if the air flow in that area is smooth or separated. Wings operate poorly in extreme turbulence.

Check the top corners of a wedge-shaped body to see if air is spilling off. The high pressure air on the top tends to spill over the sides of the body causing turbulence and added drag. To combat this, construct some *air dams.* These are flat sheet metal extensions to the sides of the body, rising several inches above the top surface. These keep the air trapped on the top surface of the body, largely eliminating spilling over the corners. The height and the effect of the air dams is determined by trial and error on the test

This test was to see if air was flowing smoothly in the area of the exit duct on the side of the body. The flow is somewhat disturbed, but in the proper direction.

Quasar Sports Prototype uses a head fairing to streamline the driver's head. This car was designed by the author, and originally had a flat tail. The need for the head fairing was determined with wool tuft tests, and it added a noticeable amount to the top speed of the car.

This McLaren is equipped with air dams running along the sides for the full length of the body. They are particularly effective at the front and near the rear wing. At these points air dams increase downforce by preventing air from spilling off the top surface of the body. On a powerful car, maximum downforce is essential to fast lap times. Louvers over wheels allow underbody air to bleed through to fill low-pressure area over wheel arches.

track. Measure straightaway speed, cornering ability, and check for air flow with the tufts. The air dams should increase downforce on the body if designed properly.

It is sometimes desired to measure actual aerodynamic lifting forces on your car. This can be done with some difficulty. It requires rigging up a device to measure the deflection of a car's suspension at speed. What has been used in the past is a stiff piece of wire attached to the upper suspension arm on a racing sports car. The wire is marked off with strips of colored tape, and it passes vertically through a small hole drilled in the fender. The driver can see the wire sticking up through the fender, and at speed he observes if there is a different amount sticking out. The amount can be converted to a force by adding weight to the car over the suspension and measuring the movement of the wire. The problem here is in being able to see the motion, plus having to sort out the motion caused by bumps in the road. A very smooth surface helps a lot. There are other displacement measuring devices, but basically they all share the same problems. The suspension movement must be averaged during observation to remove the up and down motion caused by bumps.

You may find with all this testing that the car has lifting forces on it no matter what you do. There are some shapes such as the VW beetle that are doomed to have aerodynamic problems, and there is nothing you can do to solve them completely. The best you can do is trade the car for a better one with an improved shape. As auto makers become more aware of aerodynamics there should be fewer of these inherent-problem cars on the market. Many companies now use wind tunnels to help shape the bodies for high-speed stability and low drag.

The actual air pressure on the body can also be measured during testing. The device used is a pressure gage of a very sensitive type called a manometer. Small holes are drilled in the body, and tubes run from these holes to manometers in the cockpit. A brave observer reads the manometers during a test drive.

This is a very tricky test, best left to real experts, but it is possible if you have the determination. By the way, a single pressure measurement is of little value, so many tests must be made to determine the distribution of pressure all over the body. This is the sort of testing done in wind tunnels on aircraft to predict flight characteristics before a man ever has to get inside and take off.

There has been talk of designing cars in wind tunnels, and this can be done. A wind tunnel usually tests only a model, and careful work has to be done to relate the model test data to a full-scale car. The model is stationary in the wind tunnel and the air is moving, so this creates errors because of the ground. In a real car the ground is moving at the same speed as the air, but it isn't in the wind tunnel test. There are ways of correcting for this, but it does add more difficulty to the test. Wind tunnels can measure lift, drag, and even pressure on the body. Because this sort of testing is very expensive it is best left to the ultra-serious car designers.

USING A SKID PAD

A skid pad is used to measure steady-state cornering ability. You can measure the lateral acceleration of the car at maximum cornering speeds. It will also tell if your car understeers or oversteers and will allow you to adjust these characteristics. A skid pad will not help you with high-speed stability or acceleration and braking.

A skid pad is a flat piece of pavement with a circle painted on it. The car is driven around the circle, keeping the center of the car right on the line. By measuring the time it takes to make one lap of the circle, the lateral acceleration can be computed. To do this, you need to know the radius of the circle and the time for one lap at maximum speed. The formula used is:

$$\text{Lateral acceleration} = 1.22 \times \frac{\text{Circle radius}}{(\text{Lap time})^2}$$

In this formula the circle radius is measured in feet, and the lap time in seconds. You can measure the circle radius with a steel tape, and the lap time with a stop watch. As an example, let's assume your car is traveling around a circle of 90-foot radius and makes a lap time of 10.0 seconds. Working out the formula:

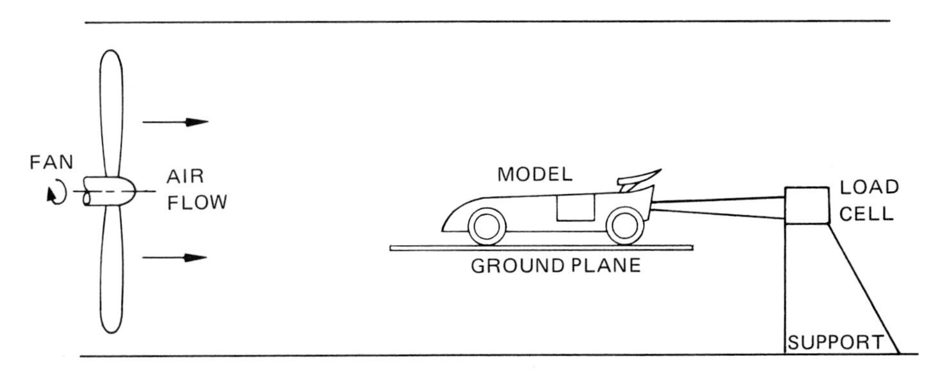

In a wind tunnel the model is stationary and the air is moving. This works the same as if the car is moving through the air except for the effect of a stationary ground surface. Complicated methods are sometimes used in wind tunnels to try to accurately reproduce the effect of the ground.
The load cell behind the model supports it and measures forces caused by the air. There is a small error due to the fact that the model has a support in the space behind it.

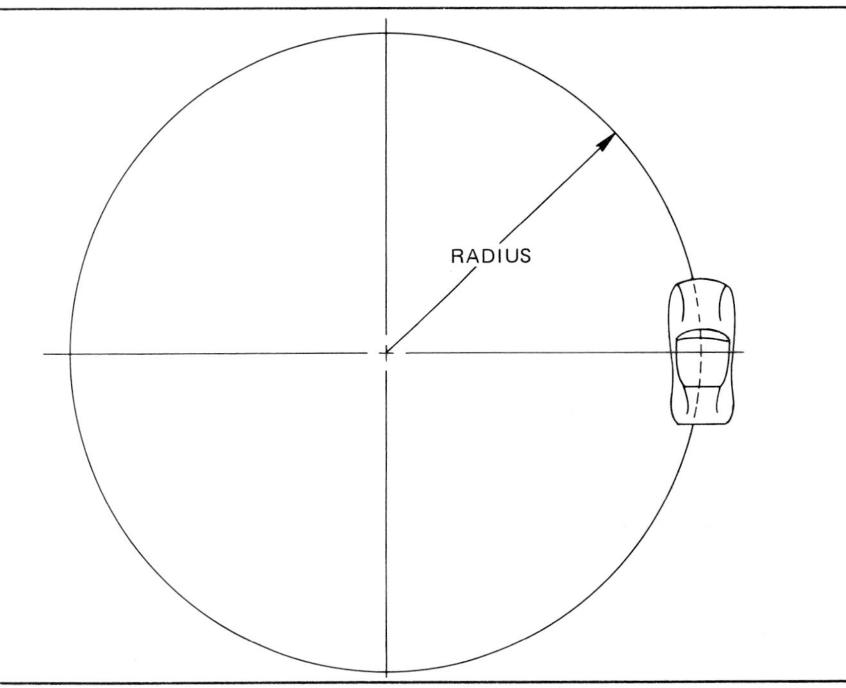

With a stop watch, measure the lap time around the skid pad circle. This is a way of measuring cornering power, and it can show you the results of suspension tuning. If you can, use skid pads of different radii to check cornering at different speeds.

$$\text{Lateral acceleration} = 1.22 \times \frac{90}{(10)^2}$$
$$= 1.22 \times 0.90$$
$$= 1.098 \text{ g's}$$

In this example the lateral acceleration is greater than 1.0 g. This is a very high figure for any car, possible only with the best racing tires plus an all-out performance chassis. Typical road cars on non-racing tires usually have a lateral acceleration around 0.7 g's.

To get an accurate value for lateral acceleration, the centerline of the car should be at the radius of the circle. It may be easier for a driver to drive around the skid pad with his inside tires on the circle. Therefore the radius used in the formula should be the radius of the painted circle plus one half the track width of the car. This is the radius to the center of the car.

At low speeds lateral acceleration equals the average grip of the tires. At higher speeds aerodynamic forces become large and grip is no longer equal to the lateral acceleration.

If you are interested in the amount of centrifugal force on your car at the limit of adhesion, multiply the lateral acceleration times the total weight of the car. This gives the centrifugal force in pounds.

In skid-pad testing, tire temperature can be a big factor. Remember the rubber temperature will change tire grip, so you must run all comparative tests at the same temperature. Thus you should have a tire pyrometer to measure tire temperature. This instrument uses a metal needle with a temperature sensor at its tip which is inserted into the hot tire-tread rubber. The probe is connected to a meter which reads temperature. Pyrometers cost from $60 to $200, but they are an essential part of suspension testing. If you are serious you should have one.

As you run consecutive laps around the skid pad the tires will get hotter and hotter. They will increase in cornering power the hotter they get, but only up to some maximum value. Any increase in temperature above this will result in a loss of traction and perhaps even destruction of the tire. Be aware of this problem and don't go too far.

If you are testing a car with racing tires you should find out what tire temperature gives the highest grip. This information is not useful for street tires, but for racing it is very important. You can sometimes change tire compounds to get a desired change in tire temperature. Cold tires are often a problem on light racing cars using wide tires. A cold day or a slick track surface makes this problem worse. With a little testing you can perhaps avoid this problem.

A very interesting item you can use at the races is the Electric Tire Preheater manufactured by Wacho Products Co. of Columbus, Ohio. This is like a high-output electric blanket specially designed to fit around a tire. It requires a 110-volt electrical supply, but perhaps you can get a small portable generator for this. This device can be used for tire testing to warm up tires without running the car. If you

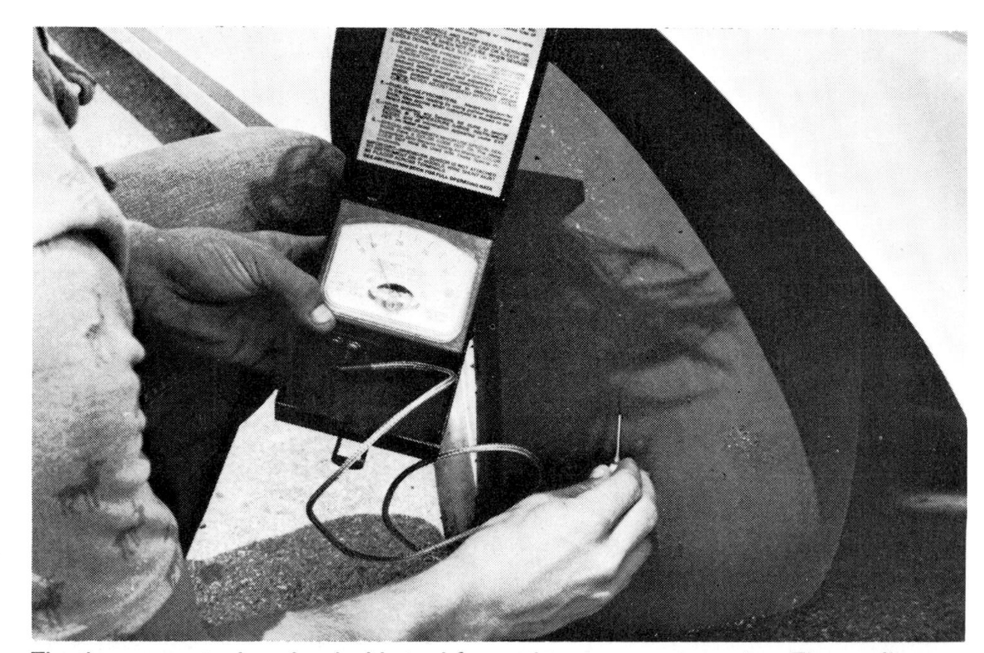

The tire pyrometer is an invaluable tool for setting up your suspension. The needle contains a temperature sensing element which registers on the meter. The needle is pushed just under the tire tread when it is still hot from a hard run on the test track. Readings must be taken within a very short time after the car stops, say 30 seconds maximum. The temperature changes sitting in the pits. When the meter reading stabilizes, move quickly to the next measuring point on the tire.

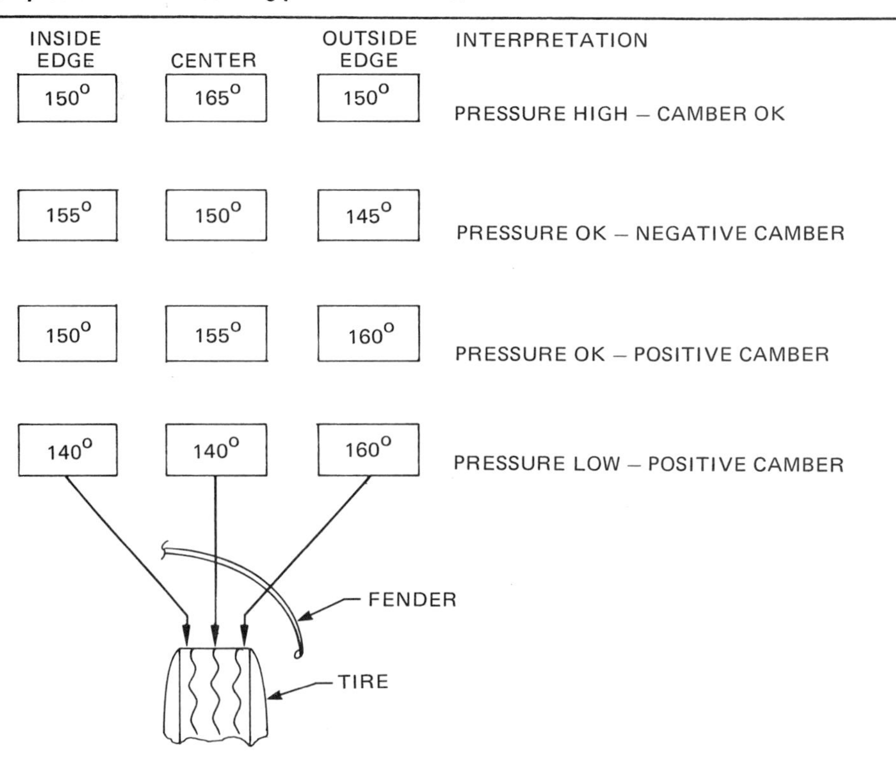

Here are some examples of pyrometer readings on a tire and how to interpret them. For correct camber the two edges of the tread should have equal temperatures. For correct tire pressure the temperature of the center should be equal to the average of the edges. Notice in the bottom row, the average of the edges is 150°, indicating low tire pressure in addition to incorrect camber.

do a lot of testing it may pay for itself in savings on tire wear. The tire preheater could be a winning trick on a cold day at the races too.

Besides tire temperatures, tire tread wear can affect the grip. If using brand-new tires, scuff them in a bit before starting to take measurements. Also stop testing when the tires get nearly to the cord. Watch their condition at all times when running a long series of tests. You should take test readings turning in both directions to make sure the tires wear evenly and that the car corners nearly the same in both directions. Naturally you will ignore this for an oval-track car.

Another side effect that can foul up testing is the engine's reaction to running in a very long corner. Watch carefully for oil starvation so you don't wipe out the lower end. If oil starvation is a problem, go back home and baffle the sump more effectively. Also be aware of possible fuel starvation or carburetor flooding due to cornering. Some carburetors, particularly the non-racing types, have a bad habit of reacting to side loads on the gasoline in the float bowl. Sometimes a change in float level will help. For tips on production carburetors see H. P. Books' Holley and Rochester Carburetor books.

A skid pad is used to determine suspension settings for maximum cornering power. One of the basic settings to vary is camber. Tires corner best with nearly zero camber, but this is affected by suspension geometry and the amount of body roll. Remember that body roll can be changed by changing the roll stiffness, so you may have to go back and readjust camber after changing anti-roll bars. It takes a lot of work to get the perfect combination. Also keep in mind that static camber setting for maximum cornering power will vary with the traction available at the tires, because body roll is affected by tire grip. Thus, adjustments have to be made at each race depending on the slipperiness of the track surface. The usual practice at the race is to leave the static camber alone and vary the anti-roll bars to provide the same amount of body roll that was determined to work best on the skid pad.

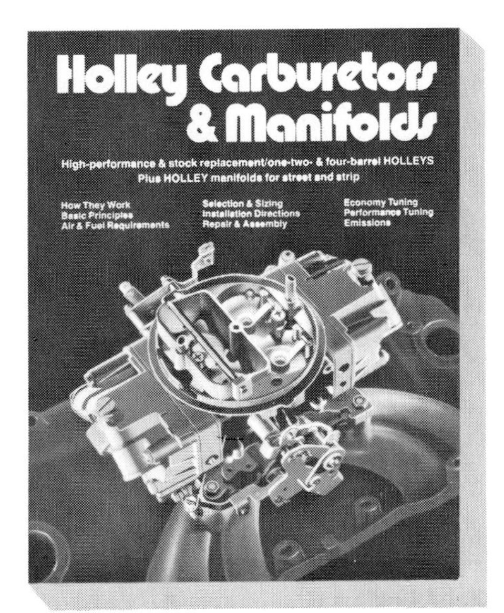

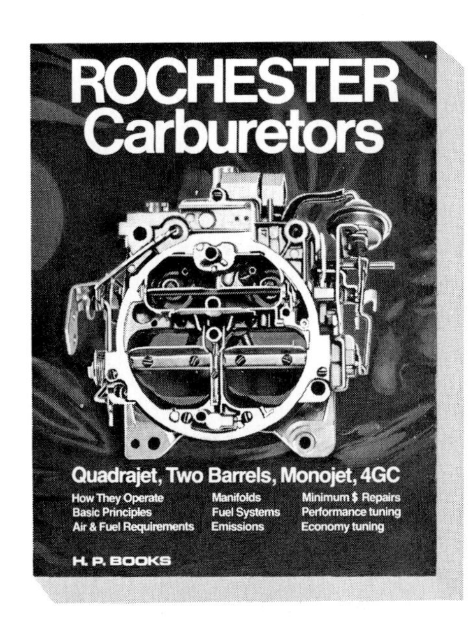

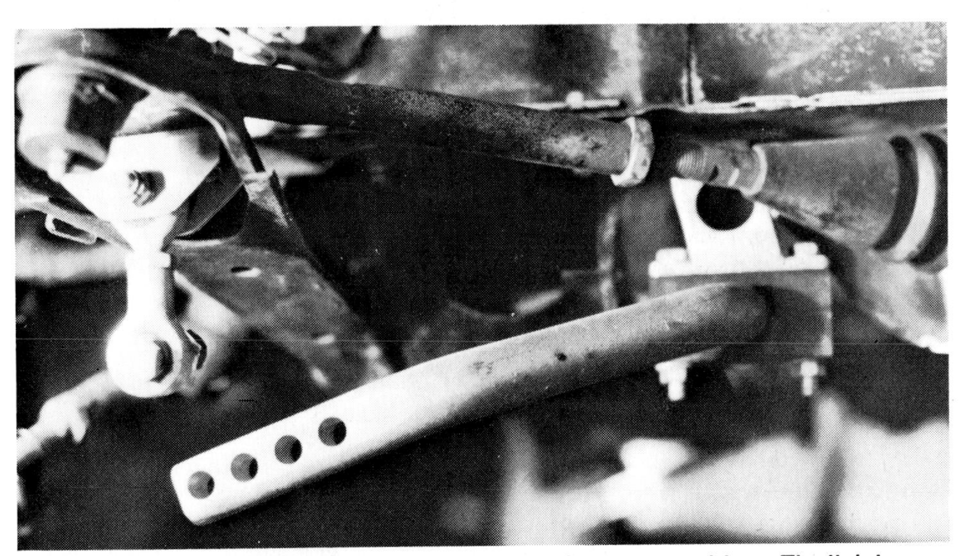

The anti-roll bar on Bill Swan's Datsun has four adjustment positions. The link is unbolted and placed in a different hole on the bar to change the stiffness. The hole closest to the end of the bar gives the lowest roll stiffness.

To set camber at the optimum angle, you need a tire pyrometer. After a hard run around the track take two readings on each tire. One on the inside edge of the tread and one on the outside edge, each about an inch from the edge of the tread.

This Spitfire is set up with a lot of negative camber on the front. It's a slalom racer, so understeer must be avoided at any cost. This much negative camber reduces braking traction and would be very poor for street driving. For slalom use, it results in fast lap times.

For optimum camber these two temperatures should be equal. If the outside edge of the tire is hotter it needs more negative camber. If the inside edge of the tire is hotter there is too much negative camber.

The camber in a corner is affected by the static setting, the suspension geometry, and the amount of body roll. If you have to set the camber at more than one degree negative to get the temperatures right, consider reducing the body roll. You can do this by increasing the overall roll stiffness of the car and perhaps get better performance. An excessive amount of negative camber reduces braking and acceleration traction and wears out tires faster. For road use, an excessive amount of camber can cause wandering on bumps and ridges in the road.

As mentioned earlier, the biggest problem with camber will occur on a car with swing-axle suspension. These cars tend to change camber a great deal in a turn, and thus need a lot of initial negative camber. The best suspension setting for a swing-axle car is a very high amount of roll stiffness plus about two degrees of negative camber. Some experimenting will help you come up with the best compromise. Any other type of suspension would probably work poorly with that much camber. A Z-bar, as described on page 144, counteracts jacking effects as are common with swing axles. High initial negative-camber angles are used to contribute to anti-jacking by reducing roll-center height. This reduces the camber thrust moment.

The best camber setting for running straight ahead is zero, because this puts the tire flat on the road. However, some compromise is necessary. You will have to juggle roll stiffness and camber settings to get the best settings. High roll stiffness is best on smooth tracks, but bumps demand softer anti-roll bars and higher initial camber setting. It is not uncommon for racing cars to be set up differently for each track. For street driving find a compromise setting and stick with it.

The best tire pressure can also be determined on a skid pad. Be aware that warming up the tire changes the pressure, so always check the pressure at the same temperature you are testing at. The pyrometer should show the center of the tire tread at nearly the same temperature as the edges when the pressure is near its optimum value. Write these temperature readings in your notebook, as they will be useful when you don't have a skid pad available. Also make a note of how much the tire pressure increases from dead cold to best operating temperature. At a race you generally will have to set the pressure cold and let it increase to its optimum value during the race as the tires warm up. This is an extremely important bit of information to obtain before your first race.

When you start skid-pad tests to adjust steer characteristics, be careful to make adjustments that do not reduce the maximum cornering power. Driver feel and maximum cornering are not the same thing, so make the changes that give you the best of both.

At times adjusting characteristics can be a real puzzle, because a change may do different things to different cars. The following table is a basic guide, but watch out for exceptions.

TO REDUCE UNDERSTEER

Increase weight transfer at rear by increasing rear roll stiffness.
Reduce weight transfer on front by reducing front roll stiffness.
Reduce front toe-in.
Increase aerodynamic downforce on the front tires.
Reduce aerodynamic downforce on rear tires.
Use wider front tires.

TO REDUCE OVERSTEER

Reduce weight transfer at rear by reducing rear roll stiffness.
Increase weight transfer on front by increasing front roll stiffness.
Increase front toe-in.
Reduce aerodynamic downforce on front tires.
Increase aerodynamic downforce on rear tires.
Use wider rear tires.

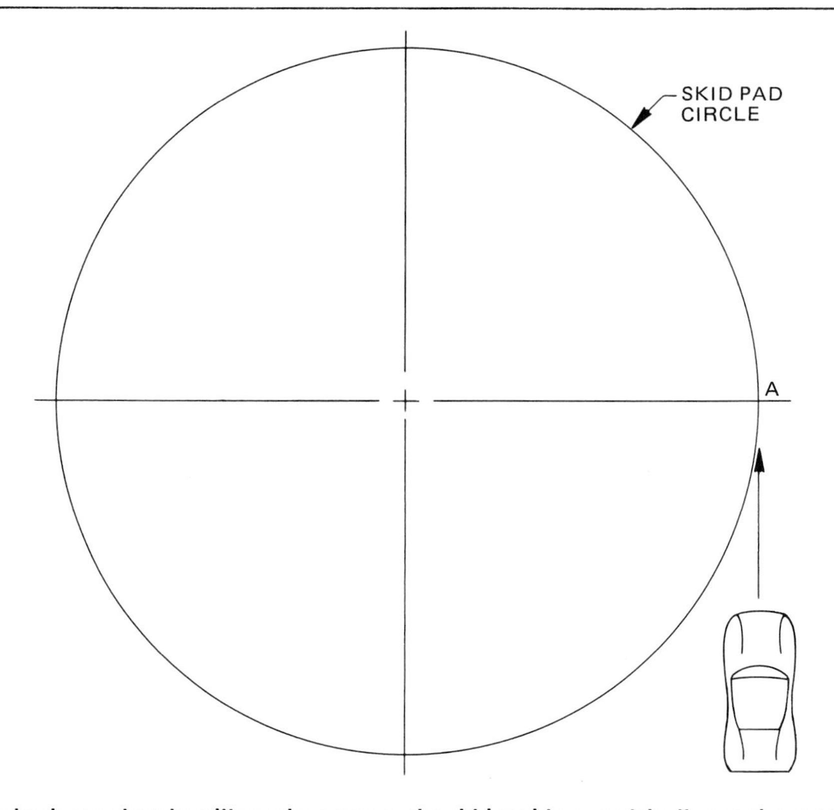

To check transient handling take a run at the skid pad in a straight line at the maximum cornering speed for the skid pad. Turn the wheels at point "A" just as you reach the skid pad circle. Make suspension adjustments to give the most controllable transient handling.

Driver is locking the front tires in a braking test at the Goodyear San Angelo Texas proving ground. Testing brakes in this manner can quickly destroy the tires, so take it easy. For more information on brake testing, design, installation, setup and servicing, get HPBooks' *Brake Handbook.* (Photo courtesy Goodyear Tire & Rubber Co.)

Transient handling characteristics can also be tested on the skid pad if you have enough room to approach the pad at speed. First determine the RPM or speed that your car runs around the skid pad at its limit of adhesion. Then to test the transient handling, run at the skid pad at exactly that speed or RPM in a path that will take you tangent to the skid pad circle. As you reach the circle turn the car into the skid pad path without changing throttle position. You can vary the procedure if you wish to test the condition of braking while or before entering the turn. Exit from the turn can also be tested by reversing the procedure.

Transient handling may be quite different from steady-state handling, as it is affected by the changing forces on the car. Again methodical trial-and-error testing technique is required to find the answers to your problems. To adjust transient handling characteristics you will find toe-in settings quite useful, as the car will respond to a very small adjustment. Certain characteristics may be designed into the car and nothing can be done about it. A car with a large weight or a long wheelbase tends to understeer when entering a corner because of its resistance to being turned. Also some cars have built-in roll steer, and this will cause transient handling to be different from steady state.

Sometimes very stiff shocks and a low roll stiffness combine to give strange transient handling problems. This is due to the fact that the shocks develop forces that resist body lean, but only when the body lean is changing rapidly. Under steady-state cornering the shocks have little effect.

Aerodynamic devices can also be tested on the skid pad, but only if it is a large one allowing high speeds. The skid pad will allow you to balance the downforce between the front and rear for best handling, but it will not tell you the best compromise between drag and downforce. This will have to be determined by measuring lap times at the actual track. Skid pad tests can give you the downforce settings that are the maximum possible, because that will give the fastest speed around the circle. Also the effect of downforce on camber can be studied on the skid pad, and compromises can be established.

BRAKE TESTING

Most cars have dual braking systems, and on many racing cars the balance between the front and rear brakes can be adjusted. You will want to make special tests for brake balance before racing the car, to get the best possible braking. If you happen to have a road car with adjustable brake balance, the test procedure is exactly the same as for a racing machine.

Find a straight piece of pavement for braking tests. It is important that the pavement not be dirty, wet, or have reduced traction compared to the surface you will ordinarily be doing hard braking on. This is an important factor of brake-balance adjustment.

Balance will only be correct for a certain value of tire grip, and it will be wrong for any other grip. Thus if the traction changes such as by running on slick pavement, you must change the brake balance again. The reason for this is that weight transfer in braking depends on the traction of the tires. The more traction, the greater the weight transfer. In a condition of high

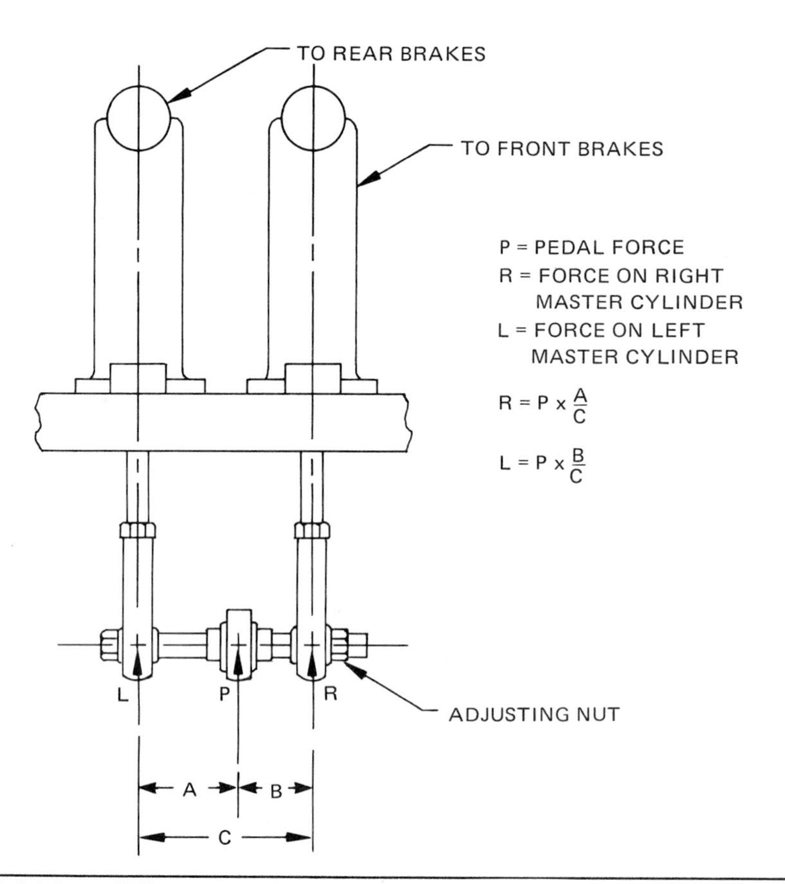

Figure 27/Pedal force "P" acts on a pivot point in the center of the balance bar, usually a spherical bearing as shown. By moving the adjusting nut the pivot point can be moved closer to one master cylinder and farther away from the other. To increase the force "R", increase distance "A". If the right-hand master cylinder goes to the front brakes, this will give more braking force on the front wheels and less on the rear.

The balance bar is mounted in the brake pedal and has a pivoting clevis on each side. To adjust it the jam nuts are loosened, and the shaft of the balance bar is turned with a screwdriver. The shaft is screwed towards the master cylinder that requires additional braking force.

weight transfer, the front brakes must get more downward force and the rear brakes less. The object of the game is to get all four tires to reach their limit of adhesion together.

It's a help to use an observer in running brake-balance tests. Sometimes it is hard for the driver to tell if the front or rear wheels are locking up, and an observer can help. If you have more than one observer, station them on both sides of the car in case one side locks up first. Run down the test track and brake as hard as possible. If you get no wheel locking, try again harder until a pair of wheels lock. Careful you don't flat-spot the tires. Just get an indication of locking without going clear through the rubber. Whichever end of the car locks up first has too much braking force. so remove some from that end of the car.

Brake balance is adjusted in three ways, two of which are easy. The hard way is to change brake-lining material. Brake linings are made in various degrees of *hardness,* which affects the coefficient of friction of the lining material. Soft linings have lots of friction, but tend to wear out faster. Also soft linings on drum brakes are more prone to fade. If you have a means of quickly changing the linings, such as on disk brakes with quick-change pads, it is possible to vary the brake balance by changing hardness of the lining material. Usually this is only done to get large changes, or if one end of the car wears the pads out much faster than the other end. In general, you want to use the softest linings that don't wear out too fast or fade.

The mechanical means of adjusting brake balance is with a *balance bar.* This is a beam between the two master cylinders which proportions the force between them. Usually the balance bar is mounted on the shaft of the brake pedal. If it is adjustable, the balance bar can be shifted from side to side so the pedal force is applied more to one master cylinder and less to the other one. The closer the pedal-force point is to a cylinder, the more force there is on that cylinder. This is illustrated in Figure 27.

Some cars have balance bars that are not quickly adjustable. However, the brake balance can still be changed by making a balance bar with slightly different dimensions, and replacing the balance bar on the car with the new one. Some balance bars can be adjusted with washers or shims to get an adjustment.

A hydraulic means of adjusting brake balance is with an adjustable pressure-limiting valve in the brake line. These have a spring-loaded relief valve in them, which pops open at a given pressure in the hydraulic line. This valve limits the pressure to one pair of brakes, and allows the pressure in the other pair to go on up as the driver pushes harder on the pedal. Remember that this valve will vary the total force required at the pedal, so you will be changing two things at once. However, once the right balance is obtained, pedal force can be adjusted by switching the entire set of linings on the car to set with a different hardness.

The most critical aspect of braking performance is *brake fade*—a partial or complete loss of braking power caused by excess heat. The term was first applied to drum brakes back in the days when all cars used them. Brake fade on drum brakes is a reduction of friction between the lining and drum. The driver feels this as an increase in pedal force required to stop the car. It can be very scary, as if someone is pouring oil on the brake linings. This increase in pedal force sometimes is accompanied by an increase in pedal travel—in extreme cases using up all the available travel. Brake fade may cause a complete loss of stopping power.

Excess heat causes fade in drum brakes. The loss of friction can be traced to vaporization of the binder in the lining material, which emits hot gas between the lining and the drum. This acts to lubricate the brake, just as high-pressure oil lubricates the bearings of your crankshaft. The high-pressure gas acts as a wedge to separate the lining and drum surfaces. The driver feels this as an increase in pedal force, sometimes to the point where the car will hardly slow down at all. Cures for this include using harder lining material, metallic linings, and drilling small holes through the drum to let the gas out. Improving the brake cooling obviously does a great deal of good, and this is what you should work on in a test session.

Another cause of brake fade with drum brakes is expansion of the brake drum due to heat. Because one side of the drum is open and the other side closed, the two sides do not expand an equal amount. There is more expansion at the open end of the drum, so the drum flares out or becomes bell-shaped.

The friction surfaces in the drum which were parallel to each other and to the brake shoes when cold, are no longer parallel when hot. The total area of contact between shoes and drum is reduced and of course braking is reduced.

Both effects combined—vaporization of the plastic binder in the shoe material plus bell-shaped drums—result in a drastic decrease in available braking force.

Disk brakes do not become bell-shaped, but excessive heat causes problems with them too. If the temperature gets high enough, the fluid boils in the caliper causing a mushy-pedal feel. If fluid boiling is severe it can result in the pedal going to the floor and a complete loss of braking power.

Fluid boiling can be cured by using higher-boiling-point fluid or improving cooling of the brake calipers. As with drum brakes, the cooling air flow is the thing that is proved out in testing.

To reduce the fluid-boiling problem, you should replace old fluid with new. Brake fluid picks up moisture with age, and lowers the boiling point. A good-quality brake fluid boils at 550 degrees F. or higher. Make sure that the fluid you use is from a fresh can. Be sure that the fluid is compatible with the brake system, as recommended by the car manufacturer. Certain fluids react with the rubber seals in the brakes, causing problems. If you have a racing car, the brake fluid should be bled out of the calipers before every race and should be completely changed each season. This along with good ducting should prevent fluid boiling.

A better cure is converting to silicone-base brake fluid. Silicone fluid has the advantage of maintaining a higher boiling point—over 750°F—because it doesn't absorb moisture. When converting to silicone fluid, the brake system must be flushed with alcohol to purge the old fluid and moisture.

Brakes can be made to fade with a series of hard accelerations to speed and quick stops. A number of these runs in rapid succession will heat the brakes to the point of fade. Be careful not to run completely out of brakes when they fade, as this can cause some real problems. If

Late-model stock car has brake ducts fabricated from aluminum sheet and heater ducting. Cooling air is directed to the center of the brake rotor and is pumped out the vent holes in the rotor as it rotates. This not only cools the rotor, but the caliper as well to prevent fluid from boiling. (Photo by Tom Monroe.)

the test strip is restricted in length be very careful.

After making the brakes fade and noting the number of runs required to do it, make a change to improve the situation. You may try brake scoops or ducting to bring in cooling air, or various types of brake lining material. Brake fluid is another item that changes the fluid boiling point with disk brakes. Make your change, wait for the brakes to cool down, and run the test again. You may wear out the linings in this sort of testing, but you will learn a lot.

If you are doing serious brake testing you may want to know the actual temperature of the brake caliper or drum. In this way you can check the effectiveness of brake cooling devices without having to make the brakes fade. A reduced temperature will tell you that a change is improving brake cooling.

There are a number of devices on the market to measure the temperature of the brake. One really good one for brake testing is the *Tempilabel* from Omega Engineering, Inc. Of Stamford, Connecticut. These are small paper stick-on labels that you attach directly to the outside of the brake drum or caliper. Each label has white dots on it that turn black when they reach the indicated temperature. The dots stay black so the readings can be taken after the car has come to rest. *Tempilabels* cost several dollars each, and they are extremely accurate. It costs a few bucks for a complete test session, but the results should be well worth it. Having brakes when the competition doesn't can be all you need to win the big race. Having brakes last all the way to the bottom of a long curving downhill mountain road is also reassuring.

SAMPLE SUSPENSION TEST PLAN

1. Warm up and course familiarization—drive car at moderate speed, observe course conditions, establish initial handling characteristics.

2. Camber—take tire temperature readings on each tire, adjust camber to give equal temperature between inside and outside edges of tire tread.

3. Tire pressure—take tire temperature readings on each tire, adjust tire pressure to give even temperature across tread.

4. Ride height—check for maximum suspension travel. Adjust ride height if necessary to prevent bottoming. Reduce ride height if possible to improve cornering speed. Reset camber to optimum settings.

5. Roll stiffness

   A. Measure roll angle using Polaroid camera to photograph car in a tight turn. Adjust total roll stiffness if necessary to get optimum roll angle. Readjust camber to find a new optimum setting.

   B. Front-to-rear roll-stiffness ratio—adjust as necessary to give desired steer characteristics.

6. Shock settings—adjust shocks to get best setting. Pay attention to performance on bumps and transient handling.

7. Toe—adjust toe setting to give desired steer characteristics and stability. Toe should be as close to zero as possible for minimum drag and tire wear. Check settings on both front and rear suspension if possible.

8. Caster—change caster if necessary to give desired compromise between high speed stability and steering effort.

9. Aerodynamics

   A. Change total downforce to give fastest lap times.

   B. Change front-to-rear downforce ratio to give fastest lap times and desired steer characteristics. Note that steer characteristics change with speed.

10. Brakes

   A. Adjust brake balance.

   B. Try harder or softer brake linings as desired.

   C. Check for fade and experiment with air ducts if required.

11. Tires—try other sizes or compounds. Optimize all suspension settings again if necessary to improve lap times.

12. Springs—try alternate spring rates. Check suspension travel and adjust ride height if required. Check roll stiffness and camber and adjust if necessary.

13. Wheels—try alternate rim width to establish optimum size for your tires.

14. Weight Jacking—if used in your type of racing try weight jacking to improve acceleration or cornering.

15. Ride balance—check by disconnecting all four shock absorbers. Next drive the car over an undulating surface and observe which end of the car rises faster and to what degree. The suspension frequency should be nearly the same at both front *and* rear. If the out-of-balance is very noticeable, a spring change should be made by softening the fast end or stiffening the slow end. Note that this method will not work if the car uses multiple-leaf springs because the interleaf friction acts as a shock absorber. Proper ride balance for multi-leaf springs can be determined only through calculations. Refer to page 141 for more details.

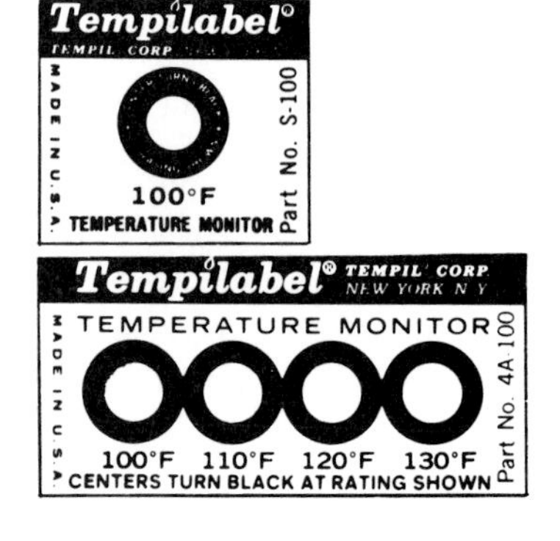

This little stick-on label can accurately measure the peak temperature of your brakes. The Tempilabel is available from Omega Engineering listed in the appendix. The white dots turn black at the temperature indicated on the label. By trying various brake scoop designs, the best one can quickly be determined. Tempilabels are available to measure temperatures from the 100° F range shown all the way up to 500° F. For braking testing, get a variety of temperatures from 250° to 500° to find out roughly how hot your calipers get. If the brakes are really hot they can get to near 500° F on the caliper. The exact temperature will vary with the location on the caliper and how hard the brakes are used during the test.

This Fiat Abarth started life as a fun little road car that gave lively performance, good handling, and reasonable practicality. After owner Chuck Thompson's modifications it was a fire-breathing monster, impossible to drive on the road, but capable of winning top-time-of-the-day at slaloms. This little car with its Formula One slicks and rock solid suspension consistently beats Corvettes and Cobras due to its superior handling.

If you are serious about getting better handling from your car, you will be interested in chassis modifications. This can be a simple job such as installing a set of heavy-duty shocks, or a very complex project such as setting up a stock car for Grand National racing. Whatever your plan, you need to obtain the right components for the job and apply them correctly. This way you get the best handling for each dollar spent.

When modifying your car for better handling you will make compromises and trade-offs. A production car is set up for the average driver under average driving conditions. Compromises have been made in handling to give the best possible ride, minimum initial cost, and a car that handles in a safe and predictable manner with any sort of skilled or incompetent driver behind the wheel. The result is usually not the best handling car for fast driving. Thus if you modify the chassis for improved handling you will be giving up some of the previously mentioned qualities to get better traction, steer characteristics more suitable for spirited or precision driving, and a car that is more fun to drive. It will most likely ride stiffer; it will surely cost money; it may not be suitable for your grandmother to drive in the rain; but it can be a very safe car in proper hands. You must be prepared for compromises and have some knowledge of how far you wish to go.

For racing use, the compromise on modifications is usually more clear cut. You will try to get the best possible racing performance within the limits of your budget. Racing results are measured with the stop watch, and you don't worry much about riding comfort. It is easier to modify a car for all-out racing than it is to get the perfect compromise for all-purpose street driving. Don't belittle the efforts of the designers of family sedans; they have a tough job to do.

TIRES

The most important items on the chassis are the tires. For better handling on road or track you need the best tires available. The best tires are always being replaced by better ones, so don't feel bad about rapid obsolescence. This is particularly true in racing tires, where it is unusual to go a full year without some major change in tires.

A modern racing tire and a radial street tire are shown for size comparison. Both tires fit a Datsun 510 sedan and use a 13" diameter wheel. The racing tire will give about 50% more traction than the street tire on dry pavement, but the street tire is superior in the wet. Unless you are racing in the dry, stay away from the slicks.

RACING TIRES

For racing use the choice is fairly straightforward. There are only a few manufacturers of racing tires, and for your particular type and class of racing the choice of tires may be very limited. Engineering data on racing tires, such as slip angle curves or coefficients of friction, is not available. Apparently either the tire companies don't have the information or it is a closely guarded secret. Anyway you won't be able to make the sort of decision you would like. If you are rich enough, the best way to select tires is to buy a set of everything that might work and test them. This is what the professionals do. Short of that, see what the winning cars are using and buy those tires. Watch the winners so you can see a trend if one develops.

You may be able to get others to cooperate with you on a tire test session. Work with someone who has tires that will fit your car and share the test results with him. You will both be one jump ahead of the other competitors.

The largest manufacturer of racing tires in this country is Goodyear. Racing tires are also available from Firestone, Dunlop of England and Bridgestone of Japan, and a number of smaller domestic manufacturers make racing tires for specialized types of racing. Some better known ones are McCreary, a manufacturer of oval-track tires; and M & H, famous for drag slicks. Each type of racing has its own requirements, and these racing-tire manufacturers know all the tricks. Technical data on tires should be obtained from the manufacturer.

For improving the overall handling of a car, road-racing tires are the trick. They dramatically improve cornering adhesion and also traction in acceleration and braking. Road-racing tires have been used on the street at times, but it is a foolish practice. They are not only illegal, they are actually unsafe for highway use. The reason for this is due to the specialized nature of the tire. A modern road-racing tire has a very light casing, totally inadequate for long life in a highway environment. The sidewalls are thin, with no protection for curb impact or weathering in the sun and the smog. The tread is slick on a dry-weather tire, giving virtually no traction in the wet. In addition the rubber compound is designed for a specific type of car and usually a short racing life, so they wear out rapidly on the street. You are much better off with a set of high-quality street tires, and they will cost you less.

Road-racing tires are designed for a particular weight and class of car, considering the car speed, weight, and restrictions on wheels. Some tires will work only on wide rims, while others are specially designed to fit a smaller rim size. Sidewall stiffness and rubber compound of the tire are designed to suit the weight and speed of the car. Consult the tire manufacturer before buying tires to be sure you are correctly applying them.

Tires for unrestricted wheel sizes have lighter casing construction than those for limited rim widths. The sidewalls are much stiffer for use on narrow wheels, to limit the lateral deflection of the tire. If you ignore the sidewall construction of the tire, you may have the tire on the wrong size wheel. If you have an unusual application, you can usually tell just by inspecting the tire what type of construction it has.

Many road-racing tires are used for slaloms, a use for which they were not designed. In a slalom the run is so short that the tires don't have time to heat up to their proper working temperatures. Also tire wear is not usually a factor due to the short time on the course. You will gain a performance advantage if you use a tire designed for a car lighter than your slalom car. Thus you will be overloading the tire, but it will tend to get hot quicker and should improve your times. You may have to run a little higher air pressure in the tire to compensate for the excess forces, but this is subject to experiment

This shows the range of racing tires offered by Goodyear in 1970. Today's range is just as varied, but with slick treads replacing most of the treaded designs shown. This large company makes specialized racing tires for road racing, oval track, off-road, drags, motorcycles, dirt tracks, stock cars, and speed record vehicles.(Photo courtesy Goodyear Tire & Rubber Co.)

This Mini is set up for slalom competition. Front tires are designed for the front of a Formula 1 car, which has less front-end weight than the Mini. These tires on a Mini would overheat in road race, but in slaloms they heat up quickly during the short timed lap. Mini tires with hard compound never get up to operating temperature in a slalom.

with the car. Stick with the recommended rim width for the tire to get maximum cornering power.

ROAD TIRES

For a road car the tire market is literally flooded with various types and sizes. There are a number of big companies with large product lines, and there are also some smaller specialty firms making high-performance tires. The first thing to do is weed out most of the economy tires. You usually get what you pay for in a tire, and the real cheapies are not what you want for fast driving. All a cheap tire has to do is hold up the car for sedate driving. It doesn't need outstanding adhesion or be able to take high speeds.

The first thing to consider is the construction of the tire. You have no doubt heard of bias, belted, and radial tires in the ads on TV. This refers to the type of construction used in the casing. The conventional tire construction that has been around for years is the bias ply. This refers to the direction that the cords run in the carcass of the tire. Bias-ply tires have the cords criss-crossed across the tire from one bead to the other. The cords make an angle of between 32 and 40 degrees with the centerline of the tread. These tires are available in all sizes and in the narrow widths are not the best tires for handling. They will usually be the cheapest type of tires you can buy, and there is nothing wrong with them for average use. Some bias-ply tires are wide and low, and these offer handling advantages because of their size.

A radial tire uses a cord angle of 90 degrees. That is, the cords run from one bead to the other directly across the tread. In addition, a radial tire has a *belt* under the tread surface, and can be made of various materials including fiberglass and steel mesh. The characteristic of a radial tire is *less* tread distortion under load and *more* sidewall distortion. Radial tires look a bit underinflated because of the high sidewall deflection under the weight of the car.

Because radial-tire sidewalls are more flexible than a bias-ply tire, the time the radial tire requires to develop a slip angle when acted on by a side force is different than the bias tire. Thus it is a bad practice to mix the two types of tires on the same car. If you use a mixture, the car may wander when crossing ridges in the road, have various instabilities, and may have drastic steer characteristics. When changing to radials, save up enough money to buy all four!

A radial tire has a number of important advantages over a bias-ply tire. The design allows the tread to work independently from the sidewalls, and the tread makes better contact with the road under all conditions. The result is a lower slip angle for any given cornering force, greater traction in both dry and wet, and better wear characteristics.

If you put radials on a car which previously used bias-ply tires, the steering seems quicker and more responsive. This is a result of lower slip angles. There will be less tendency for the car to be disturbed when crossing a ridge in the road at a shallow angle. The limit of adhesion will be higher with radials, but the slip angle at the limit will be less. Thus radial tires will give less warning than bias-ply

This is Firestone's Super Sports Wide Oval passenger tire. It is bias construction, two-ply nylon cord. This is a wide-tread tire, but shares the stiffness characteristics of all bias-ply tires. Notice the cord angles under the tread.

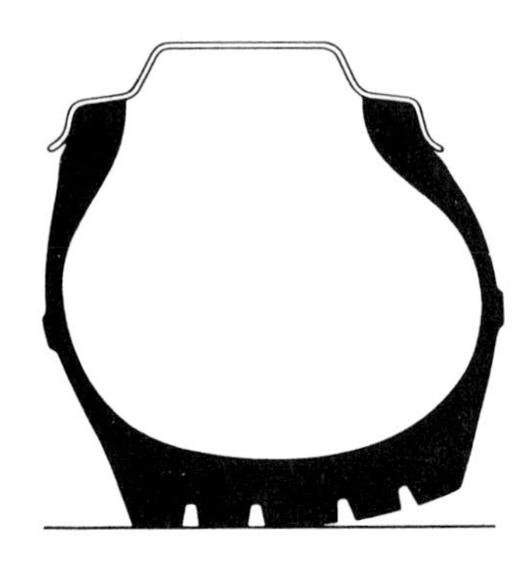

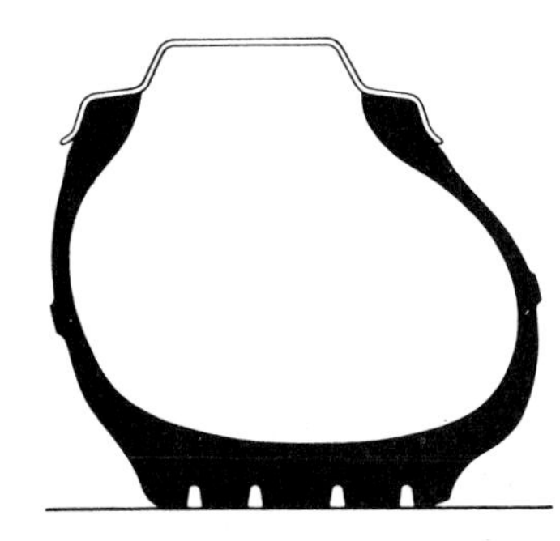

These drawings from Pirelli illustrate the difference between the bias-ply and the radial tire in a corner. The bias-ply tire, top, loses contact with the road as both tread and sidewalls distort under a cornering load. The radial tire, bottom, flexes mostly in the sidewall area keeping the tread flat on the road. The result is better adhesion for the radial.

Here's a close-up of the interior construction of Firestone's Wide Oval Radial passenger tire. It has two radial body plies and four belt plies. The belt material is criss-crossed for maximum lateral stiffness, and the 90-degree cord angle of the sidewalls allows flexibility.

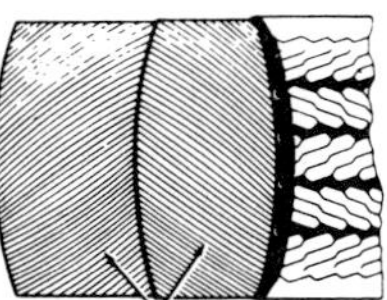

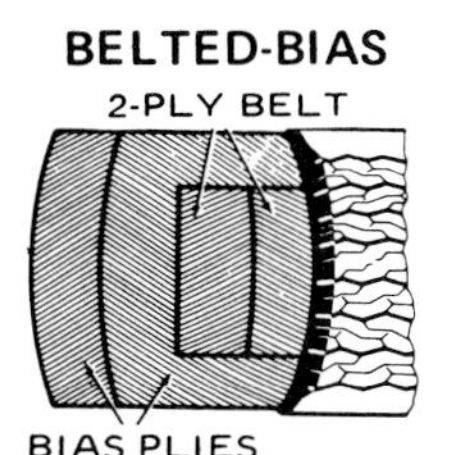

RADIAL-PLY
4-PLY BELT
RADIAL PLIES

Drawing courtesy PV4 magazine

The bias-belted or belted-bias tire is a cross between the bias-ply tire and the radial. The sidewalls are similar to the bias-ply tire, but a belt under the tread is added. This gives tire characteristics that are a compromise between the other two types.

tires, and will take more driver concentration when driving very hard. Radial tires will wear a lot longer than the bias-ply tires, and thus may cost less in the long run in spite of their usually higher initial cost.

If your car was not equipped with radials as standard equipment, the ride may seem a bit more harsh at low speeds, and there may be an increase in road noise. This characteristic varies a lot with the car, and on some cars it is not noticeable. You will have less tread distortion with radials, which reduces rolling drag of the tires. You can expect a small increase in top speed and a noticeable improvement in gas mileage, although mainly at lower speeds.

If you check tire manufacturer's recommendations you will most likely find that a radial is safe at a higher speed than a bias-ply tire. This is due to the cooler-running of the radial because of less tread flexure.

All these characteristics make the radial tire the best for a high-performance road car, and this is why more new cars every year are equipped with radials as standard equipment. You will find radial tires in all sizes for just about any car. They are available in a wide variety of tread designs and are made of various materials by the different manufacturers. It is hard to draw specific conclusions on which brand to buy from all this variety of information, so talking with owners of radial tires on a car like yours may be the best guide.

Another type of tire construction is the bias-belted tire. This design is a cross between a conventional bias-ply tire and a true radial and has some of the characteristics of each. The tire uses bias-ply cord angles and is also equipped with a belt under the tread similar to a radial. These tires are a compromise between the two types and perhaps have an application for modest street driving. The advantages of a bias-belted tire over a bias-ply tire are not as pronounced as a radial, and likewise any disadvantages are not as severe.

TIRE SIZE SELECTION

Once you have decided on a basic tire construction you have to select the proper tire size for your car. The important dimensions are outside diameter, tread width, maximum section width, and wheel rim width. Also of some importance is the tire weight.

The tire industry has recently introduced new size designations for tires, and some of these give you little or no direct information on the important dimensions of the tire. The only way to be sure is carry a steel tape in your pocket when shopping for tires, and measure each one. The weight can also be measured with a small spring scale once you have arrived at a size you want.

The outside diameter of the tire determines the ride height of the car and also the effective overall gear ratio. A lower tire will reduce the ride height, give you more clearance around the tire, and improve the acceleration in each gear. Also your engine will be turning more RPM at cruising speed and fuel economy will suffer. In addition the speedometer will read fast if the tire is smaller than stock.

The effect of a change in tire diameter on gearing is the ratio of the diameters. New diameter divided by old diameter is an indication of how much top speed will change, neglecting air-resistance effects.

The same ratio is an indication of the change in acceleration. If you reduce tire diameter by 10%, acceleration should increase by about 10%, but top speed will be reduced. There is some "small print" which you should read carefully. In most automotive competition, the game is to get from one place to another in the shortest time.

The car which is first into the turn is the one that got there in the shortest time. Time from one point to another varies according to the square root of acceleration. The square root of 10% is about 3%. Of course the engine is turning about 10% faster and once it passes the power-peak it ain't accelerating any more.

Obviously the change in ride height is about half the change in diameter of the tire.

The first step in selecting tires is to choose the outside diameter, considering the ride height and gear ratio that will result. Once the outside diameter is chosen, you can measure clearances and see how wide a tire will fit.

For best traction and cornering power you want the widest possible tread. However, there are limits to this as follows:

You reach the limit of available widths on the market.

The tire won't fit under the car.

You don't have the right wheels for those wide tires.

You wish to improve handling in wet weather and give up some in the dry.

You want lighter tires than the widest ones available, to improve the ride and handling on rough roads.

The first step in selecting the tread width is to see what is available in the outside diameter you selected. Tread width varies from one manufacturer to the next, so you must measure it directly off the tires for comparison. Go to the tire dealer with a steel tape and a notebook and take your own measurements. It's OK to measure unmounted tires as long as the tread is reasonably flat. The tread width doesn't change a noticeable amount with air in the tire, but it may change flatness.

When measuring tread width note the tire size. You will see all sorts of confusing numbers and letters, but they can be useful to determine the dimensions of your new tires. Figure 28 gives the dimensions of each tire size. The design rim is not necessarily the best rim width to use. It is merely the rim that the maximum section width is measured on. The maximum section width changes 0.2" for each 1/2" change in rim width. Thus starting with the wheel rim and section width shown in Figure 28, the section width for any other rim width can be determined.

Let's take an example. Assume you want to put tires of 25.5" outside diameter on your car. After measuring tread widths at the dealer's you find that E60-15 is a desired tire size. In Figure 28 you find that tire has a section width of 8.7" mounted on the design rim of 6". However, you want to put these tires on 7" rims. If you increase the rim width by 1" over the design rim, the increase in section width is 0.4". Remember the section width changes 0.2" for each 1/2" change in rim width. The new section width of this tire on a 7" rim is 9.1", the original 8.7" plus the increase of 0.4".

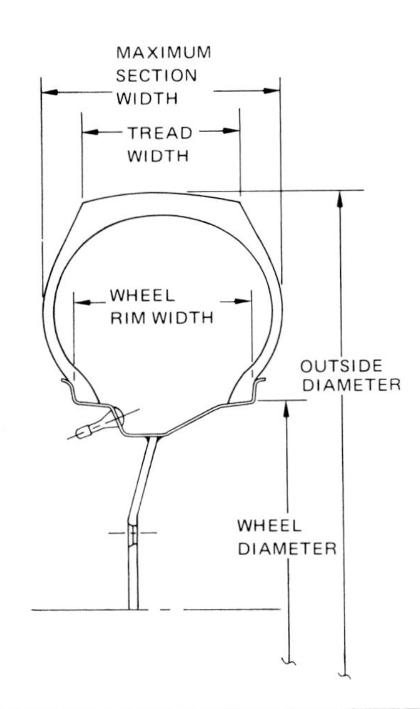

When selecting new tires for your car you need to know all the dimensions shown. The dimensions may not be standardized for a particular size tire, so it is best to actually take measurements off the tire in the dealer's store.

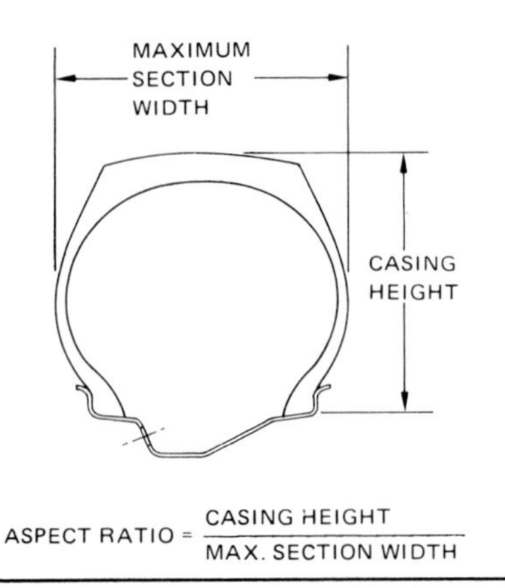

$$\text{ASPECT RATIO} = \frac{\text{CASING HEIGHT}}{\text{MAX. SECTION WIDTH}}$$

Tire aspect ratio is shown in this drawing. Low-profile tires for high-performance are those with low aspect ratios. Old-fashioned tires have larger aspect ratios.

BIAS AND RADIAL PLY TIRES

TIRE SIZE DESIGNATION	DESIGN RIM WIDTH	SECTION WIDTH	OVERALL DIAMETER
50 SERIES – BIAS PLY			
C50-13	6.50	9.40	22.46
F50-14	7.00	10.20	24.32
H50-14	8.00	11.35	25.36
M50-14	9.00	12.55	26.48
N50-14	9.00	12.85	26.84
B50-15	6.00	8.55	23.58
G50-15	7.00	10.35	25.46
H50-15	8.00	11.15	26.06
L50-15	8.00	11.65	26.76
N50-15	9.00	12.65	27.50
50 SERIES – RADIAL PLY			
BR50-13	6.50	9.15	22.06
GR50-14	8.00	10.95	24.76
GR50-15	7.00	10.35	25.46
HR50-15	8.00	11.15	26.06
LR50-15	8.00	11.65	26.76
60 SERIES – BIAS PLY			
A60-13	5.50	7.85	22.48
B60-13	6.00	8.35	22.96
C60-13	6.00	8.60	23.36
D60-13	6.50	9.05	23.74
D60-14	6.00	8.65	24.44
E60-14	6.50	9.10	24.78
F60-14	7.00	9.55	25.28
G60-14	7.00	9.85	25.80
H60-14	7.00	10.25	26.40
J60-14	7.00	10.45	26.70
L60-14	8.00	11.10	27.20
B60-15	5.50	7.80	24.38
C60-15	6.00	8.25	24.76
E60-15	6.00	8.70	25.50
F60-15	6.50	9.20	25.94
G60-15	7.00	9.70	26.44
H60-15	7.00	10.05	27.08
J60-15	7.00	10.25	27.40
L60-15	7.00	10.50	27.86
60 SERIES – RADIAL PLY			
AR60-13	5.50	7.85	22.48
BR60-13	6.00	8.35	22.96
ER60-13	6.50	9.25	24.16
AR60-14	5.50	7.70	23.16
FR60-14	6.50	9.35	25.30
GR60-14	7.00	9.85	25.80
JR60-14	7.00	10.45	26.70
LR60-14	8.00	11.10	27.22
ER60-15	6.00	8.70	25.52
FR60-15	6.50	9.20	25.94
GR60-15	6.50	9.50	26.44
HR60-15	7.00	10.05	27.08
JR60-15	7.00	10.25	27.40
LR60-15	7.00	10.50	27.86
70 SERIES – BIAS PLY			
A70-13	5.00	7.10	23.30
C70-13	5.50	7.80	24.24
D70-13	5.50	8.00	24.70
C70-14	5.00	7.45	24.90
D70-14	5.50	7.85	25.32
E70-14	5.50	8.05	25.76
F70-14	5.50	8.30	26.24
G70-14	6.00	8.75	26.82
H70-14	6.00	9.10	27.50
J70-14	6.50	9.50	27.80
L70-14	6.50	9.75	28.32
A70-15	4.50	6.60	24.68
C70-15	5.50	7.50	25.58
D70-15	5.50	7.70	26.02
E70-15	6.00	8.10	26.40
F70-15	6.00	8.35	26.92
G70-15	6.00	8.60	27.46
H70-15	6.00	8.95	28.10
J70-15	6.50	9.35	28.44
K70-15	6.50	9.40	28.68
L70-15	6.50	9.60	28.92
70 SERIES – RADIAL PLY			
AR70-13	5.00	7.15	23.22
BR70-13	5.50	7.60	23.78
CR70-13	5.50	7.85	24.16
DR70-13	5.50	8.05	24.60
CR70-14	5.50	7.65	24.92
DR70-14	5.50	7.90	25.24
ER70-14	5.50	8.10	25.68
FR70-14	6.00	8.55	26.18
GR70-14	6.00	8.85	26.66
HR70-14	6.50	9.40	27.34
JR70-14	6.50	9.55	27.74
LR70-14	6.50	9.80	28.24
BR70-15	5.00	7.10	25.14
CR70-15	5.50	7.50	25.58
DR70-15	5.50	7.75	25.94
ER70-15	5.50	7.95	26.32
FR70-15	6.00	8.40	26.84
GR70-15	6.00	8.65	27.40
HR70-15	6.50	9.20	28.04
JR70-15	6.50	9.40	28.34
KR70-15	6.50	9.50	28.54
LR70-15	6.50	9.65	28.84
MR70-15	7.00	10.15	29.24
78 SERIES – BIAS PLY			
A78-13	4.50	6.60	23.46
B78-13	5.00	7.05	24.00
C78-13	5.00	7.25	24.46
D78-13	5.50	7.70	24.84
A78-14	4.50	6.45	24.16
B78-14	4.50	6.65	24.70
C78-14	5.00	7.05	25.24
D78-14	5.00	7.35	25.50
E78-14	5.50	7.65	26.00
F78-14	5.50	7.90	26.50
G78-14	6.00	8.35	27.06
H78-14	6.00	8.70	27.76
J78-14	6.00	8.80	28.20
A78-15	4.50	6.35	24.80
B78-15	4.50	6.55	25.38
C78-15	5.00	6.95	25.82
D78-15	5.00	7.15	26.24
E78-15	5.00	7.35	26.64
F78-15	5.50	7.70	27.22
G78-15	5.50	8.05	27.68
H78-15	6.00	8.55	28.36
J78-15	6.00	8.70	28.72
L78-15	6.00	8.85	29.30
M78-15	6.50	9.35	29.68
N78-15	7.00	9.80	30.16
L84-15	6.00	8.65	29.64
78 SERIES – RADIAL PLY			
AR78-13	4.50	6.50	23.36
BR78-13	4.50	6.75	23.88
CR78-13	5.00	7.15	24.32
DR78-13	5.00	7.35	24.76
ER78-13	5.50	7.75	25.16
AR78-14	4.50	6.40	23.98
BR78-14	4.50	6.60	24.54
CR78-14	5.00	7.00	25.00
DR78-14	5.00	7.20	25.40
ER78-14	5.00	7.40	25.80
FR78-14	5.50	7.85	26.30
GR78-14	6.00	8.30	26.86
HR78-14	6.00	8.60	27.56
JR78-14	6.50	8.95	27.94
AR78-15	4.50	6.25	24.70
BR78-15	4.50	6.45	25.24
CR78-15	5.00	6.85	25.70
DR78-15	5.00	7.05	26.10
ER78-15	5.50	7.45	26.48
FR78-15	5.50	7.70	26.94
GR78-15	6.00	8.15	27.52
HR78-15	6.00	8.45	28.18
JR78-15	6.50	8.80	28.58
LR78-15	6.50	9.00	29.08
MR78-15	6.50	9.20	29.58
NR78-15	7.00	9.70	29.92

Figure 28/This table shows the outside diameter and section width for each tire size. The section width is for the tire mounted on the wheel of the design rim width shown. Manufacturers' standards allow variations of up to 7% in these dimensions, so take your own measurements just to be sure. (Courtesy of The Tire and Rim Association, Inc.)

Here is a side-by-side comparison of four Goodyear tires of different aspect ratios. The low profile 50 series tire on the right is the Goodyear Rally GT 50, a high-performance road tire. Notice the wide tread grooves in the wider tires for improved wet weather traction. (Photo courtesy Goodyear Tire & Rubber Co.)

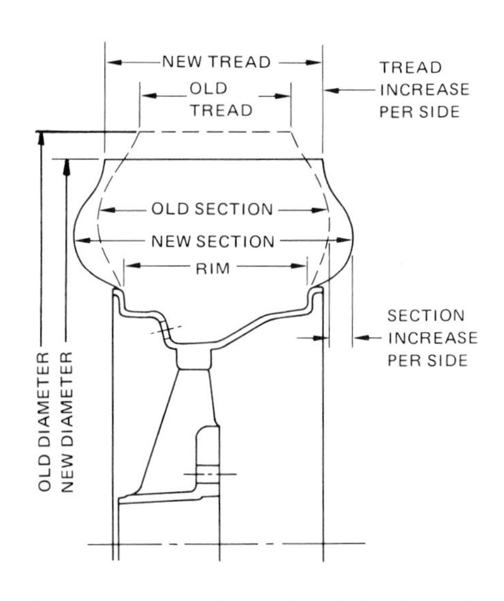

A sketch of old and new tire dimensions such as this can be helpful in visualizing where clearance problems are. The tread width increase and the section increase are the areas to check. If the increase per side is larger than the clearance with the old tires you have trouble. This sketch assumes you are not changing wheels—only the tires. With new wheels this gets more complex.

When measuring tread widths and section widths you will notice tires grouped by *aspect ratio.* The aspect ratio is the casing height divided by the maximum section width. *Low-profile tires* are those with the lowest aspect ratio. These are the widest for a given outside diameter. Older or conventional tires had aspect ratios of from .90 to .80. The newer 70 series tires have an aspect ratio of .70. The lower profile 60 series and 50 series tires have aspect ratios of .60 and .50 respectively. For high-performance driving you will be looking at tires with these lower aspect ratios.

For each tire size there are several rim widths that will work. A list of these approved rim widths is shown in Figure 29. You should use the widest possible rim on the list for maximum cornering power. However, narrower rims may be used if the tire section width on the wide rim is too big for the space under the fenders. Also you may not wish to spend money for wide wheels if your existing rims fall within the range of acceptable sizes. This is one more compromise to make.

To measure your existing wheel rim width take one off the car and lay it on a flat surface. Put a straightedge across the sidewall of the tire and measure the maximum section width of the tire. This is dimension A in Figure 30. Then measure from the straightedge to the surface of the wheel at the tire bead, dimension B in Figure 30. Next estimate dimension C, which is the metal thickness at the tire bead. For steel rims, C is about 1/8". For cast aluminum or magnesium wheels it is about 1/4". Add B to C, multiply by 2, and subtract this amount from A. That gives you the rim width. The rim width is usually measured in inches, even on foreign cars. Rim widths go in 1/2" increments as shown in Figure 29.

If you have wide enough rims already, you have a fairly easy job of measuring tire clearance. However, if you plan to buy new wheels, you can change both the rim width and the track width of the car. Read the next section on wheels *before* making your final decision.

Before putting a down payment on that new set of tires, measure the car to be sure they will fit. To keep this discussion simple, let's assume you are not changing wheels. Just the tires will be new. If you are using tires that are smaller than before in section width, tread width, and outside diameter, don't even bother to measure the car. However, the usual thing is to go wider on the tire, so check for interference.

First make a sketch of the new tire mounted on your existing wheel. Measure the tread width, maximum section width, and outside diameter, and write these dimensions on the sketch. On the same drawing put these same dimensions of your existing tires. This will help you visualize where the problems may be.

Now take a look under the car and see how close your old tire is to hitting the frame, suspension, or inner body panels. Check this on the front wheels with the steering turned to full lock in both directions. The front tires do not pivot directly around their treads but move forward and

TIRE SIZE DESIGNATION	APPROVED RIM WIDTHS
A70-13	4½, 5, 5½, 6, 6½
C70-13	5, 5½, 6, 6½, 7
D70-13	5½, 6, 6½, 7
C70-14	5, 5½, 6, 6½, 7
D70-14	5, 5½, 6, 6½, 7
E70-14	5½, 6, 6½, 7
F70-14	5½, 6, 6½, 7
G70-14	6, 6½, 7, 8
H70-14	6, 6½, 7, 8
J70-14	6, 6½, 7, 8
L70-14	6½, 7, 8, 9
A70-15	4½, 5, 5½, 6
C70-15	5, 5½, 6, 6½
D70-15	5, 5½, 6, 6½, 7
E70-15	5, 5½, 6, 6½, 7
F70-15	5½, 6, 6½, 7
G70-15	5½, 6, 6½, 7, 8
H70-15	6, 6½, 7, 8
J70-15	6, 6½, 7, 8
K70-15	6, 6½, 7, 8
L70-15	6½, 7, 8, 9
AR70-13	4½, 5, 5½, 6, 6½
BR70-13	5, 5½, 6, 6½, 7
CR70-13	5, 5½, 6, 6½, 7
DR70-13	5½, 6, 6½, 7
CR70-14	5, 5½, 6, 6½, 7
DR70-14	5, 5½, 6, 6½, 7
ER70-14	5½, 6, 6½, 7
FR70-14	5½, 6, 6½, 7
GR70-14	6, 6½, 7, 8
HR70-14	6, 6½, 7, 8
JR70-14	6½, 7, 8, 9
LR70-14	6½, 7, 8, 9
BR70-15	4½, 5, 5½, 6, 6½
CR70-15	5, 5½, 6, 6½
DR70-15	5, 5½, 6, 6½, 7
ER70-15	5, 5½, 6, 6½, 7
FR70-15	5½, 6, 6½, 7
GR70-15	5½, 6, 6½, 7, 8
HR70-15	6, 6½, 7, 8
JR70-15	6, 6½, 7, 8
KR70-15	6, 6½, 7, 8
LR70-15	6½, 7, 8, 9
MR70-15	6½, 7, 8, 9
A78-13	4½, 5, 5½, 6
B78-13	4½, 5, 5½, 6, 6½
C78-13	5, 5½, 6, 6½
D78-13	5, 5½, 6, 6½, 7
A78-14	4½, 5, 5½, 6
B78-14	4½, 5, 5½, 6
C78-14	4½, 5, 5½, 6, 6½
D78-14	5, 5½, 6, 6½
E78-14	5, 5½, 6, 6½, 7
F78-14	5, 5½, 6, 6½, 7
G78-14	5½, 6, 6½, 7
H78-14	5½, 6, 6½, 7, 8
J78-14	6, 6½, 7, 8
A78-15	4, 4½, 5, 5½
B78-15	4½, 5, 5½, 6
C78-15	4½, 5, 5½, 6
D78-15	4½, 5, 5½, 6, 6½
E78-15	5, 5½, 6, 6½
F78-15	5, 5½, 6, 6½, 7
G78-15	5½, 6, 6½, 7
H78-15	5½, 6, 6½, 7
J78-15	5½, 6, 6½, 7, 8
L78-15	6, 6½, 7, 8
M78-15	6, 6½, 7, 8
N78-15	6½, 7, 8, 9
L84-15	5½, 6, 6½, 7, 8
C50-13	6, 6½, 7, 8
F50-14	6½, 7, 8, 9
H50-14	8, 9, 10
M50-14	8, 9, 10, 11
N50-14	9, 10, 11, 12
B50-15	5½, 6, 6½, 7
G50-15	7, 8, 9
H50-15	7, 8, 9, 10
L50-15	8, 9, 10
N50-15	8, 9, 10, 11
BR50-13	6, 6½, 7, 8
GR50-14	7, 8, 9, 10
GR50-15	7, 8, 9
HR50-15	7, 8, 9, 10
LR50-15	8, 9, 10
A60-13	5, 5½, 6, 6½, 7
B60-13	5½, 6, 6½, 7
C60-13	5½, 6, 6½, 7, 8
D60-13	6, 6½, 7, 8
D60-14	5½, 6, 6½, 7, 8
E60-14	6, 6½, 7, 8
F60-14	6, 6½, 7, 8
G60-14	6½, 7, 8, 9
H60-14	6½, 7, 8, 9
J60-14	7, 8, 9
L60-14	7, 8, 9, 10
B60-15	5, 5½, 6, 6½, 7
C60-15	5½, 6, 6½, 7
E60-15	5½, 6, 6½, 7, 8
F60-15	6, 6½, 7, 8
G60-15	6, 6½, 7, 8
H60-15	6½, 7, 8, 9
J60-15	6½, 7, 8, 9
L60-15	7, 8, 9, 10
AR60-13	5, 5½, 6, 6½, 7
BR60-13	5½, 6, 6½, 7
ER60-13	6, 6½, 7, 8
AR60-14	5, 5½, 6, 6½, 7
FR60-14	6, 6½, 7, 8
GR60-14	6½, 7, 8, 9
JR60-14	7, 8, 9
LR60-14	7, 8, 9, 10
ER60-15	5½, 6, 6½, 7, 8
FR60-15	6, 6½, 7, 8
GR60-15	6, 6½, 7, 8
HR60-15	6½, 7, 8, 9
JR60-15	6½, 7, 8, 9
LR60-15	7, 8, 9, 10
AR78-13	4½, 5, 5½, 6
BR78-13	4½, 5, 5½, 6
CR78-13	4½, 5, 5½, 6, 6½
DR78-13	5, 5½, 6, 6½, 7
ER78-13	5, 5½, 6, 6½, 7
AR78-14	4, 4½, 5, 5½
BR78-14	4½, 5, 5½, 6
CR78-14	4½, 5, 5½, 6
DR78-14	5, 5½, 6, 6½
ER78-14	5, 5½, 6, 6½, 7
FR78-14	5, 5½, 6, 6½, 7
GR78-14	5½, 6, 6½, 7
HR78-14	5½, 6, 6½, 7, 8
JR78-14	6, 6½, 7, 8
AR78-15	4, 4½, 5, 5½
BR78-15	4½, 5, 5½, 6
CR78-15	4½, 5, 5½, 6
DR78-15	4½, 5, 5½, 6, 6½
ER78-15	5, 5½, 6, 6½
FR78-15	5, 5½, 6, 6½, 7
GR78-15	5, 5½, 6, 6½, 7
HR78-15	5½, 6, 6½, 7
JR78-15	5½, 6, 6½, 7, 8
LR78-15	6, 6½, 7, 8
MR78-15	6, 6½, 7, 8
NR78-15	6, 6½, 7, 8

Figure 29/This list shows rim widths approved by The Tire and Rim Association, Inc., for each modern road-tire size. The tire listed works on any of the rims shown for that size, but try to use a rim toward the upper end of the size range for maximum cornering power. The maximum section width of the tire will vary with rim width, so select rim width before measuring tire clearances.

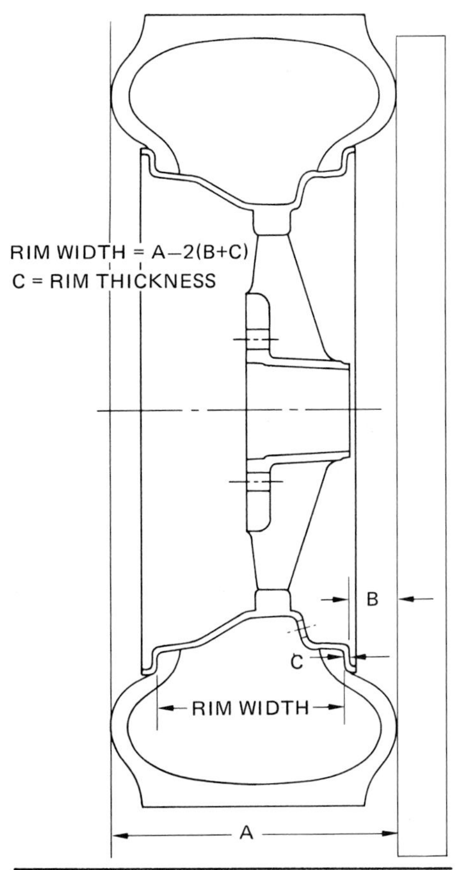

Figure 30/To find rim width of your existing wheels, remove tire and wheel from car. Place on a flat surface and measure dimensions shown. The metal thickness of the rim, "C", is about 1/8" for steel wheels and about 1/4" for mags.

backward when they are turned. Check the clearance between the tread and the fender lip to see if this motion will cause problems.

Consider the condition where the wheel travels upward into the fender and check the clearances at that point. This can be done by adding a lot of weight to the car to compress the springs. Check the clearance when the car rolls by jacking up one side of the car using a floor jack with casters so the weight is carried by the other two tires.

With all these measurements you should be able to tell how much wider you can go on either the tread or the section of the tire without hitting anything. If

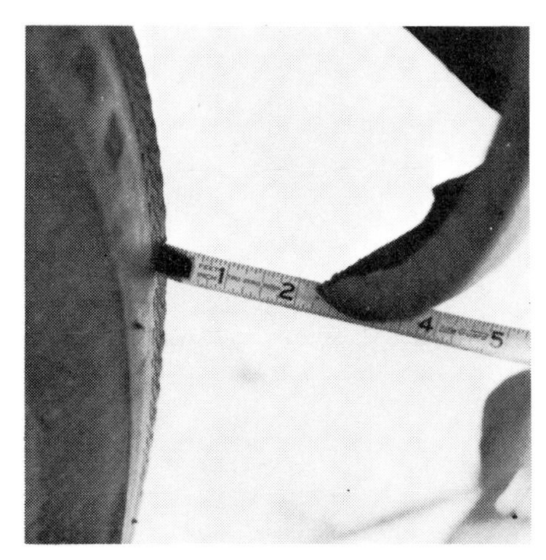

This tire could be 2" bigger on the outside edge and still clear the fender lip. Be sure to turn the wheels when measuring here, as the other side of the tire may differ in clearance. Look at all possible combinations of steering and suspension movements and find the most critical clearances with the tires. Check both inboard and outboard sides. Check the right and left sides of the car, as many cars are not symmetrical.

To fit fat racing tires under this Datsun it was necessary to build fender flares. These were done by professionals and really look like it. The cost of modifications such as this can be hundreds of dollars. But it's pretty, isn't it?

Here are some big tires on a little car, but they don't help the handling very much. The wheel rim is too narrow for these tires, curving the tread noticeably. Wide fender flares were added but the big tires will hit the underside of the flares with only about an inch of suspension travel. This is a case of poor planning.

you are not considering a large change in tire width, then perhaps your car dealer can confirm that a certain size tire will fit. If they can't be sure, do all the measuring yourself. You will be terribly disappointed if those new wide skins burn holes in the fenders every time the car rounds a turn. You are better off with narrow rubber or different fenders.

Certain modifications are possible to squeeze big tires under the car. Check with your local body shop on the cost of flaring the fenders, re-contouring the wheel well opening, or moving the inner fender panel. If the tire hits the anti-roll bar at full lock, perhaps a new bar can be built to clear. Also you may be willing to sacrifice some steering lock to get more rubber. These are all compromises that should be decided on before you buy the tires.

Even though you can gain traction with the widest tread, there are reasons for selecting a narrower tire. An important reason is wet-weather traction. A wide tire has a problem known as *hydroplaning.* Hydroplaning is where water is forced between the tire and the road, lifting the tire up on a fluid wedge. This totally destroys traction, and the car usually goes out of control. Wide tires, a light car, fast driving and deep water are the worst combination.

Besides tire width, the tread design has a great effect on a tire's tendency to hydroplane. Tread grooves give the water under the tire a place to go, breaking up the wedge of water that tends to lift the tire. Wide tread grooves offer the best rain performance. This and other effects of tread design are shown in Figure 31.

Tire weight is of some concern. Lower unsprung weight will give better cornering on bumps and the ride will be improved.

Classification	Pattern	Tread Pattern	General Purpose	Handling stability on dry	Cornering Braking on wet	Life	Riding Comfort	Winter Property
RD-20	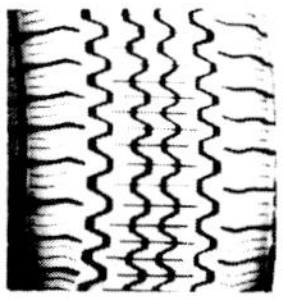	Rib	General Use	3	3	2	2	
RD-201		Block	Sporty driving	2	1	2	3	3
RD-102		Rib/Block 70 Series	High Speed driving	1	1	3		
RD-150		Rib	Soft/Quiet riding	3	3		1	
RD-170V		Rib (Steel Br)	Long Mileage for Highway Use	1	2	1		
RD-601		Rib/Block Reinforced (Radial)	Station Wagon			2		
R-19P		Block	Winter tyre for Highway driving		3	3		2
RD-701P		Block	Winter tyre for deep snow		3	3		1

Note: 1 = Best 2 = Better 3 = Good Blank = Standard

Figure 31/Here is a comparison of Bridgestone Radial tread designs. The advantages and disadvantages of each design are shown in the table. Similar tread patterns are made by other manufacturers, so this table can be used to compare other brands. Bridgestone offers tires for every purpose, as you can see from this selection. (Courtesy of Bridgestone Tire Co., Ltd.)

Engineers can observe and photograph tire behavior through a glass plate set into the test track at the Goodyear proving ground near San Angelo, TX. Photographs of tread distortion and hydroplaning are analyzed to help design tires with improved performance. (Photo courtesy of Goodyear Tire and Rubber Co.)

Typical of a custom street wheel is Cragar's Mag Master. This is a cast aluminum "slot-mag" style wheel, and is one of the least expensive of Cragar's custom wheel line. Weights range from 15 to 24 pounds, and it is available in 13", 14", and 15" diameters. Various rim widths, rear spacings, and bolt patterns are available for a wide range of street applications. (Photo courtesy Cragar Industries)

Acceleration and braking are also affected by the weight of the tires on account of their flywheel effect. Tire weight is difficult to determine without a direct measurement, so just take a fish scale or a small platform scale to the tire dealer and take your own measurements. Weight of the tire is not the most important factor, but it may help you decide which of similar tires you wish to use. Tire weight is more important on a light car than a heavy one, because it represents a larger percentage of the vehicle weight.

WHEELS

When you select tires for improved handling, often the wheels should be changed too. Most cars come stock with wheels that are the narrowest rim width which will work. For best cornering power you want a wider rim than the minimum. How wide you can go depends a lot on the tire construction. For road tires, check Figure 29 or the tire manufacturer's specs. For best handling use a rim width at least as wide as the tire tread.

You benefit from a rim up to several inches wider than the tread. For racing tires check with the tire manufacturer or select a rim at least as wide as the tread width. Usually the wider the better on rims, and you can use a rim up to 2" wider than the tread on most racing tires.

A wider rim effectively stiffens the sidewalls of the tire, and there is less deflection of the tire in a corner. This means lower slip angles and a higher limit of adhesion. You may also be able to notice a slightly harsher ride with a really wide-rim wheel.

Other reasons for obtaining new wheels besides looks are to increase the strength and stiffness of the wheel, improve brake cooling, and most important of all to lower the unsprung weight. An additional reason, sometimes vitally important, is to change the track width so big tires fit on the car. New wheels are much better than spacers for getting proper tire clearance.

Look at Figure 32 before going out to buy wheels. It shows four dimensions used to describe a wheel. One dimension shown, *rear spacing,* is related to another term called *offset.* Rear spacing and offset are two ways to describe the same thing. I'll clarify that in a minute.

If cars were driven on dry roads only, tires could be treadless like the Indy tire at the left. Goodyear engineer Bob Stella points out the tread grooves in the Eagle GT which channels water from between the tire tread and the road surface. Because it lacks a similar tread pattern, the high-speed Indy race tire can't be driven on a wet track. (Photo courtesy Goodyear Tire and Rubber Company.)

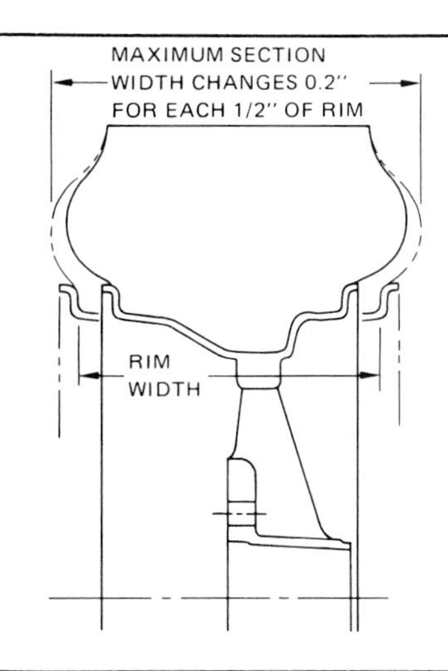

This drawing shows what happens to a tire mounted on a wider rim. The tread width doesn't change, but the sidewalls are braced better and the maximum section width increases. Wide rims are necessary to get maximum cornering power from any tire.

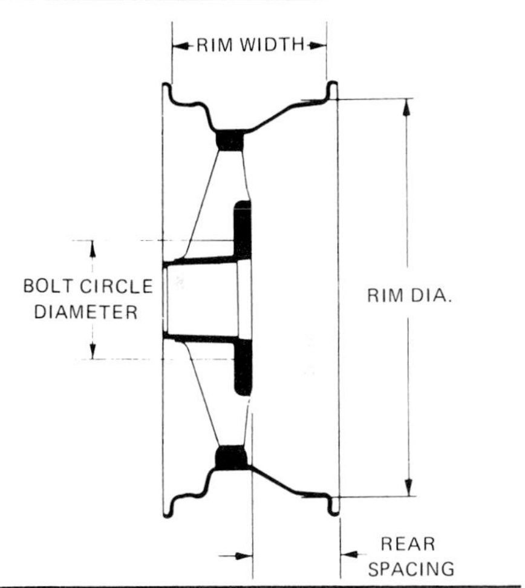

Figure 32/Rim width, diameter, rear spacing, and bolt-circle diameter are the dimensions necessary to describe a wheel. Another term related to rear spacing is *offset.*

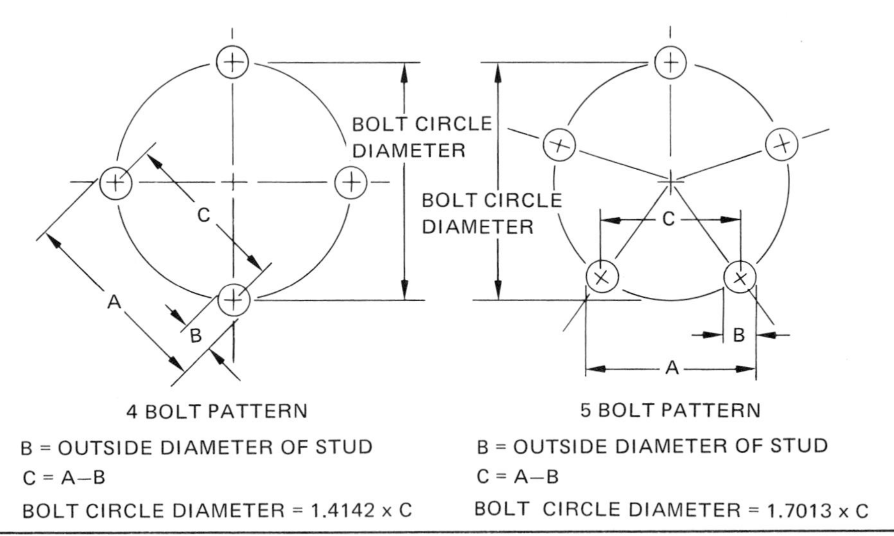

Figure 33/The bolt circle on the car should be measured before buying new wheels. The easiest way is to measure between adjacent studs with a caliper across the outsides of the two studs—dimension "A" in the drawing. Then measure the stud diameter "B", and calculate the center distance "C". Then multiply by the number shown on the drawing to compute bolt-circle diameter. Of course 4-bolt patterns can be measured directly between opposite studs.

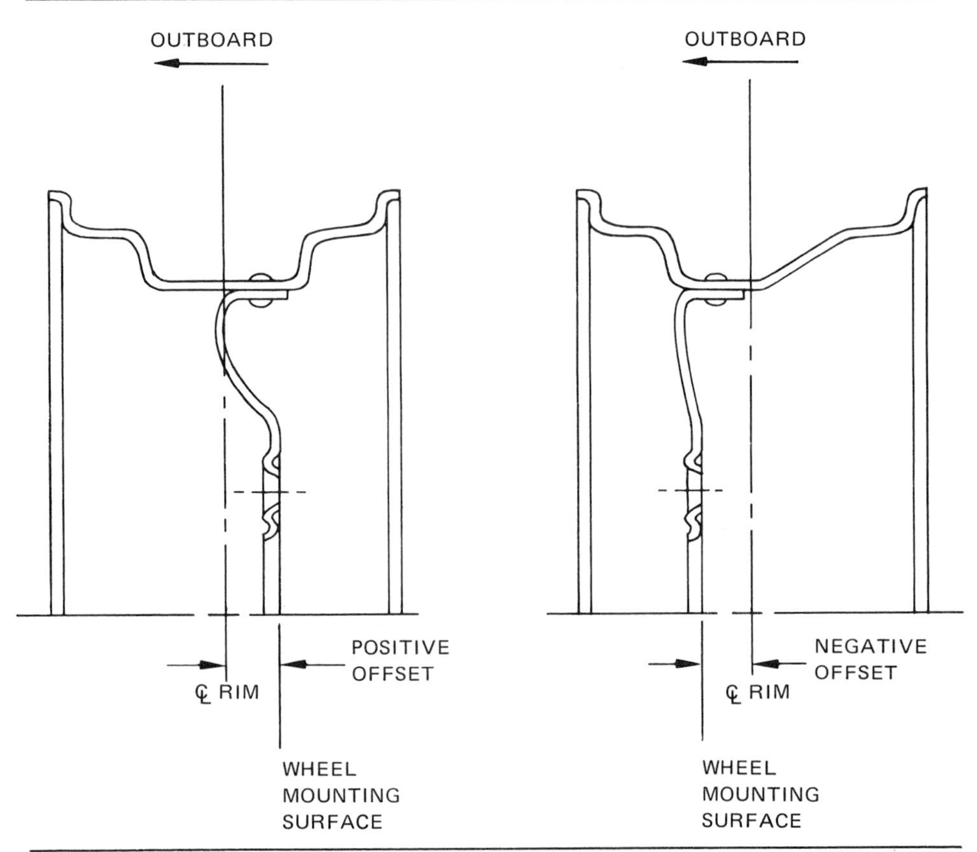

Figure 34/Positive offset results in a wider track than negative offset. A wheel with the mounting surface at the center of the rim has *zero* offset.

To select wheels for your car, first you should choose wheel diameter and rim width. These are determined by the tire you have decided to use along with rim-width recommendations made earlier in this section.

Next check the bolt pattern on your car. It can be measured as shown in Figure 33. If you have a metric bolt pattern, measure it in inches and multiply by 25.4 to convert to millimeters. Also check the stud diameter and threads so you can buy the proper lug nuts with the new wheels. If you don't have a thread gage, check with your car dealer for the proper thread diameter and pitch. Many foreign cars use metric threads, even if the bolt pattern is inches. Also some cars use left-hand threads on one side of the car. Take a good look at the threads, and don't assume anything.

Now let's get to the matter of offset and rear spacing. Rear spacing is the distance between the mounting surface of the wheel and the *inboard edge* of the rim, as you saw in Figure 32. It's plain, simple, and easy to measure.

Offset is the distance between the mounting surface of the wheel and the *center* of the rim, as shown in Figure 34. From a chassis-engineering standpoint, offset is handy to use because it tells you directly what a change in wheels will do to the track width of the car. Track is measured between the center of one rim or tire and the center of the other. If you change wheel offset, you are moving the center of the tire and obviously you are changing track.

Wheels can be built so the center of the rim is outboard of the wheel mounting surface, called *positive offset.* Positive offset makes the track wider.

Wheels built so the center of the rim is inboard of the wheel-mounting surface, called *negative offset,* make the track narrower.

I prefer to work with *offset* numbers and then convert these numbers to *rear spacing* if necessary to compare manufacturer's wheel specs. This discussion proceeds the same way—the effects of offset, followed by the method of converting the dimensions to rear spacing.

TYPICAL BOLT PATTERNS FOR WHEELS

Make	Year and Model	No. Bolts	Bolt Circle Diameter
AMC	60-75 Ambassador, AMX, Gremlin, Hornet, Javelin, Matador	5	4-1/2 in
Buick	65-75 Skylark, Special GS300-400, Apollo, Century	5	4-3/4 in
	65-75 Le Sabre, Centurion, Electra, Riviera, Wildcat	5	5 in
	75 Astre, Skyhawk	4	4 in
Cadillac	60-75 All models	5	5 in
Chevrolet	71-75 Vega	4	4 in
	60-64 Chevy II, Corvair	4	4-1/2 in
	57-70 Caprice, Impala, Bel Air	5	4-3/4 in
	70-74 Monte Carlo	5	4-3/4 in
	65-69 Corvair	5	4-3/4 in
	64-75 Chevelle, Chevy II, Camaro, G-10 Van, Nova, El Camino	5	4-3/4 in
	71-75 Full size models, Caprice, Impala, Bel Air, G-20 Van	5	5 in
	75 Monza	4	4 in
Corvette	54-75 All models	5	4-3/4 in
Chrysler	60-75 All models except Imperial	5	4-1/2 in
Dodge	71-75 Colt	4	4-1/2 in
	71-75 Dart, Demon, Lancer, Swinger with drum brakes	5	4 in
	71-75 Dart, Demon, Lancer, Swinger with disk brakes	5	4-1/2 in
	60-75 Full size models	5	4-1/2 in
Ford	71-75 Pinto	4	4-1/4 in
	74-75 Mustang II	4	4-1/4 in
	70-75 Maverick	4	4-1/2 in
	60-67 Falcon, Mustang	4	4-1/2 in
	49-72 Full size models	5	4-1/2 in
	71-75 Maverick	5	4-1/2 in
	49-75 Bronco, Pick-up, E200 Van	5	5-1/2 in
	64-75 Falcon, Fairlane, Mustang V-8, Torino, Ranchero, Van	5	4-1/2 in
	73-75 Full size models	5	5 in
International	72-75 Scout, Travelall	5	5-1/2 in
Lincoln	70-72 Continental	5	4-1/2 in
	69-72 Mark III, IV	5	4-1/2 in
	65-69 All models	5	5 in
	73-75 Continental, Mark IV	5	5 in
Mercury	71-75 Capri	4	4-1/4 in
	61-75 Comet	4	4-1/2 in
	64-75 Comet, Cougar, Cyclone, Montego	5	4-1/2 in
	61-72 Monterey, Montclair, Park Lane, Marquis	5	4-1/2 in
	73-75 Full size models	5	5 in
Oldsmobile	64-75 Cutlass, F-85, 442, Omega	5	4-3/4 in
	64-75 Full size models	5	5 in
Plymouth	60-75 Belvedere, Fury, GTX, Roadrunner, Satellite	5	4-1/2 in
	70-75 Barracuda	5	4-1/2 in
	73-75 Duster, Scamp, Valiant with disk brakes	5	4-1/2 in
	60-75 Duster, Scamp, Valiant with drum brakes	5	4 in
Pontiac	69-75 Grand Prix	5	4-3/4 in
	64-75 Firebird, GTO, Grand Am, Tempest, Ventura	5	4-3/4 in
	65-75 Bonneville, Catalina, Grandville, Star Chief	5	5 in
Foreign Cars	Austin Healey Sprite	4	4 in
	Alfa Romeo	4	108mm
	BMW 1600, 2002	4	100mm
	Fiat	4	98mm
	Datsun 510, 240Z, 260Z 280Z	4	4-1/2 in
	Ford (English)	4	4-1/4 in
	Lotus Elan, Europa, Seven	4	3-3/4 in
	Mazda	4	100mm
	MG B	4	4-1/2 in
	MG Midget	4	4 in
	Opel 1900, Manta, GT	4	100mm
	Porsche 914 (4 cylinder)	4	130mm
	Porsche 914-6, 911, others	5	130mm
	Toyota 1600 Corolla, Corona Mk II, Celica	4	4-1/2 in
	Triumph Spitfire	4	3-3/4 in
	Volvo 144, 122, P1800	5	4-1/2 in
	VW 68-74 sedans	4	130mm
	VW 48-67	5	205mm

This table shows bolt patterns of some domestic and foreign cars. To be sure, *always* measure the bolt pattern on your car before buying other wheels. Don't assume that wheels with *identical* bolt patterns will interchange. Brake clearance, center hole, and tire clearance may prevent it.

Changing wheel offset changes the track width of the car. Changing rim width or tire width has no effect on the track. Adding positive offset makes the track wider. If your wheels are negative offset to begin with, adding positive offset means *reducing* the amount of negative offset.

When installing wider tires and wheels it is normal to use wheels with more positive offset. Besides widening the track this creates more clearance between the inside of the tire and the car chassis. Then if the tires touch the outer fender lips, the fenders can be flared to clear. It is impossible to use wide tires when there isn't enough room on the inboard side of the tire.

Rear wheels on this March Formula car have positive offset to reduce loads on the wheel center and wheel bearings during hard cornering. This is explained in Figure 35.

Front wheels of March Formula car have negative offset. This puts steering pivot axis near center of tire for easier steering and better handling.

Adding positive offset to wheels can make a car corner faster because the resulting wider track reduces weight transfer. However, there are some side effects. On the front of the car, increased positive offset increases the distance between the steering pivots and the center of the tire. This makes the car steer harder. It also increases the loads on the steering system which increases the steering-linkage deflection, and this may require extra toe-in to compensate. All these things are undesirable, so if possible *keep the front offset near stock.*

At the rear, increased positive offset is not as bad. There usually is more tire clearance problem in the rear, requiring positive offset. If the car has independent rear suspension, increased track may cause more toe change during acceleration and braking which will require a change in rear toe setting. This is usually not important enough to notice in street driving, but could cause a problem on a racing car.

Changing offset affects wheel-bearing loads. On some cars the wheel bearings are so marginal in strength that an increase in positive offset causes them to fail prematurely.

A road car is designed with wheel offset to minimize loads on the bearings in straight-line driving, but in a corner the wheel bearings get higher loads. In most cases a positive-offset wheel will increase the loads in straight-line driving, but reduce the loads in a corner. This is shown in Figure 35. The best bet for street driving is to keep the offset as close to stock as possible. If necessary go ahead with *small* changes in offset and don't worry about bearings. It only becomes an intolerable problem with a large change in wheel offset, say more than 2 inches.

Before ordering new wheels you must determine what offset you need. Start by measuring your existing wheels. Figure 36 shows how this is done. Remove the wheel from the car and lay it down on a smooth floor with the outboard side of the wheel down. Lay a straightedge across the tire sidewall and measure to the floor and to the wheel mounting surface as shown. Then calculate the offset. The formula will also tell you if the offset is positive or negative. If a tire is not on the

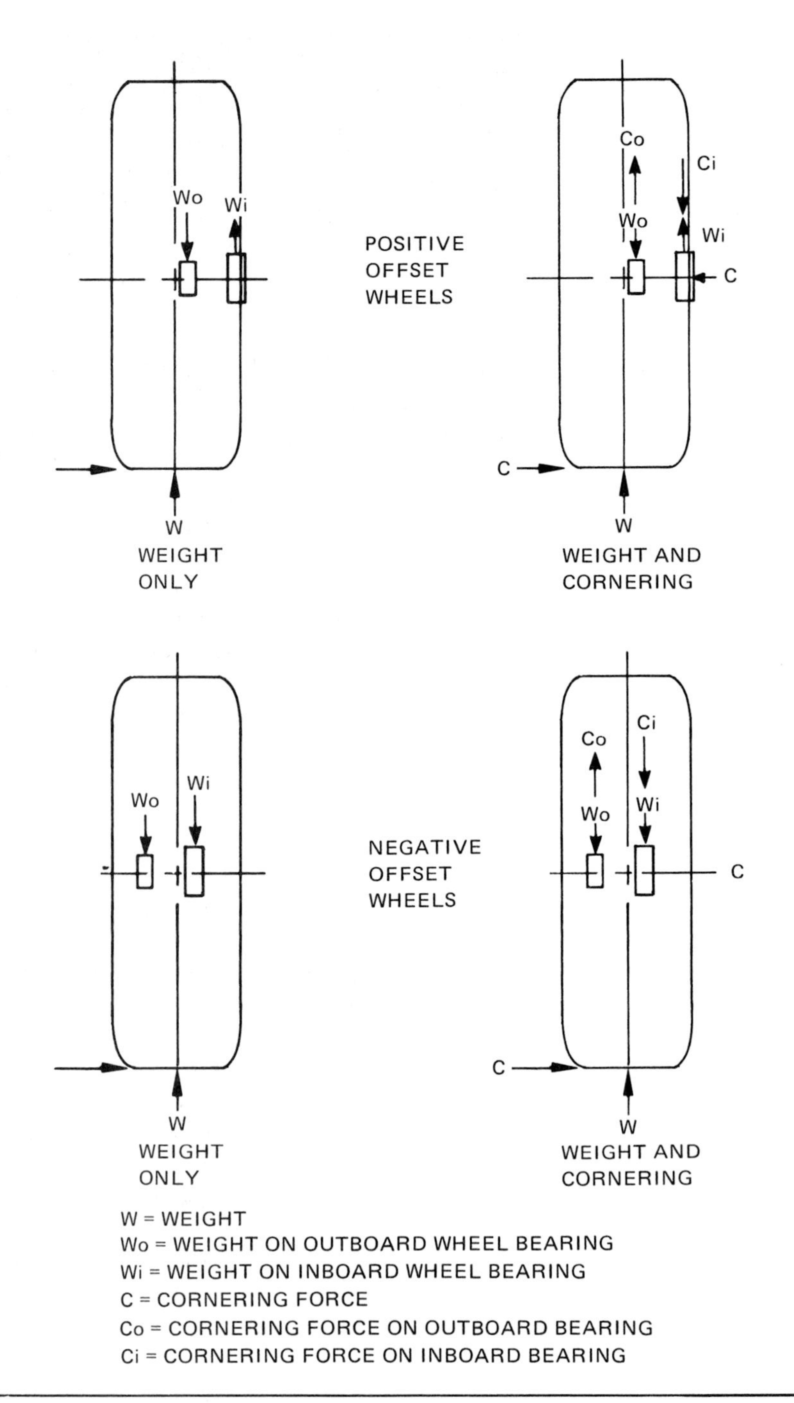

Figure 35/Wheel bearings carry weight of vehicle, "W", during straight-line running. They also carry cornering forces, "C", in turns.
With negative offset, weight loading is well distributed between inner and outer bearings. However, cornering loads upset this balance by increasing force on the inside bearing and decreasing force on the outside bearing. This makes cornering much harder on the inside bearing than straight-line driving.
With positive offset, the situation is reversed. In straight-line driving, bearing loads are uneven but the addition of cornering forces tends to even them out. Thus cornering loads are less severe than those of straight-line driving.
In these sketches, offsets are exaggerated for clarity. Small changes in offset don't affect bearing loads enough to worry about. In any event, wheel-bearing life will probably be affected more by your driving than anything else.

wheel, use the same method but measure to the edges of the wheel.

Now go to the car and measure clearances around your existing tires. Check all positions of suspension and steering movement. You may have to jack up the car on one side to simulate body roll. Put a lot of weight on the car to make the suspension approach full bump. If you can't make the suspension bottom out, remove the springs. On leaf-spring cars, use a turnbuckle in place of the shock and compress the spring by tightening the turnbuckle. This is a lot of trouble, but if you want to use the widest rubber it must be done.

Because stock offset will give the best overall results for street use, first check clearances without changing offset. This means that additional tire and rim width is equally divided between each side of the wheel center. If your new tires have a section width 4" wider than your existing ones, this means the clearance will be reduced by 2" on both the inboard and outboard sides of the tire.

Pay particular attention to clearances on the inboard side of the wheel and tire. If you widen the wheel rim width *without* changing offset, the inboard edge of the rim will be closer to the steering linkage and suspension. If it turns out that there is not sufficient wheel or tire clearance on the inboard side, determine how far the wheel should be moved outwards to clear. You can imagine putting spacers inside the wheels to do this. Remember this is with your *new* rims and tires.

After figuring out how far to move the new rims outboard to clear, double check the outboard side of the tire for clearance. Remember you need clearance inside the fender for one half the increased tire width *plus* the amount you move the wheel outboard.

The amount you move the wheel outboard is the increased positive offset. The added positive offset plus the offset of your existing wheels will give the offset of your new wheels.

Some wheel manufacturers use the term *rear spacing* rather than offset to describe their wheels. Rear spacing is the distance from the wheel mounting surface to the inboard edge of the wheel, shown in Figure 32.

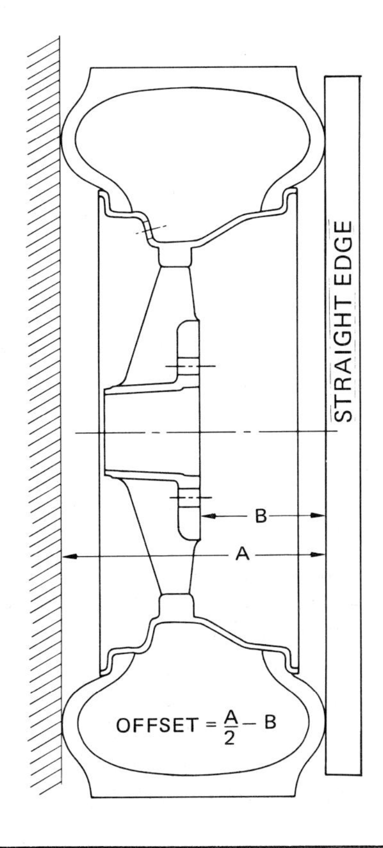

Figure 36/You can measure offset with or without the tire mounted on the wheel. Place wheel with the outboard side down and take dimensions "A" and "B". Offset is calculated with the formula shown. If "B" is larger than "A/2" the offset is negative. If tire is not on wheel, take measurements to edge of rim rather than tire.

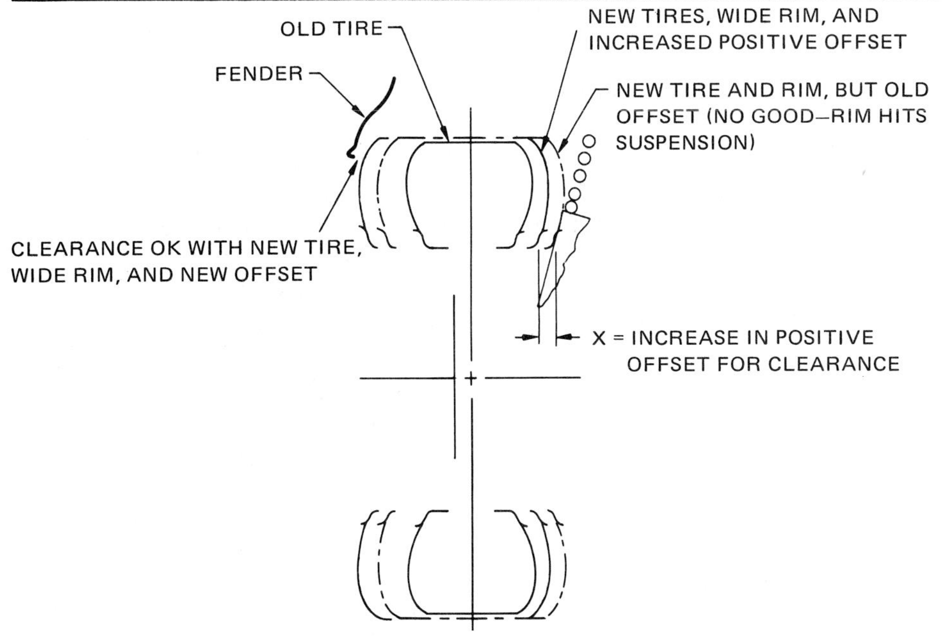

Make a sketch like this to see what offset your new wheels should have. First draw the old tire and the position of the closest parts such as the fender and the suspension. Then draw the new tire mounted on the new wider rim, but without changing offset. This means the new tire is centered on the old tire, with the increased width divided equally between both sides. If there is interference on the inboard side as shown, move the wheel outboard enough to clear. This is distance "X" shown above. The new wheel offset is found by adding the increased positive offset "X" to the old offset. If spacers were used, "X" would be the required spacer thickness under a wheel with the same offset as your old wheels.

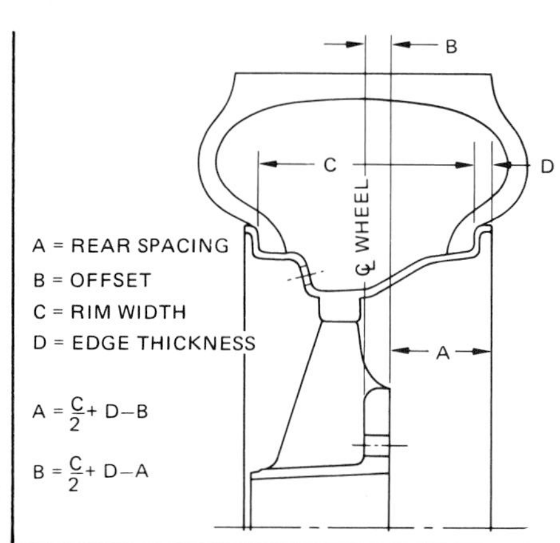

Figure 37/If you know offset you can calculate rear spacing and vice versa. In these formulas if the offset is negative be sure to put in a negative number for "B". If you don't change rear spacing when buying new wheels, the new wheels cannot hit the suspension on the inboard side, but watch out for tire clearance problems.

Rear spacing of this custom wheel can be measured directly as shown. This wheel has the tire pushed off the rim, so the measurement can be taken by resting the 2 x 4 on the edge of the wheel and measuring to the mounting surface. If the tire were inflated, the rear spacing is determined by measuring from the tire sidewall to the mounting surface of the wheel, and then subtract the distance from the tire sidewall to the edge of the wheel. Wearing necktie adds a little class to the operation.

Rear spacing is easy to measure directly off the wheel. There are no positive and negative values, so there is less confusion than with offset. However, some care must be used to describe your new wheels using rear spacing measurements.

If you have already figured out your wheel offset, you can determine the rear spacing as shown in Figure 37. Notice that you need the distance from the tire bead to the edge of the wheel to do it. If you have a tire mounted on the rim this isn't easy, but it can be measured with several dimensions as shown. The typical steel rim has an edge thickness of about 1/2 inch.

A safe way to specify new wheels is to keep the rear spacing unchanged and just widen the rim. This puts the inboard edge of the *rim* the same distance away from the suspension as before, so it can never interfere with anything. You still have to check *tire* clearance on the inboard side, but usually there is no problem. This method increases the positive offset by half of the rim-width increase. On many cars this gives too wide a track, causing the tires to run into the fenders or other problems. Don't let wheel dealers sell you on this method for selecting new rims. It's safe for them, but you may not be satisfied with the results. Always check clearances on both sides before putting your money down.

There is one way to increase the positive offset of wheels without buying new ones. This is use of spacers under the wheels. Spacers are to be used only as a last resort, and then only for small changes in offset, say less than 1/2 inch. Spacers are often marginal in accuracy and if not used properly they can be dangerous.

Spacers can cause the wheels or lug studs to fail. Particular attention must be paid to the lug nuts when using spacers. Never add spacers without changing wheel studs and lug nuts to the strongest possible design.

If spacers are used under steel wheels be careful that they make contact over the same area on the wheel as before. Inadequate contact surface on the wheel can cause the wheel to flex during hard cornering leading to wheel failure. Longer studs should be used with spacers so the lug nuts get full engagement with the threads. Studs must have an *unthreaded* shank in contact with the holes in the spacer. Spacer holes should fit close to the shank of the studs. If the spacer has big sloppy holes or contacts threads, the studs can fail due to bending. NEVER NEVER use a stack of washers as a spacer, *instant disaster can occur!*

If spacers are used under light-alloy wheels longer studs can be used or long-shank lug nuts. The best bet if using the long lug nuts is to bolt the spacer to the wheel and match-ream the lug holes in the spacer for a perfect match. A close fit with the shank of the lug nut is essential for adequate strength. Cragar Industries and other wheel manufacturers offer a selection of long-shank lug nuts for cast "mag" wheels.

Beware of adapters to change one bolt pattern to another. These are heavy and often are too weak for really hard driving. It is best to invest in the proper wheels rather than take a chance.

WHEEL TYPES

There are almost as many types of custom wheels as there are tires. These products can be broken down into two major categories, street wheels and racing wheels. Just as with tires, they should be used only where intended. There may be little difference in appearance but the two types of wheels are designed for totally different jobs. A racing wheel is designed to be as light as possible for the type of car and the type of racing. Thus it may have a very small safety factor, and may fail if used on too heavy a car. In addition, a racing wheel must be very stiff under a cornering load unless it is only used for straight-line racing.

A street wheel is designed for a different environment. It must have a large enough safety factor to go practically forever without failure under all sorts of conditions including overloading, impacts with curbs, severe and repeated bumps, and corrosion. A street wheel is not necessarily designed for long service in cornering at racing speeds, nor is it necessarily as light or stiff as it could be. It is, however, always designed to a price, even the most expensive one on the market.

Racing Wheels—Other than the heavy stock-car steel wheels and some spoked off-road steel wheels, there are three types of racing-wheel construction, all using lightweight aluminum or magnesium alloys. The oldest type is the cast wheel, usually made of magnesium. In the mid-60's I invented a new type of racing wheel, the "monocoque" sheet-metal wheel. This type wheel consists of two spun or stamped sheet-aluminum halves, bolted together with a circular pattern of small bolts near the rim. The center of each wheel is reinforced with a solid spacer in the lug-bolt area which is sandwiched between the two halves.

In recent years a new type of "modular" wheel has been developed. It uses thinner-gage spun or stamped sheet-metal rim halves which are bolted to a cast-aluminum or magnesium center. These wheels are very light and very expensive. You can easily spend over

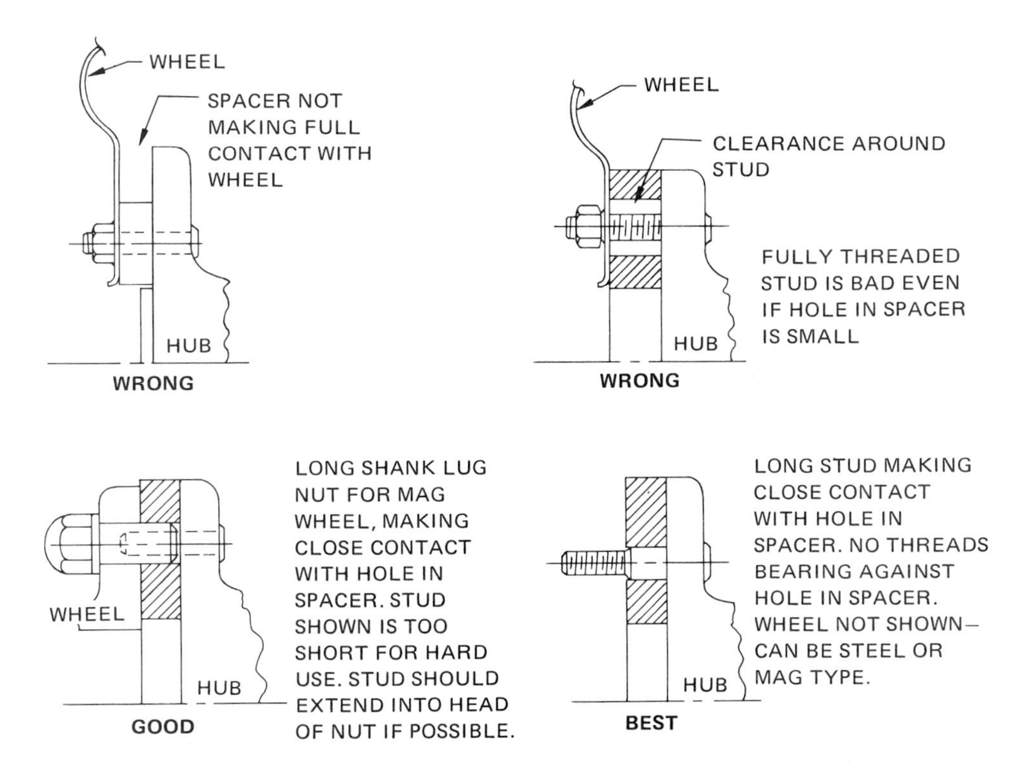

If poverty or a taste for adventure causes you to try spacers under your existing wheels, check these ways to do it. The top two methods will quickly satisfy the adventurous spirit. The bottom two are a reasonable accommodation to a dwindling bank balance.

Modular racing wheels are a development from the traditional cast-magnesium wheels and the monocoque spun wheels. Modular wheels incorporate sheet-metal rims, which offer high-impact strength, bolted to a cast, ventilated center. A damaged rim can be replaced at less cost than that of a new wheel.

$1,000 on one set of these wheels, so they are a luxury on anything but an all-out competition machine.

Different types of racing wheels have specific advantages, depending on the type of racing you do. The cast-magnesium wheel was first developed for its great lateral stiffness and strong cast center section, not to mention its light weight. In addition, vents can be cast in to accommodate air flow for brake cooling. Cast wheels are very accurate because they are machined on a lathe. They have no joints to cause tolerance or fit problems. So, if you want highly accurate wheels having good brake-cooling caracteristics, use one-piece cast wheels.

The disadvantage of cast wheels in rough use is their brittleness. Castings tend to shatter if hit, usually ruining the wheel. Another consideration is the rim is not changeable as it is with modular or monocoque wheels. So if you envision any need to change rim widths or your car will see rough going, stay away from cast wheels.

The monocoque wheel was developed to overcome the disadvantages of the cast wheel. I developed this type of wheel at Chassis Engineering Company, first for road racing, then later for off-road and drag-racing applications. The wheels are assembled from two halves, so many different widths can be made from a limited number of manufacturing tools. This makes it reasonably cheap to produce custom sizes for just about any car. Also, the solid center-section allows any bolt pattern to be drilled in the wheels. The monocoque wheels are as light as cast-magnesium wheels and have a much higher impact strength. The rims bend rather than break if you smash into a rock, as in off-road racing. A bent rim can usually be straightened and used again. And in the worst case, only half the wheel needs replacing.

The disadvantage of monocoque wheels is a lack of large holes in the center for brake cooling. Some of these wheels have been made with cooling holes, but they are never as big as on a cast wheel. If you need brake cooling, you must do it with scoops and ducts, rather than expecting help from the wheel fanning air over the brake.

Minilite Sport wheels are similar to their famous magnesium racing wheels but are cast in aluminum.

This is a cast magnesium wheel designed for racing only. It is made for a Formula car weighing about 1000 pounds, with a racing tire inflated to about 20 psi. This is a very light wheel, ideally suited for its intended use. It would not be strong enough for a heavy car, and would not last long on the street. Keep the intended use in mind when selecting wheels.

Probably the best type wheel is the modern "modular" wheel, which combines the best features of monocoque and cast wheels. These are produced by a number of companies: all use sheet-metal-rim halves bolted to a cast center. Modular wheels offer the impact strength of the monocoques, plus the brake-cooling holes possible with cast centers. Rims can also be replaced if damaged, or rim widths can be changed.

Where to Get them—One of the top manufacturers of cast-magnesium racing wheels is Minilite of England. This brand, marketed in the USA by Linkport, Inc., has an outstanding reputation in the road-racing field. They have a wide range of sizes and offsets, and are designed for most popular sports cars and sedans. Minilites are popular because of their reliability in racing use. The quality and strength means that the wheels are neither the lightest or the lowest cost, but they do offer good value for the money. Another plus is they are one of the few magnesium racing wheels that are suitable for the street.

Another manufacturer of high-quality cast racing wheels is LeGrand Race Cars.

The LeGrand wheels were originally designed to be used on their extensive line of Formula and Sports Racing cars, but they have expanded their line to fit many light cars. They can be obtained for Formula Ford, Super Vee, and other cars of similar weight.

Monocoque wheels are produced in large numbers by Cragar Industries and by Centerline. The Cragar wheels are known as Super Trick, and they are very popular in drag racing and off-road competition. They range in size from skinny front wheels for rails to the giant 16-inch diameter rear wheels. In addition to conventional bolt-on wheels, Cragar also offers a spindle-mount front wheel for dragsters which saves the weight of a separate wheel hub. The hub and bearings are bolted between the two halves.

The Centerline wheels differ from the other monocoque wheels as they are stamped rather than spun. This method of forming has cost advantages over spinning, but the tooling is more expensive. Thus, stamped wheels cannot be made in as many sizes as spun wheels for the same tooling cost, but the manufacturing costs for the sizes that are made will be less. Centerline wheels use rivets rather than bolts to join the rim halves together, and thus the replaceable rim advantage is lost unless you leave it up to the factory.

The highest quality monocoque wheels in my opinion are made at the Monocoque Wheel Company in Santee, California. This little shop is a descendant of my original monocoque-wheel manufacturing firm, Chassis Engineering Company. Monocoque Wheel Co. has developed the manufacturing technique to a fine art, and emphasize high quality and close tolerances. They use high-quality alloy-steel aircraft bolts in the rim, and they make sure each wheel runs true before leaving the shop. Monocoque Wheel Co. specialties in custom sizes for drag racing cars, and can turn out new wheels in a very short time.

Modular wheels are being produced by a number of firms, both here and abroad. Many of them are similar in appearance, and it is difficult to tell the difference without reading the lable on the wheel.

The Crager *Super Trick* wheel has dominated the field of drag racing due to its light weight and wide range of sizes. Pictured here is an ultra-light front wheel for drag racing and a large rear wheel. Weights range from 9 to 17 pounds, and virtually any size is available. The wheels are also made for off-road racing, where their ability to absorb impact overloading without cracking makes them a superior type of wheel.

Of the foreign brands, BBS is perhaps the best known, having been in the business for a number of years. BBS wheels are distributed by Intermag of Berkeley, California.

A high-quality small shop building modular wheels in this country is Jongbloed Modular Wheels of Santa Ana, California. They were one of the first to develop this design as an outgrowth of the monocoque road-racing wheels. Jongbloed can supply special sizes to fit most cars and they are very familiar with road-racing applications.

Street Wheels—Countless varieties of street wheels are on the market today. Prices and styles cover a wide range, and are as confusing as the marketplace for tires. There are several things to remember. Quality doesn't come cheap, so you generally get what you pay for in a custom wheel. Any reputable street wheel manufacturer should have a wheel that is **SEMA** (Specialty Equipment Manufacturers Association) approved. This is your assurance that the wheel is designed to pass a maximum-load endurance test.

This sticker is an indication that the wheel has passed the severe SEMA test. There are some good wheels that have not been SEMA approved. If a wheel has not been SEMA approved, ask the manufacturer for *equivalent* evidence of safety on the street. Some street wheels are guaranteed against failure, and this is evidence of quality.

The stock steel wheel is reasonably light and reliable for street use. However, the center is too thin for hard cornering, and tends to flex. Less flexibility is one big advantage of cast-aluminum custom wheels.

This test is so severe many standard-equipment wheels will not pass.

When shopping, you will be looking for a wheel that fits your car, has the proper rim width and offset, and is the lightest weight you can get. Carry a steel tape and a scale with you when shopping for wheels so you can compare and measure. Also you may choose a certain type of wheel construction because of its structural properties.

Custom street wheels are made of steel or aluminum. They can be cast, stamped, forged, or flame-cut. They are available painted, polished, or chromed. Some have brake-cooling holes and some do not. Some are a lot lighter than others. I will try to sort out the main differences here.

The common street wheel is made of sheet steel. It has a rolled rim and a stamped-steel center, welded or riveted in place. These are the lowest-cost wheels, usually fairly light too. The disadvantage is usually a lack of lateral stiffness. Some original-equipment wheels are weak and will not take the stresses of road racing for a long period. If you have steel wheels on your car and wish to widen the rim you may find other steel wheels in a junk yard or at a dealer that will bolt right on. Look for a car with the same bolt pattern and then compare offset, rim width, and other clearance dimensions to see if they will work. Sometimes the centerhole in the wheel is the wrong diameter and won't clear the hub. Other times the wheel shape may not clear your brakes, particularly if your car is equipped with disks. Check all these important details before putting your money down.

You may also have a wheel specialty shop widen your rims. This is done by cutting the rim apart and welding in a band or a new rim half. This sort of thing is not as popular as it once was because complete wide rims are becoming more available. The split-and-welded rim is not as strong as the original wheel, and failures have been known to happen with poor jobs. If you go this route check the quality of the work, and ask how the fabricator will back up the job. If possible, welding a complete new rim on your centers works a lot better. Steel wheel shops may be able to do this and give you just the right offset. Do not assume this will be

This VW has steel wheels with widened rims which were split and welded by a specialty shop. This type of wide wheel is usually fairly cheap and light, but the strength is strictly according to the skill of the welder. For rugged use, one-piece rims are preferred.

This wheel has a steel rim welded to a steel center cut from heavy plate. Even though the spokes lighten the center considerably it is a rather heavy wheel compared to cast aluminum types. Notice how the spokes are staggered at the weld points to provide added lateral stiffness to the center.

the cheapest way to go—custom work sometimes comes high.

A similar wheel uses a steel rim with a flame-cut or punched-out steel plate center. These wheels are popular for rugged off-road use, because of the much stiffer center. The heavy center plate is made lighter by using holes or a spoke pattern, but some of these wheels are still the heaviest available. They also are among the strongest in sheer impact strength. Take a good look at these steel wheels if your car gets very rugged use.

The majority of custom street wheels are made of aluminum. Most of them are cast, but some are made by an exotic forging process. A forged aluminum wheel will have the very best strength properties, and is highly desirable if both hard road use and occasional racing are intended. The optional spoked-alloy wheels offered by Porsche are made by forging. These are as light as a good racing wheel and are suitable for street driving also—a rare combination in any wheel. Unfortunately Porsche wheels don't fit anything but a Porsche, but don't overlook them if you are using a Porsche bolt pattern on a special car.

One-piece cast aluminum wheels for the street are made by a number of reputable companies including Cragar, American, Ansen, Rocket, Keystone, ET, and Minilite.

Assuming these have all passed the SEMA test or the equivalent, they are all adequate in strength. There are spoke types and slot-mag types, mostly only a difference in appearance. The spoke type may offer slightly better brake cooling in a marginal situation, while the more solid slot-mag center may be a bit stronger

A number of street wheels use a cast-aluminum center with a steel rim. The strength of this type wheel is sometimes questionable at the attachment between the center and the rim. Some quaint schemes have been tried on the cheaper brands, so stick with something that looks like quality. Again stay away from any wheel that has not passed SEMA. The steel rim offers an advantage in mounting tires and curb impact, because the rim is rugged. It may be the best compromise if *all* other factors are equal.

These Porsche wheels are made from aluminum by forging. They are nearly as light as the lightest racing wheels and are strong enough for unlimited street use. These are probably the best all-around wheels you can buy but unfortunately they only fit a Porsche. The price is as high as the quality. You get what you pay for.

The Crager S/S is a high quality two-piece custom wheel using an aluminum center and a steel rim, available in a wide range of rear spacings. The center can be mounted at various locations on the rim. S/S wheels are available in 13", 14", and 15" diameters, widths from 4" to 10". (Photo courtesy Cragar Industries)

The Cragar Mach 8 is a one piece "mag" type wheel. Actually made of cast aluminum, this wheel ranges in weight from 18 to 25 pounds; sizes in 14" and 15" diameters and 6, 7, and 9-inch rim widths. Rear spacing is 3 3/4 inches on most models, which is suitable for many street applications. (Photo courtesy Cragar Industries)

SPRINGS

The basic characteristic of springs which makes them useful is the fact that they are springy. An uncompressed coil spring exerts no force as long as it is uncompressed. The more you compress it, the more it pushes back.

If it takes a force of 100 pounds to compress a coil spring one inch, then when compressed that much it can support a weight of 100 pounds. The behavior of coil springs is that it will require an additional 100 pounds to compress the spring another inch. Now, 200 pounds is resting on the spring and it has compressed or deflected a total of 2 inches

If one corner of a car weighs 400 pounds and that corner is resting on this spring, it will compress 4 inches when the car is sitting still. If the car is driven and hits a bump, the spring should compress still more—perhaps another 4 inches before the suspension hits the bump stop on the frame. That's a total of 8 inches of spring travel, with the spring force ranging from zero to 800 pounds. Actually most coil springs still have some compression even when the suspension is fully extended—they are compressed to install

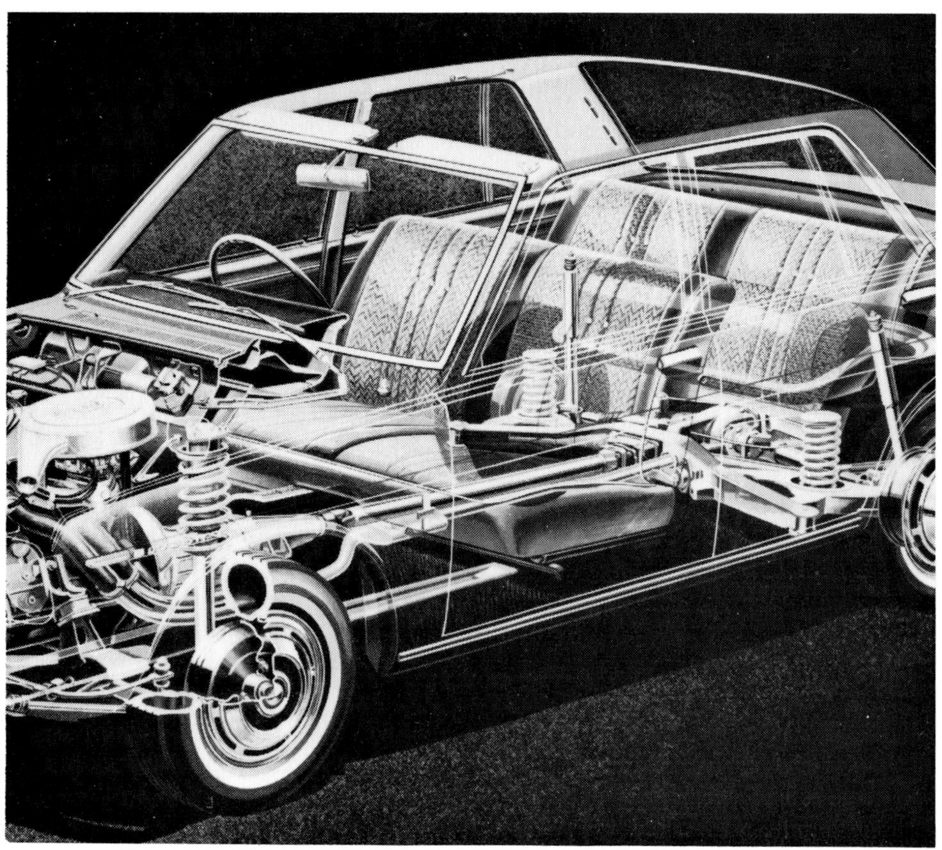

There's a spring and shock at each corner of a car. Spring stiffness has a big effect on ride and handling. Car is a Datsun 510.

them in the car—so the actual range of spring forces may be from 200 to 1000 pounds in this example.

A suspension travel of 8 inches is common on passenger cars but not suitable for most racing cars. Modified springs are sometimes helpful in improving the handling of a car. Generally spring modifications are for one of two reasons: to change the stiffness of the suspension, or to change the ride height of the car. Because springs control so many factors in car handling, what you may think is a simple change can result in complex handling differences.

The basic purpose is to absorb bumps and irregularities in the pavement. Springs provide a reasonable degree of comfort for the occupants of the car, and they help maintain contact between tires and road. The wrong springs can reduce both ride quality and handling on bumpy surfaces.

In general, soft springs give the best ride and also the best traction on rough surfaces. The disadvantage of soft springs is a lot of up-and-down motion of the car on its suspension, resulting in inferior handling under certain conditions. For example soft springs will allow more nose dive under braking with greater camber change on an independent front suspension and a net loss of braking traction.

On banked corners, soft springs can allow the car to ride the bump stops, a very serious condition leading to poor handling. If large aerodynamic forces are acting on the car, soft springs will allow a large change of ride height with change in speed, and this can cause entirely different handling characteristics between low-speed and high-speed driving. Even though body roll in corners can be controlled by anti-roll bars, the springs are the only effective limit to vertical motion of the car on its suspension. A special case is the Z-bar described later in this section.

The usual modification to springs is to stiffen them for less vertical motion. This has the side effect of increasing roll stiffness of the suspension, so the car will have less roll in the turns. Increasing the spring stiffness has definite limits for both ride and handling. For racing use, the springs should be fairly stiff, so this modification may be just what you need. However, for drag racing, stiffening the springs will result in less lifting of the nose and thus less weight transfer to the rear tires. If only one end of the car is stiffened and not the other, the resulting change in relative roll stiffness may alter the steer characteristics. It may also cause uncomfortable fore-and-aft pitching motions on certain surfaces.

In every case, the ride will become harder with stiffer springs, and you will need stiffer shocks to control the springs properly.

No matter how you go about selecting springs for your car, you must eventually judge the result by a criterion common to all people and all cars—*how does it feel?*

Everybody has experienced the soft wallowing undulations of a large passenger car designed for a boulevard ride, and many have felt the harsh jiggling vertical motions of a stiffly sprung performance car.

The difference is how fast the body bounces or *oscillates* up and down on the suspension springs. Shock absorbers or

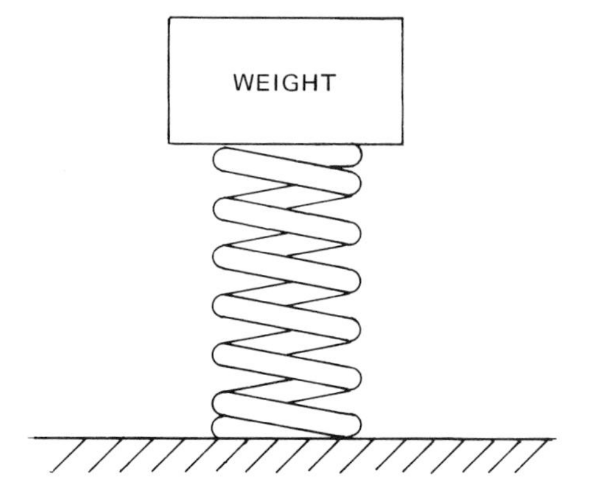

If set into motion, the weight will bounce up and down on that spring for a long time. The bouncing will be at a steady rate called the natural frequency. Cars do the same thing.

dampers tend to mask the effect of vertical body movements because they oppose those movements. For analysis, we imagine that the shocks are disconnected, leaving only the two essentials which determine the rate of vertical oscillation. These are the weight of the chassis and the stiffness of the springs.

A natural characteristic of any weight associated with a spring is vibration or oscillation at some definite rate. This is done by a pendulum—where the springiness is contributed by gravity—or a tuning fork which has some weight and some springiness in each half. It is also a property of a weight hanging from a spring, or sitting on a spring. In all such mechanical systems, if the weight is moved away from its rest position and then released, it will oscillate back and forth, or up and down, at some steady rate or frequency.

You can probably see intuitively or by experiment that more weight tends to make the system oscillate at a lower *natural frequency* and a stiffer spring tends to make it vibrate faster.

Weight is a commonly used expression, but *stiffness* is not. If you test some coil springs for stiffness by compressing them between your hands, you will conclude that the springs which are less stiff compress more with the same force applied. Spring stiffness is the force applied to the spring divided by the amount the spring compresses or deflects as a result of that force. If force is measured in pounds and deflection in inches, spring stiffness is expressed in *pounds per inch.*

A common symbol used for spring stiffness is K, called *spring constant,* or *spring rate.*

$$K = \frac{\text{Force}}{\text{Deflection}}$$

The chassis of a car sits on springs. Disregarding the effect of shock absorbers, every bump in the road causes the chassis to bounce on its springs at some natural frequency determined by the weight of the chassis and the stiffness of the springs.

Because the stiffness of the front springs is not usually the same as the rear springs, and because the weight of the front end of the car is not usually the same as the weight at the rear, most cars have two natural frequencies of vertical oscillation—one at the front and a different rate at the rear.

Natural frequency is expressed in cycles per second, where a cycle is one complete up-and-down motion of the car.

Imagine you are sitting in a car without shocks and someone bounces the springs. If the natural frequency is very low and the motion large, say 3 inches, you would not feel like you were in a car at all. Instead the slow up-and-down motion would resemble that of a boat gently rising and falling with the waves. The ride would possibly make you seasick.

If we increase the natural frequency of the suspension to 1 cycle per second

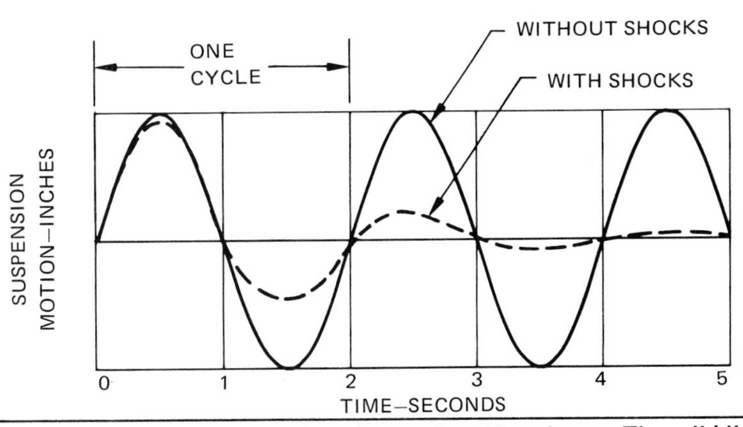

Here is a graph of suspension motion after a wheel hits a bump. The solid line shows the motion without any shocks, the car continuing to bounce up and down a number of times. The dotted line shows how shock absorbers cause the motion to disappear very quickly. The natural frequency can be determined from this graph by seeing how many cycles occur in one second. In this example the suspension natural frequency is 1/2 cycle per second—which will make you seasick.

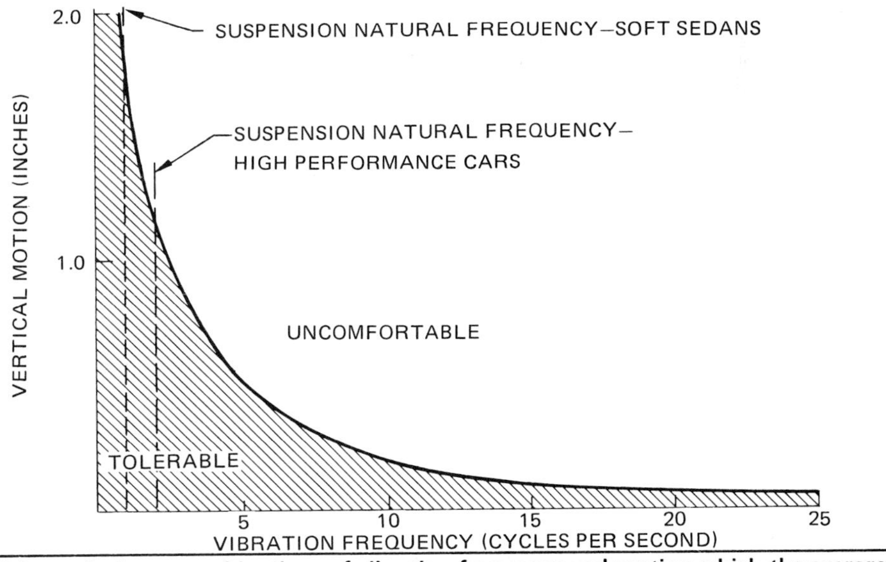

This graph shows combinations of vibration frequency and motion which the average person finds tolerable. Above the curve is the discomfort zone, which obviously is to be avoided. A typical sedan has a low-natural-frequency soft suspension so vertical motions of over 1 inch are not uncomfortable. A sports car or racing car requires less lean and variation in ride height, so a higher natural-frequency suspension is used. The stiffer racing suspension means less comfort. Typical suspension frequencies are shown as dashed lines on the graph.

and reduce the vertical motion, the ride would begin to feel better, more like a car. Increasing the natural frequency to 2 cycles per second would result in a jiggly ride like a truck or a racing car. A further increase to 10 cycles per second would blur your vision and vibrate your body to the point of total discomfort, like riding on a jackhammer. A car with no springs would have a natural frequency of about 10 cps. The car would be bouncing only on the tires.

There is a comfort zone which the human body likes best, and this is what most cars are designed for. Going far outside the limits results in a very uncomfortable ride.

If you plan to change the springs for better handling, you should first determine the natural frequency of your car's suspension. Then you can make an intelligent change that won't be a blind guess. To calculate natural frequency, you must find both the vertical stiffness of the suspension and the *sprung weight*—meaning the weight sitting on the springs.

The *suspension vertical stiffness* —also called *wheel rate*—is the amount of vertical force at the wheel necessary to move the suspension one inch vertically.

You can find wheel rate by direct measurement. Add weight to the car so its effect on each wheel is known. If you place the weight directly between a pair of wheels, it will divide equally between the wheels. Then measure the vertical and aft after the weight is applied to eliminate scrub. Then measure the vertical movement of the suspension. The vertical stiffness is the added weight per wheel divided by the resulting suspension vertical movement.

If you place a 400-pound weight centered over the rear axle, it adds a force of 200 pounds to each rear wheel. If this makes the body of the car move one inch closer to the ground, the suspension vertical stiffness at *each* wheel is 200 pounds per inch.

Because of friction in the suspension, it is often difficult to do this job accurately. The best way is to remove the shocks and place a scale under each tire to accurately measure the change in weight on the wheel. If you like calculations, you can determine the suspension vertical stiffness by the following formula:

Here's how to measure the vertical suspension stiffness of a small racing car. A person of known weight climbs on the car, centering his weight between the two tires. The change in ride height from unloaded to loaded is measured with the scale along the side of the car at the front-wheel centerline. Be sure to bounce the car several times and roll it back and forth to get rid of the effect of friction and tread change every time you make a load change when checking ride height. Friction is particularly bad with leaf springs.

$$\text{Suspension vertical stiffness per wheel} = \frac{\text{Spring stiffness}}{(\text{Mechanical advantage})^2}$$

Some important ideas have been concealed in the last few paragraphs and they may not be obvious. Let's spend a minute on them.

First, the stiffness of a suspension is not the stiffness of the spring all by itself. The important stiffness is that "seen" by the wheel or the body, which may be different on account of leverage.

Also, suspension stiffness is the same whether "seen" by the chassis or the wheel. This is another way of saying that moving the chassis down 2 inches is the same as moving the wheel up 2 inches. It is often more convenient to consider wheel movements as you will see in just a minute.

Finally, there must be a reason that the mechanical advantage term is *squared* in the equation above for suspension vertical stiffness. You can tune chassis without ever knowing why, but if you are intrigued by such things, let's piece it together.

Mechanical advantage is another way of saying *leverage.* Imagine a front suspension using A-arms, with the wheel at the end of the A-arm and the coil spring exactly halfway between the wheel and the suspension pivot on the frame. When the wheel moves 2 inches, the coil spring will compress only one inch because of leverage or mechanical advantage.

The trick here is to recognize that a change in leverage changes *both* the amount of *force* at the end of the lever and the amount of *movement* at the end of the lever. Because spring stiffness is force divided by movement, using a lever to operate a spring causes a curious happening.

Let's assume two cases and calculate some forces and movements. Suppose the stiffness of the spring itself, measured all alone, is 800 pounds per inch. With the coil spring halfway between pivot and wheel, move the wheel upward one inch. The spring compresses only half an inch. If spring rate is 800 pounds per inch, compressing it by half an inch causes a force at the spring of 400 pounds. But because of the 2:1 leverage at the wheel, the force at the wheel is only 200 pounds.

As far as the *wheel* is concerned, the stiffness caused by the spring located somewhere else is still force at the wheel divided by deflection at the wheel. Two hundred pounds force divided by one inch of wheel movement is a stiffness at the wheel of 200 pounds per inch. Because that's what the wheel "sees," it is called *wheel rate.*

Now let's double the leverage so it is 4:1. When the wheel moves upward one inch as before, the spring will compress only 1/4 inch. That *same* spring will exert a force of 200 pounds when compressed only 1/4 inch because the rate is still 800 pounds per inch. The force of 200 pounds at the spring will become only 50 pounds at the wheel because the leverage is now 4:1.

What stiffness does the wheel experience? A force of 50 pounds divided by a movement of one inch is a wheel rate of 50 pounds per inch.

Doubling the leverage reduced the wheel rate by a factor of 4.

The mechanical advantage of a suspension is the amount of vertical movement at the tire to move the spring one inch. Movement of the spring is measured as shown in Figure 38: along the centerline of a coil spring, vertically for a leaf spring, and in the direction of driving-arm motion for a torsion bar. The mechanical advantage can be measured or calculated. A direct measurement is usually easiest. Just move the suspension vertically by jacking up the car slightly. Measure the motion of the spring. Mechanical advantage is the vertical motion of the body divided by the movement of the spring.

When measuring mechanical advantage, do not use large motions or take measurements near the end of suspension travel. Mechanical advantage often changes with suspension motion, and is only significant when measured at or near the ride height position. Simple calculations for natural frequency assume a constant mechanical advantage, and this is nearly correct for small suspension movements near the ride height. Natural frequency calculations of variable-stiffness suspensions are too complex for this book.

After you have measured or calculated the suspension vertical stiffness, you must determine the sprung weight. Often this is approximated by subtracting an estimated unsprung weight from the total weight per wheel. A more accurate unsprung weight figure can be obtained with a direct measurement. Place the car on jack stands under the frame, thus supporting all the sprung weight. Then disconnect the rebound limiting devices, usually the shocks or the leaf springs, and weigh the unsprung weight by placing a scale under each tire. The sprung weight at each wheel is the total weight on each wheel minus the measured unsprung weight. There is a small error in this procedure unless you remember to add in the contribution to unsprung weight of any moving parts disconnected from the suspension. For clarification refer to page 27.

Finally you can calculate natural frequency as follows:

$$\text{Natural frequency} = 3.13\sqrt{\frac{\text{Suspension vertical stiffness per wheel}}{\text{Sprung weight per wheel}}}$$

Another method for finding the natural frequency is to use the following formula:

$$\text{Natural Frequency} = \frac{3.133}{\sqrt{\text{Static Deflection}}}$$

Here the natural frequency is measured in cycles per second, and static deflection is measured in inches. The term *static deflection* is the vertical distance the car body has to be raised to take the load completely off the springs, starting at the ride height.

It is possible to measure the static deflection of your suspension and avoid calculating sprung weight and vertical-suspension stiffness. You must disconnect the shocks or other rebound straps from the car so the load can come completely off the springs when the body is raised. Then measure the ride height of the car, the distance from some reference point on the frame to the ground. Next jack up the car slowly until the load is completely off the springs. The spring should rattle around but just barely. Stop there and take another ride-height measurement. Subtract the first reading from this number to get the static deflection. This method is limited to coil springs or other designs where static deflection can be easily measured. If it cannot, use the formula on page 139.

In this formula the suspension vertical stiffness is measured in pounds per inch. The sprung weight is in pounds. The

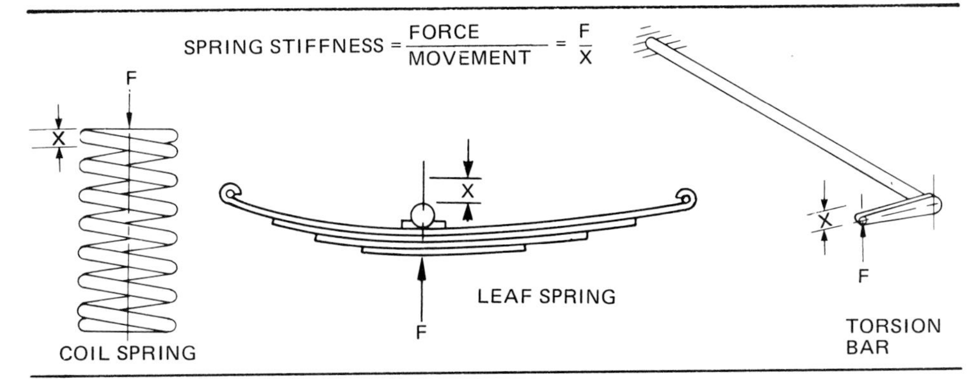

Figure 38/Spring stiffness is the amount of force applied to the spring divided by the movement caused by that force. The directions of forces and spring movements are as shown. Measure both force and movement in the normal direction of spring motion. Spring stiffness is expressed in pounds per inch.

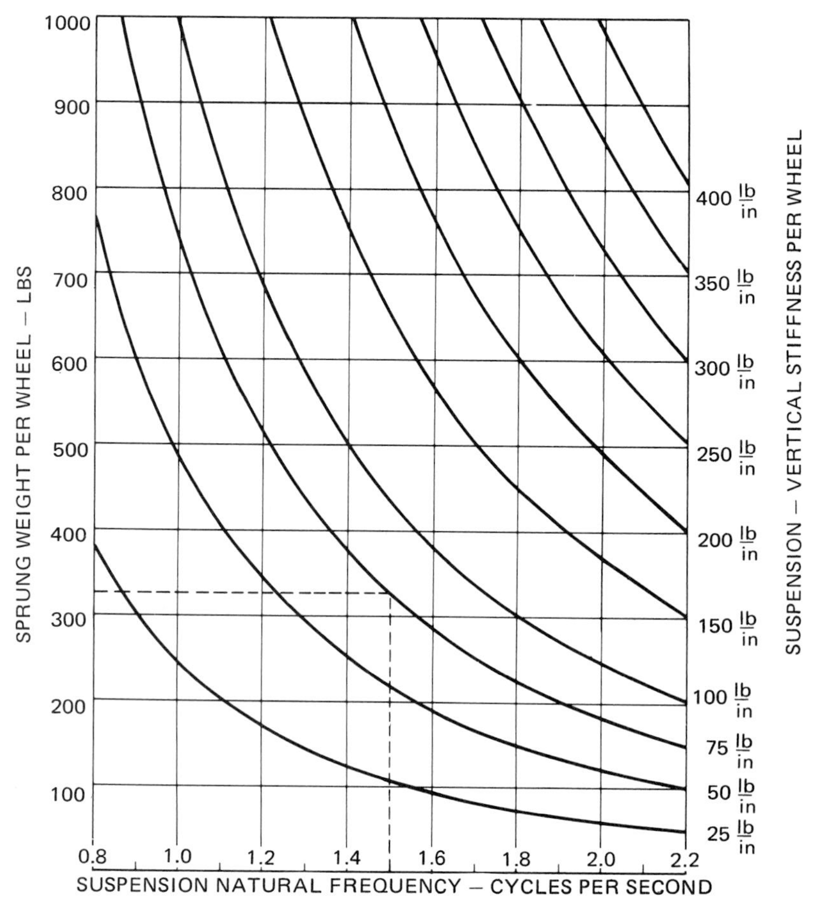

Figure 39/If you know vertical suspension stiffness and sprung weight, you can find the natural frequency from this graph. Measure the vertical suspension stiffness per wheel and sprung weight per wheel as described in the text.

Use the graph as shown by dotted lines. With a suspension vertical stiffness per wheel of 75 pounds per inch and sprung weight per wheel of 325 pounds, the natural frequency of oscillation will be 1.5 cycles per second.

natural frequency is cycles per second.

To avoid arithmetic you can determine the natural frequency from Figure 39. Find the curve with your suspension vertical stiffness. Draw a horizontal line at the sprung weight per wheel, and note the point where this horizontal line intersects the curve. Then draw a vertical line down from this intersection point and read off the suspension natural frequency on the bottom of the graph.

Figure 39 is useful if you wish to change springs. You can select a natural frequency and see what vertical suspension stiffness is required with a certain amount of sprung weight. The new spring stiffness can be determined once you know the desired suspension vertical stiffness:

Spring stiffness =
(Suspension vertical stiffness per wheel)
x (Mechanical advantage)^2

There is another interesting fact that can be determined from Figure 39. Note that if you reduce the sprung weight of a car the natural frequency increases. This happens when you lighten a car for racing, so sometimes there is no need to change the stock springs. By lightening the car you get an effect similar to stiffening the springs.

The problem with this is: As sprung weight is reduced, the ride height goes up, resulting in a c.g. height increase. This may result in more weight transfer and less cornering power. Therefore, the free height of the springs should be less to maintain or lower the c.g. Here again, you must do everything carefully and in steps so you don't end up with the wrong ride height.

For example, if you take the body off your VW and convert it into a light dune-buggy, the sprung weight is reduced. If you leave the springs the same, the natural frequency goes up, and what was once a soft-riding sedan now has the ride of a racing car. Change the springs if you are making a large change in car weight and don't want the ride to change.

The only real comparison of suspension stiffness is natural frequency. A suspension with high natural frequency has a firm ride and a low natural frequency rides softly. Because sprung weight of the car enters into the calculation of natural frequency, changing sprung weight has the same effect as changing the spring stiffness.

The natural frequency of the suspension should fall in the range of one to two cycles per second. The lower limit is for soft-riding sedans and the upper limit is for racing cars. You can make a compromise, depending on how back-breaking you think you want the ride to be.

Traction on bumps will be better with softer springs, but smooth-pavement handling will be better with the stiffer springs. Decide these things before you start to modify springs.

Pitching is an uncomfortable fore-and-aft motion of the car, caused by one end of the car moving up on the suspension while the other end of the car is moving down.

To minimize pitching motions, the front suspension should have a lower frequency than the rear suspension. Then the front takes longer for the rise and fall after hitting the bump, and the rear suspension "catches up" with the front. When the front and rear suspensions move simultaneously there is no pitching.

Pitching can be largely eliminated this way, but only at one road speed. Select the speed at which you would like the best ride, and design the springs to that condition. There will be some pitching at other speeds. For a road car 55 MPH is a typical cruising speed. For a racing car use the average speed of the typical race lap.

Knowing the car's wheelbase and the average speed, you can compute the time it takes for the rear wheels to hit a bump after the front wheels have hit it. The formula is as follows:

$$\text{Time to travel wheelbase} = \frac{.0568 \text{ Wheelbase}}{\text{Car speed}}$$

where time is in seconds, wheelbase is in inches, car speed is in miles per hour.

The time for one full cycle of motion must be determined for both the front and rear suspension. This time is called the *natural period,* and is equal to one divided by the natural frequency. The front natural period should be greater than the rear natural period by the time it takes for the car to travel its wheelbase length. Thus if you select the natural period for the rear, you then calculate the natural period for the front by merely adding the wheelbase travel time to it. One divided by the natural period is the natural frequency. To summarize;

$$\text{Natural Period} = \frac{1}{\text{Natural Frequency}}$$

Front natural period—Rear natural period = Time to travel wheelbase

$$\text{Natural Frequency} = \frac{1}{\text{Natural Period}}$$

when natural frequency is in cycles per second; natural period is in seconds per cycle.

As an example let's calculate the natural frequencies for a small racing car such as a Super Vee. Assume the wheelbase is 90 inches and the average speed for this class of car is 100 miles per hour. The time it takes for the car to travel its wheelbase is:

Time to travel wheelbase =

$$\frac{.0568 \times 90 \text{ inches}}{100 \text{ miles per hour}} = .0511 \text{ seconds}$$

For a desired rear suspension natural frequency of 1.8 cycles per second the natural period becomes:

Natural period =

$$\frac{1}{1.8} = .5556 \text{ seconds}$$

Adding the time to travel the wheelbase gives the natural period of the front suspension.

Natural period of front suspension =

$$.5556 + .0511 = .6067 \text{ seconds per cycle}$$

Natural frequency of front suspension =

$$\frac{1}{.6067} = 1.65 \text{ cycles per second}$$

Besides suspension stiffness, springs are modified to change the ride height of the car. Usually you are trying to lower the CG. If you have leaf springs, they can be re-arched or have the eyes reversed by a spring shop. Changing the arch only changes the ride height, not the stiffness. If you change the number of leaves, that will change both ride height and stiffness. You can stiffen a leaf spring by adding more leaves. A spring shop can make them for you or you can disassemble some leaf springs for parts. Juggle leaves around until you get the stiffness you like. Adding a long leaf will increase stiffness more than adding a short leaf.

For drag racing you may wish to raise the ride height and CG for better weight transfer by installing long shackles on the rear leaf springs. Before doing this, you should do some planning. The amount of ride-height increase for a particular shackle change is shown in the accompanying drawing. The ride height is not raised the

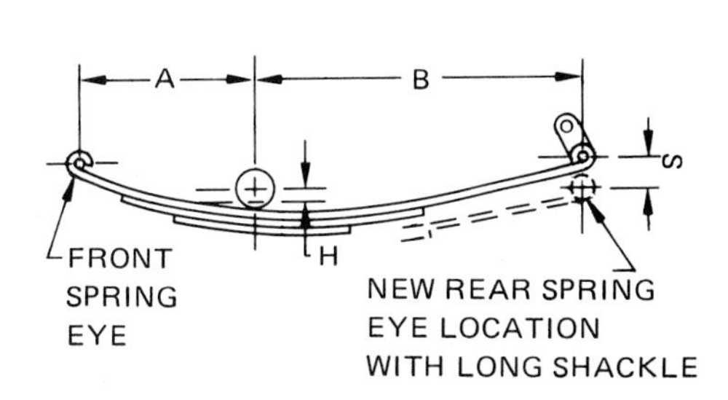

$$H = \frac{S \times A}{A + B}$$

This approximate formula is good for springs which are nearly horizontal and for shackles which are nearly vertical.

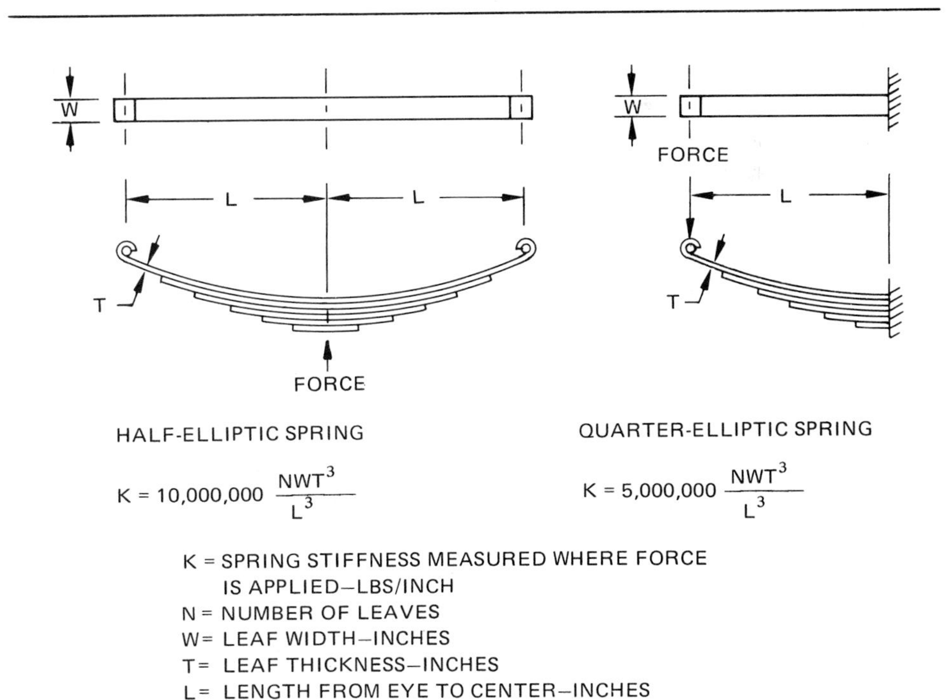

$$K = 10{,}000{,}000 \frac{NWT^3}{L^3} \qquad K = 5{,}000{,}000 \frac{NWT^3}{L^3}$$

K = SPRING STIFFNESS MEASURED WHERE FORCE IS APPLIED—LBS/INCH
N = NUMBER OF LEAVES
W = LEAF WIDTH—INCHES
T = LEAF THICKNESS—INCHES
L = LENGTH FROM EYE TO CENTER—INCHES

For steel leaf springs these formulas give spring stiffness. The half-elliptic spring must have the load in the center as shown. All leaves must be of constant thickness for these formulas to apply. Even so, the formulas are only approximate, and the spring should be tested to be sure of stiffness.

same amount as the shackle is changed. Usually, it is about half as much. A long shackle at the rear of the spring will change roll steer, along with raising the car. To allow adjusting roll steer (roll axis), which may be necessary when you change the ride height and CG, consider fabricating a new bracket to allow moving the front spring eye mounting.

On a coil spring of a certain wire size and coil diameter, stiffness is determined by the number of coils. Thus cutting the spring increases the stiffness and reduces ride height. If you only want to reduce ride height the spring can be shortened without changing its stiffness. A spring shop can do this for you, or you can try it at home. To do it yourself you need a spring compressor. This is a long large-diameter bolt and two thick metal plates. The plates should have a hole in the center for the bolt. Clamp the spring between the plates and compress it by tightening the bolt. Make sure it doesn't slip or you could kill yourself!

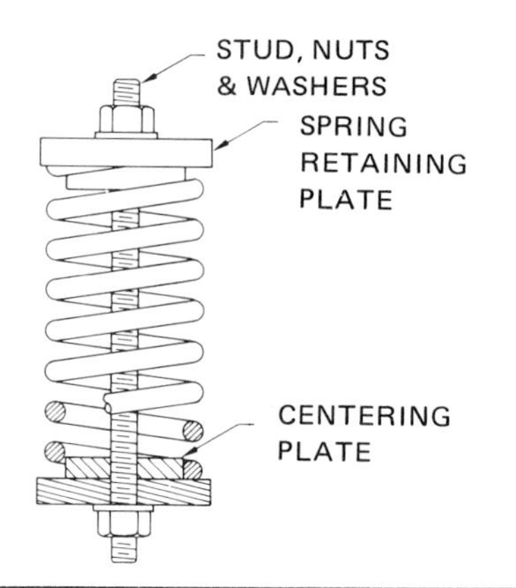

Here is a simple spring compressor that can be used for shortening a coil spring in an oven. The spring retaining plates should be thick steel with a clearance hole for the stud. The centering plates can be other material such as aluminum or plywood. The centering plates are necessary to keep the spring from slipping off center and bending the stud. If this happens the stud may break and the flying parts could easily kill someone! When designing this tool use *very* strong parts, particularly for big springs!

With the spring collapsed, put it in an oven at about 400 degrees F. Leave it in for a few minutes and take it out. Let it cool and remove the spring from the compressor. It will now be shorter by some amount. With luck it is not too short, and some additional time in the oven will make it more nearly correct. This trial-and-error process is used on the other spring. Any differences can be made up with metal shims when you install the springs on the car. It is disaster to try shortening springs by heating them with a torch—you will probably ruin the spring.

If you want to put stiffer springs on your car as well as lower it, these two changes tend to work against each other. A stiffer spring compresses less from zero load to its normal riding position. Thus cutting off the end of a coil spring will not lower the car as much as you removed, because the spring also got stiffer when you cut it.

If mathematics is your bag, you can compute the stiffness of a coil spring. The formula is:

$$K = \frac{W^4 G}{8ND^3}$$

where:

- K = stiffness of the spring in lbs. per inch
- W = diameter of the spring wire, in inches
- G = 12,000,000 for steel springs
- N = number of active coils (number of free coils + 1/2)
- D = diameter of the coil measured to center of wire, in inches

Notice how sensitive spring stiffness is to the diameter of the wire. If you double the diameter of the wire you multiply the stiffness of the spring by 16. It is also quite sensitive to coil diameter—a larger coil results in a softer spring. Note the direct relationship between the number of active coils and the stiffness. If you cut off half the active coils, spring stiffness is doubled. Cutting off active coils or heating and collapsing the end coils solid is the way to change the stiffness of a coil spring. All you can do is make it stiffer, never softer. Buy a new spring if you require a softer one.

If you are checking in junk yards for different springs, bring your measuring tools and electronic calculator to figure stiffness. Don't go by the looks of the spring or you will be fooled. Remember a tiny difference in wire diameter makes a huge difference in stiffness, so measure

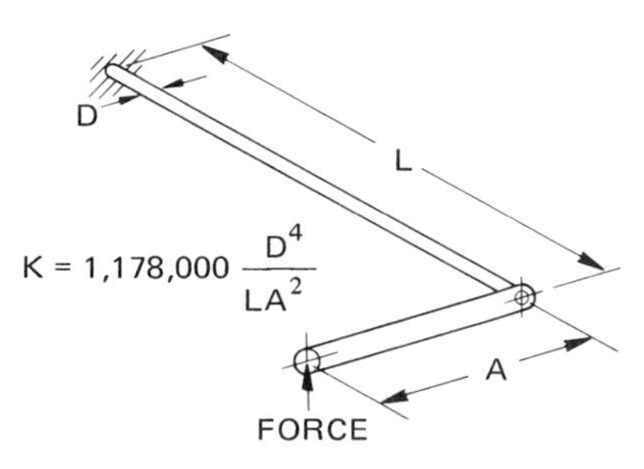

D = BAR DIAMETER—INCHES
L = BAR LENGTH—INCHES
A = LEVER ARM LENGTH—INCHES
K = SPRING STIFFNESS MEASURED WHERE FORCE IS APPLIED—LBS/INCH

The stiffness of a torsion-bar spring can be computed by this formula. It applies to steel torsion bars with dimensions as shown. Bars that are not solid round sections are more complicated to compute. By changing length of the lever arm the stiffness of the torsion bar can be adjusted.

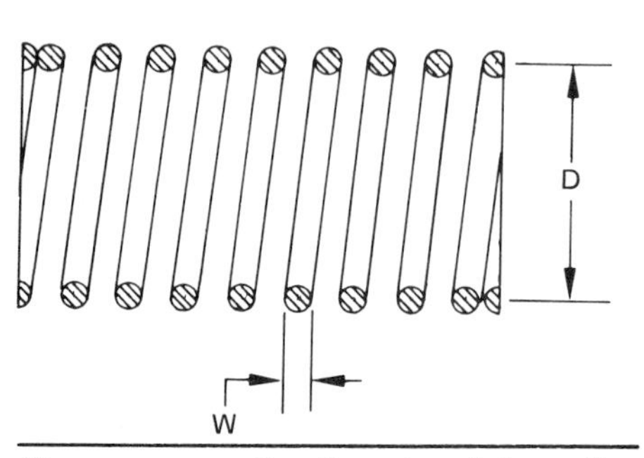

If you measure the diameter of the coil and the wire diameter as shown, you can calculate the spring stiffness using the formula in the text. The coil diameter "D" is measured to the center of the wire. An easy way to find the diameter "D" is to measure the outside diameter of the coil and subtract the wire diameter "W". Use very accurate measurements of both "D" and "W", as a small error in measuring will result in a large error in the calculated spring stiffness.

it with a micrometer. The number of active coils is the number of free coils plus one half coil. The free coils are the ones free to move, usually all but the two end coils.

In addition to stiffness you must also consider the inside diameter, end treatment and direction the coils are wound. The inside diameter should be at least as big, end treatments should be the same and the coils should be wound in the same direction as the springs being replaced to ensure proper seating. To prevent sagging occurring as a result of overstressing, wire size in the new springs should be at least as big as those replaced.

There are four types of end treatments you are likely to encounter—open, squared, square and ground and pig-tailed. It is not a good practice to use the wrong type of end because the spring may become unseated and disaster would likely result. Determining the direction a coil spring is wound: Look at the end of the spring and follow the wire around in a clockwise direction. If the wire goes away from you it is a right-hand wind and vice-versa.

You can measure spring stiffness with a bathroom scale and a steel tape. Measure the free length of the spring resting on the bathroom scale and note the scale reading. Now push the spring and measure its deflection while reading the increase in weight on the scale. The spring stiffness is the increase in scale reading divided by the spring compression. If you are strong enough to deflect it one full inch the arithmetic is simple.

If your car requires stiffer springs, one way to do this is to bolt on helper springs. There are a large number of helper springs on the market, both coil and leaf types. Some bolt on in addition to the existing springs, and some come with heavy-duty shocks as a single unit. All of these increase the suspension vertical stiffness, and thus increase the natural frequency. You won't get a huge increase in stiffness with any of these springs, so the car will still be reasonably comfortable.

One type of helper spring has adjustable stiffness. These are the air helper springs such as those marketed by the Air Lift Company. The more the air pressure the greater the stiffness. Thus you can tune the stiffness to get the ride you want. Air Lifts also change the ride height of the vehicle when the pressure is changed, so once the right spring stiffness is obtained you will have to adjust the ride height to the desired value.

The relationship of spring load and spring deflection is shown in Figure 40 for Air Lifts at 40 PSI pressure. Spring stiffness increases with increasing spring deflection. This is a *rising rate* suspension system, which increases suspension vertical stiffness with suspension movement toward full bump. Rising-rate suspension is highly desirable, giving a soft ride at the ride-height position, but increasing stiffness when approaching the bump stops. This allows more severe bumps to be taken without hitting the stops, but does not sacrifice ride comfort. Rising-rate suspension is particularly valuable on racing cars, which have limited suspension travel because of low ride height.

On cars meant for street use, stiff springs usually are *not* one of the most useful modifications. You can get more handling improvement with stiffer anti-roll bars and the ride will not suffer nearly as much as modifying the springs. You may consider lowering the car with the springs, but this modification is best done by using smaller-diameter tires. This way you still have full suspension travel. On a road car a softer ride is more desir-

Stiffness of a small coil spring can be measured with a ruler and a bathroom scale. Just lean on the spring and measure the increase in scale reading and the deflection of the spring. Spring stiffness is the scale reading increase divided by spring deflection.

Here's a very handy tool for measuring stiffness of big springs. This Wacho portable coil-spring checker allows quick and accurate measuring of coil spring stiffness. This expensive tool is a must for a serious racer, particularly if experimenting with a variety of coil springs. (Photo courtesy of Wacho Products Co.)

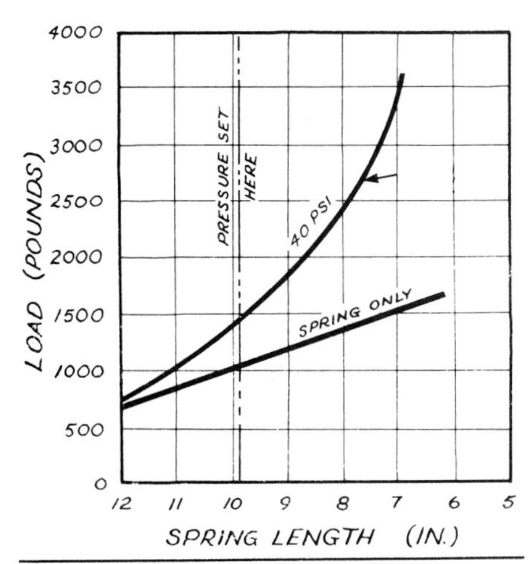

Figure 40/This shows the relationship be-between spring load and spring deflection for a suspension with and without Air Lifts. For a leaf or coil spring only, the curve is a straight line. This indicates the spring stiffness remains constant. The Air Lifts, however, have an increasing spring stiffness as the spring is compressed. Note the increasing steepness of the curve. This gives a *rising-rate* suspension.

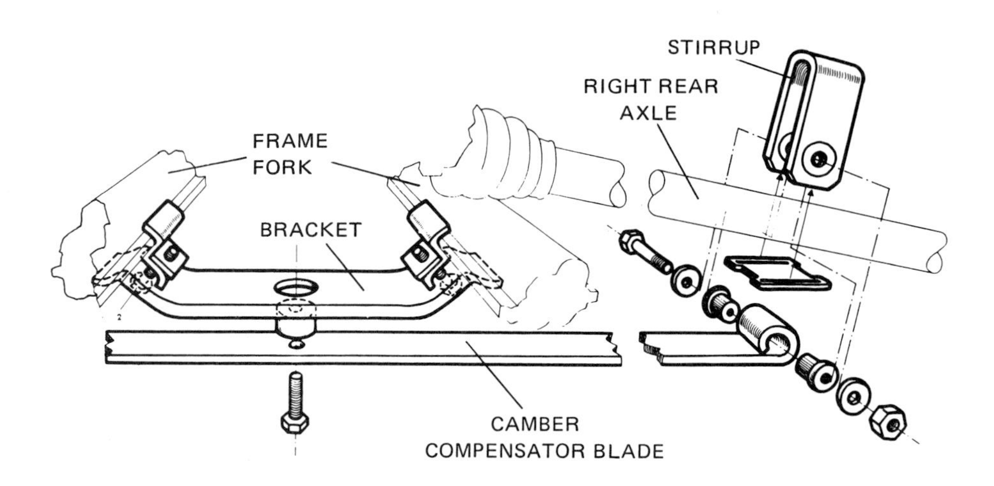

The camber compensator looks like a transverse leaf spring bolted under a VW swing axle suspension. This device opposes going into tail-up attitude in the turn. Reduction of jacking effect helps corerning and stability of a VW. If you have an early swing axle VW this is an essential bolt-on modification, even if you only drive to the grocery store. The safety and stability improvements are well worth it.

The Z-bar on this racer is mounted above the rear suspension with one end of the bar attached to each swing axle through pivoted links. To change stiffness, the links can be clamped at different positions on the bar. The bar is attached to the frame with bearings, similar to the attachment of an anti-roll bar. However, a Z-bar is twisted only when both wheels rise or fall together. Thus it has no effect on roll stiffness but increases the vertical stiffness of the suspension. It reduces the swing-axle jacking effect.

able than on a racing car, and the lower road-speeds mean aerodynamic forces are very small. All in all, modified springs are not very helpful except in highly-modified cars. A word to the wise: Don't change the springs unless you have a really good reason.

If you have a car with swing-axle rear suspension there is a bolt-on leaf spring that will help the handling. This is called a camber compensator. The device mounts under the trans-axle on a swing-axle suspension and connects with straps to both axles. It looks like a transverse leaf spring, but it acts only when both wheels droop together. This reduces the jacking effect during cornering that makes a swing-axle rear suspension oversteer. This type of device is almost essential for maintaining moderate camber angles during cornering with a swing axle.

A similar device is a Z-bar. This is somewhat like an anti-roll bar, but Z-shaped in contour. It is twisted only when both wheels of the car move in the same direction. In this way it acts something like a spring, but with a difference. The Z-bar does not add anything to roll stiffness, unlike the suspension springs. When the car leans it offers no resistance. Also the Z-bar resists both bump and droop motions of the suspension. In the droop direction this is similar to the action of a Camber Compensator. The Z-bar has the advantage of being adjustable in stiffness and preload, assuming the ends are properly designed. The Z-bar is standard equipment on Formula Vee road racing cars, which are required to use VW swing axle suspension. A Z-bar is a desirable item for street use too, if your car has swing-axle suspension.

A Z-bar or a Camber Compensator reduces the jacking effect of a swing-axle suspension. This increases cornering power due to smaller camber angles in a turn. If you set your car up with negative camber to compensate for jacking effect, adding a Z-bar or Camber Compensator means you can use less negative camber. Thus braking and stability will improve. For street driving these devices tend to reduce oversteer and high-speed instability, usually a problem on a car with swing-axle rear suspension. This is probably the best item you can buy to improve the handling of a swing-axle suspension system.

SHOCK ABSORBERS

Many sedans and sports cars can be helped by better-quality shock absorbers. On most production cars, the manufacturer uses the softest and cheapest shocks to lower manufacturing cost and to get the softest ride under non-spirited driving conditions. For hard driving, most original equipment shocks simply will not do the job.

Without shocks a car would bounce uncontrollably on the springs. The car would hit the bump and droop stops often, and the ride would be a constant series of bounces. The tires would have little adhesion on turns because they would be bouncing up and down on the road surface. Shocks are used on all automotive suspensions to damp out these wild oscillations. The shocks do not hold up the car, but they do absorb much of the energy when the wheels hit a bump in the road. All a shock does is convert the energy into heat, thus resisting unwanted motion.

If the suspension tries to move rapidly, the shock resists this with a high force, caused internally by the shock piston trying to ram oil through a tiny valve. Adjustable shocks change the size of this valve by an external knob or adjuster.

The heat generated inside a shock warms up the oil that it contains. As the oil gets hot, it thins out just as the oil does in your engine. Thin oil offers less resistance to being pushed through a small hole, so the shock loses its damping force. This is *shock fade,* and it can be a real problem under certain conditions. To combat fade, the shock must be larger and contain more oil. On some cars, dual shocks are used for this purpose. In any event, a good-quality shock will be bigger and heavier than a cheap light-duty shock.

Another problem with shocks is cavitation or foaming of the oil. Gas bubbles are formed inside the shock and it offers little resistance to movement of the suspension. High-quality shocks are designed to have less tendency toward foaming, both by their design and by the fact that

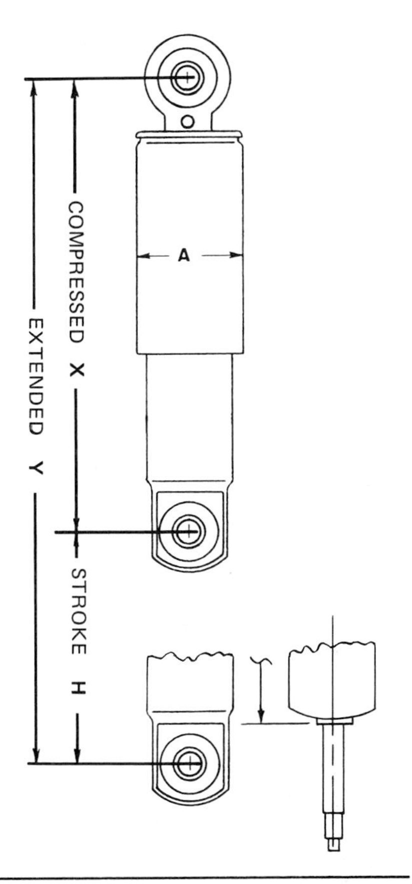

The important dimensions of a shock are extended length, compressed length, stroke, and body diameter. The length is measured center to center on the eyes as shown. On shocks with studs, measure to the shoulder at the base of the stud. For special shock applications be sure to have all the above dimensions handy.

they tend to run cooler.

Of course the brute strength of the shock is important if you are planning to do hard driving. Heavy-duty shocks have larger-diameter rods and bigger parts on the ends for increased strength. Also the design of the seals is better on the more expensive shocks. Leakage of fluid past the seals is one source of shock failure.

The amount of force a shock develops for a given speed of compression or expansion of the shock is its *damping* or *stiffness.* When people talk about stiff shocks they are referring to shocks that offer a great amount of resistance to motion. At very slow speeds there is only the friction of the moving parts inside the shock, as the oil offers almost no resistance. The force increases roughly four times for

The racing Koni shock has a separate adjustment for bump and rebound. The bump or compression stroke adjustment is the knob on the body of the shock. The rebound or extension stroke adjustment is made by inserting a pin into the small holes just under the upper shock eye. Adjustment procedure is given in Chapter 3.

each doubling of shock-absorber speed of movement.

For street use the force is much less on the bump stroke than on the rebound stroke. When the wheel rises on a bump it encounters a rather small resistance. A large force on the upward motion of the wheel would jar the car and cause a poor ride. The major part of the resistance to motion is when the wheel is being pushed back toward the road by the spring.

The stiffness of a shock must be matched to the stiffness of the springs. Because springs are normally dependent on weight of the car, heavier cars require stiffer shocks. Thus shocks for a Buick will be much too stiff for a VW, assuming they are mounted in a similar manner on the suspension.

Shock manufacturers seldom supply any engineering data about their products. You would like to know the relationship of force to motion when designing the suspension on a special car. Figure 41 shows the sort of information you would like if you could get it. Since you usually can't, you must resort to trial and error or use of adjustable shocks on special cars.

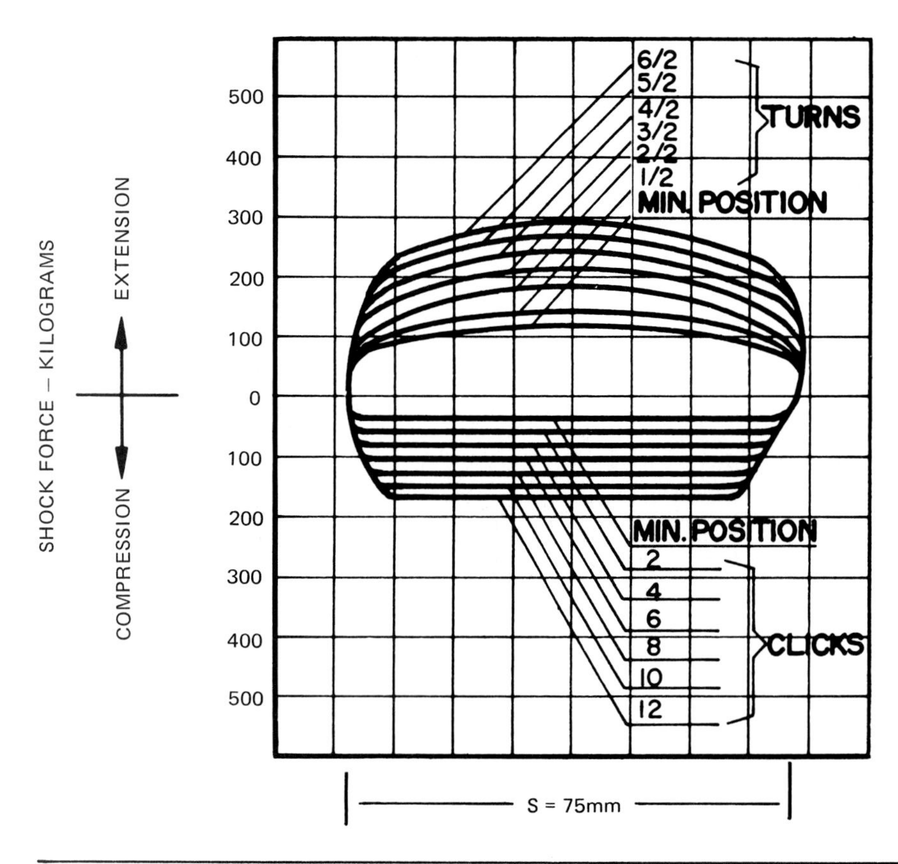

Figure 41/This graph shows the relationship between force and stroke for a double-adjustable Koni shock. The various curves show how the shock adjustments vary the force separately in the extension and compression strokes. This test is run at a particular shock velocity. If the shock were tested at a higher velocity all forces would be higher. (Courtesy Kensington Products Corp.)

For production cars shock manufacturers have designed proper shocks, and all you have to do is install them. There are a number of quality brands, all more expensive than the minimum price you could pay. Stay away from the cheapest ones, and stick with heavy-duty or high-quality shocks. Monroe, Gabriel, Koni, Armstrong, and Bilstein are all well-known companies with a large product line. Some of these companies also build racing units and can help you select the proper shocks for your car.

Unlike racing wheels and tires, racing shocks can be used on the street. However, there are several disadvantages, the major one being cost. The best racing shocks such as the aluminum Konis used on road-racing and drag cars cost over $200.00 each. Racing shocks usually do not have rubber end bushings, but instead use spherical bearings with metal races. This gives better shock action on small bumps, but the metal fittings do transmit more road noise and vibration from the suspension. Also many racing shocks have closer to a 50-50 damping force ratio between the bump and rebound stroke. This will make the car ride stiffer. In addition, some racing shocks do not have dust covers and thus don't have as much protection against damage to the precision-ground shaft. If the shaft of the shock is nicked or corroded, the seal will leak oil and the shock will soon be useless.

An advantage of many racing shocks is that they are externally adjustable. Some such as the racing Konis have separate adjustments for the bump and rebound stroke. On racing cars, the shocks can be tuned to give the maximum adhesion on whatever course you are running. Because the temperature of the shock may vary with the course and the temperature of the day, adjustable shocks are desirable for the ultimate in suspension tuning.

Adjustment of the shock is largely a trial-and-error process that you do at the track as described in Chapter 3. A rough adjustment can be done in your garage by bouncing the suspension. Start with the shock set on full soft and see how many times the car bounces up and down before it comes to a stop. It usually will be more than one complete up-and-down cycle. Then increase the shock stiffness until the car only goes down and back to rest. Additional stiffness will probably not be necessary, but tests should always be made at the track to confirm the proper setting.

Bilstein shocks are used on some racing cars such as Porsche. They are designed for the particular racing car and are not externally adjustable. Bilstein shocks are rather unique in their design and deserve special mention. They have a pressurized gas chamber inside which acts against the oil at all times. The result of this is a resistance to foaming of the oil, because gas bubbles do not form as readily when liquid is under pressure. This is the same theory behind the pressurized cooling system in your car. The boiling point is raised by pressure. The gas pressure shock also runs cooler because the working cylinder is exposed to the air and not insulated as is the reserve chamber on a conventional shock absorber.

Bilstein shocks have special seals to withstand the high pressure of the oil. Thus they will not leak and will work perfectly well if mounted upside down. On most cars this saves unsprung weight, because the heavier oil-filled body of the shock is stationary and the lighter rod and piston move with the suspension. If you try this with any conventional shock it will not work. Air will be in a portion of the shock normally filled with oil.

The gas pressure of 25 atmospheres inside the Bilstein tends to push the shock into the open position at all times. Thus it is always fully extended when not mounted on the car. According to Bilstein, this force is approximately 40 pounds, and it remains nearly constant through the travel of the shock. This extra force acts in addition to that of the springs and may make a small difference in ride height on a light car.

Bilsteins have had remarkable success in off-road racing, apparently because of their resistance to fade and foaming. Some models are made specially for racing. The importer is happy to assist a race car owner select proper shocks for his car. Contact Bilstein of America for engineering assistance if you need it. Be sure to include all information on your car such as weight, type of suspension, and the use of the car. Also the mechanical dimensions of the shock must be specified. Measure a shock from eye to eye extended and collapsed, and give the dimensions of the mounting bolts. It pays to get expert help on special applications.

There are some shocks that serve a dual purpose: to damp the suspension and also to change the spring stiffness. These are called overload shocks, and they increase the stiffness of the suspension. There are two types, those using coil springs mounted on the shock and those using air pressure in the shock. The air type can be adjusted for ride height with an air valve outside the car. The main purpose of these shocks is to raise the ride height and provide stiffer springs for a car or truck that has to carry extra weight. However, because they also stiffen the springs and are good heavy-duty shocks, they can also improve the handling of some cars. If you install these shocks you will increase the ride height and the spring stiffness at the same time, so be sure this is what you need before you buy them.

Just because a Monroe Load Leveler or a Gabriel Load Carrier looks like a racing coil-shock unit, don't assume it will work as the only suspension spring on a special racing car. These overload shocks use a shock that is much too stiff for the spring that is on it, because it is meant to go on a car that already has springs. The spring will have to be much stiffer if it is to be the only spring on the suspension. Overload springs generally are about 60 pounds per inch spring stiffness, and this is only satisfactory on a very light car. On such a car the shocks would be too stiff, so they just won't work for that purpose.

One advantage of the Bilstein shock is that it mounts upside down. Thus the unsprung portion of the shock is the relatively light piston and rod. This saves a small amount of unsprung weight over a shock which mounts the other way.

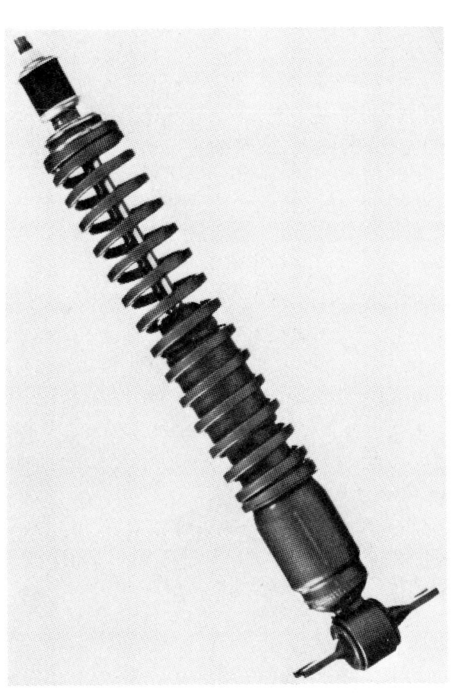

This Monroe *Load-Leveler* is both a heavy-duty shock absorber and an auxiliary coil spring which adds to the suspension stiffness. Although it has both a spring and a shock, these units are not intended to serve as the entire suspension. (Photo courtesy Monroe Auto Equipment Company)

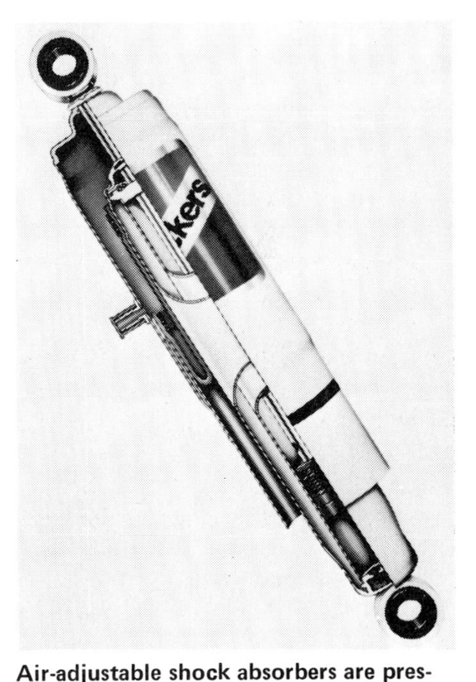

Air-adjustable shock absorbers are pressurized air in a cavity inside the shock to add stiffness to the suspension—just as an added coil spring does. This cross-section of a Gabriel *Hi-Jacker* shows the air valve and a flexible diaphragm which converts the top portion of the shock body into an air chamber. (Photo courtesy Maremont Corporation)

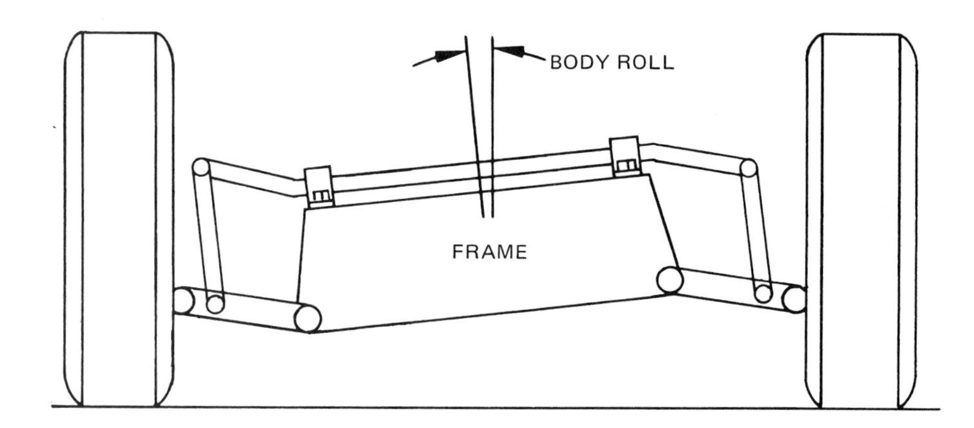

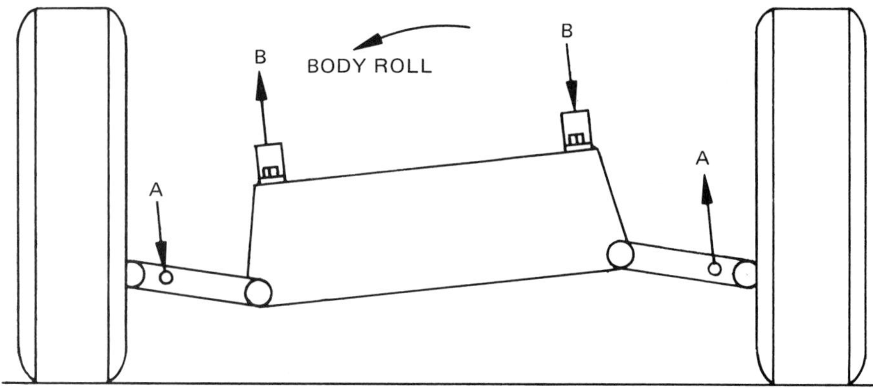

The top drawing shows how an anti-roll bar is twisted when the body rolls in a turn. This creates forces at the four points where the bar is attached to the vehicle. The forces are shown in the bottom drawing. Forces A on the suspension increase weight transfer to the outside tire. Forces B on the frame resist body roll. The effect is a reduction of body roll and an increase in weight transfer at the end of the chassis which has the anti-roll bar. Because the total weight transfer due to centrifugal force is not changed, the opposite end of the chassis has reduced weight transfer.

MGA competing in a slalom could really use an anti-roll bar. The car leans too much, causing positive camber on the front tires. This loses traction on the front tires and causes understeer. Besides cornering faster, the car would be more fun to drive if it had less roll.

This sedan is burning the inside tire trying to accelerate through a right hand turn. A stiffer front anti-roll bar would cure this problem. For street or racing this is a modification that really helps.

ANTI-ROLL BARS

Installing heavier anti-roll bars will give you more improvement in handling per dollar spent than almost anything else you can do. Most production cars use an anti-roll bar, usually in the front, but they are a compromise for average driving conditions. This means the anti-roll bar is stiff enough to limit roll to acceptable values if the car is driven conservatively. If hard driving is the normal operating condition for the car, typical production anti-roll bars simply are not stiff enough. They allow so much roll that on a car with independent suspension the tires may operate at a positive camber angle in a turn. In addition, a car that rolls less in a turn is more fun to drive and has better transient response.

Most anti-roll bars are used on the front suspension, except on front-wheel-drive cars. It adds more roll stiffness to the front, which reduces rear suspension weight transfer in a turn. This delays or eliminates lifting one of the driving wheels, and may create an understeering tendency by increasing the loading on the outside front tire.

If you try to improve your car with a stiffer front anti-roll bar you will definitely get less body roll in a turn. However, if you go to extremes, front-wheel lifting may be the result.

The steer characteristics resulting from a stiffer front anti-roll bar are unpredictable. On cars with similar front and rear suspension designs the tendency is for a stiffer front anti-roll bar to cause more understeer, due to increased weight transfer in the front. On cars with independent front suspension and solid rear axle, the tendency is often toward less understeer, particularly if the car was designed to roll a great deal in the corners. The reason is, the stiffer anti-roll bar prevents excessive positive camber from occurring on the front tires in a turn. This effect is sometimes stronger than the extra weight transfer, and the result is less understeer. Usually this result is strongest on cars which inherently understeer a great deal, so a larger front anti-roll bar is almost always a help to the handling with no other changes.

If you install an anti-roll bar on the rear, or if you stiffen an existing one, the effect is to create oversteer. This is *not* a desirable modification unless the car under steers very heavily under all conditions. One reason for making this modification, however, would be if you wanted to race the car in slaloms. Ultra-tight cornering requires the car to oversteer to help it around the turns. A typical car understeers a lot in very tight turns, so a rear anti-roll bar may be just the thing for this type of racing. Be careful with the car on the street with slalom suspension, because it may be dangerously unstable on high-speed corners. Oversteer usually gets less controllable as speed increases.

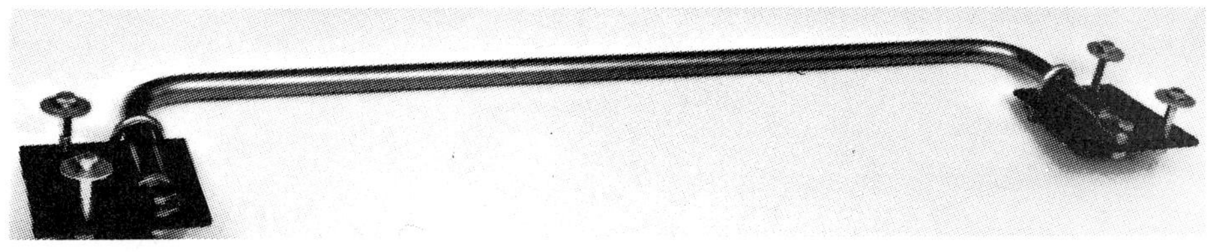

This is a rear anti-roll-bar kit from Addco Industries. On rear-wheel-drive cars, a rear anti-roll bar is used with a much heavier bar in front. On front-wheel-drive cars, a rear bar may be used alone or with a light front bar.

On front-wheel-drive cars, the large bar goes on the rear, not the front. Because the driving wheels are at the front of the car, you want the increased roll stiffness at the *rear*. This improves traction at the driving wheels and reduces understeer which front-wheel-drive setups always seem to have. An anti-roll bar or a larger one would be used at the front of a front-wheel drive car *only* if the rear wheel lifts during steady-state cornering, and if the car leans too much. So use a front bar with caution unless you like heavy understeer on your front-wheel driver.

Anti-roll bars are designed so they twist when the car rolls in a turn. A vertical motion of both wheels does not put any loads in the anti-roll bar. Thus it only affects the roll stiffness and not the vertical-suspension stiffness. However, when only one wheel of the car hits a bump, the anti-roll bar is twisted and adds to the vertical-suspension stiffness. This tends to make the ride somewhat stiffer, but not anywhere near as much as stiffening the springs. Thus soft springs and stiff anti-roll bars are a good combination for a reasonable ride plus flat cornering.

The stiffness of anti-roll bars is usually chosen to limit the roll angle to a certain amount. For racing, the ride is not as important as for a road car, so common practice is to limit the body roll to 2 degrees or less. This can be considerably more on a road car and not be unacceptable. It all depends on your personal preference in the compromise between ride and handling. My suggestion is limiting the car to 4 degrees roll at maximum cornering power for street use.

Exact calculation of how stiff to make an anti-roll bar is very complex. The easiest way is to test a heavier bar from the car manufacturer or from an accessory

Photo on top shows a stock Datsun cornering at 25 miles per hour. The bottom photo shows the same car cornering at the same speed after installation of an Addco anti-roll bar kit. The roll angle has been reduced from 6 1/2° to 4°. (Photos courtesy of Addco Industries)

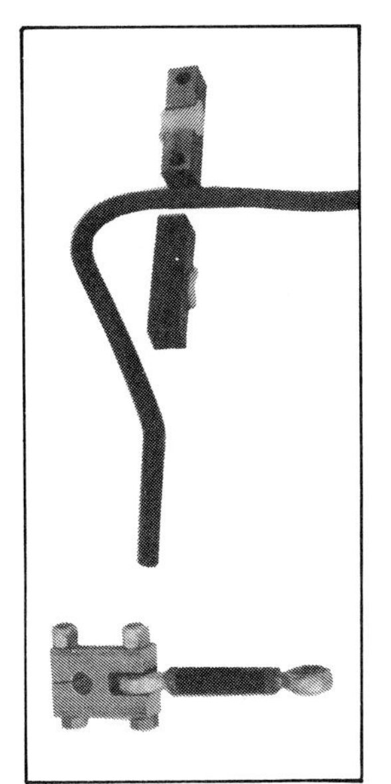

These sway-bars are modified production Chevette units. Solid bearings with Delrin inserts replace the production rubber bushings at the frame. The rear bar shown on the left uses an infinitely adjustable block with a weld-on round extension. The front bar uses an extension with a series of positive locating holes. This is due to the tight clearances at the front suspension.

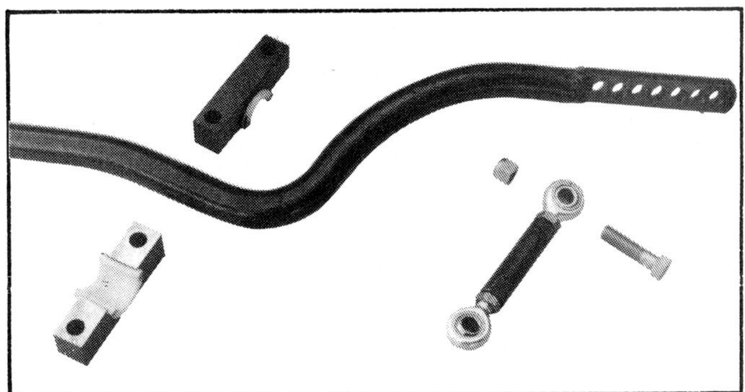

supplier and simply install it on your car. These bolt-on kits are usually engineered for reasonable performance on the street or on the track, and you can save yourself the trouble of selecting the stiffness of the bar.

First determine what the car does in a turn. The best way to find the roll angle is to use a camera. Station the photographer in a position where he can shoot a head-on shot of the car as it rounds a corner. A skid pad or a large parking lot may be a good spot to conduct this test. Drive the car as hard as possible in a turn and photograph it at maximum lean. Measure the angle of lean off the photograph with a protractor. Take a number of photos to be sure you are getting a true head-on shot.

Now install a stiffer anti-roll bar and take photos again. Comparison of the photos will show the difference in roll angle with two different anti-roll bars.

Another means of evaluating the effect of a stiffer anti-roll bar would be to take the bar off completely and take photos of the car cornering. This is not a good thing to try unless the car is a sports car, because excessive body roll on some sedans can cause them to overturn.

The stiffness of an anti-roll bar is determined as shown in Figure 42.

As with a coil spring, the diameter of the bar is very important. In addition to the stiffness of the bar itself, the mechanical advantage of the suspension affects the overall roll stiffness caused by an anti-roll bar. To make the bar most effective, the end of the bar must be as close as possible to the tire. If the end of the bar is mounted on the suspension upright, the mechanical advantage is nearly one. Most anti-roll bars are mounted with a mechanical advantage of more than one. As with the springs, the effective vertical stiffness at the tire is equal to its vertical stiffness of the anti-roll bar divided by the mechanical advantage squared.

If you cannot find an anti-roll bar kit for your car perhaps you can find a stiffer bar at a wrecking yard. With the formula of Figure 42 you can calculate the stiffness of a new anti-roll bar based on its dimensions. Look for a stiffer bar off a station wagon to use on your sedan. Many station wagons use stiffer anti-roll bars because of their high gross weight.

Replacing rubber anti-roll-bar bushings with solid bushings at the frame and suspension can have more of an effect than a bigger bar. In an actual test of anti-roll-bar stiffness, a 0.80-inch diameter bar using solid bushings gave the same effect as a 1.00-inch rubber-mounted bar.

Making an anti-roll bar adjustable requires a modification to most production bars. The loops at the bar's ends must be cut off to allow the installation of an adjusting clamp as shown on pages 149 and 151. The clamps can then be positioned on the bar to change the suspension's roll stiffness.

It is also possible to build a bar at home if you have a welding torch. Use 4130 steel for bars up to 1 inch and 4340 steel for larger diameters. These chrome-molybdenum alloy steels should be heat treated after bending.

First lay out a pattern for your bar by bending some stiff wire or by cutting a pattern on a piece of plywood. Mark the center of your piece of bar stock and also mark the center on your pattern. Start with the bends near the center first and work to the ends of the bar, keeping the bar centered on the pattern as closely as possible.

To bend steel bar it is necessary to heat it to a yellow color. Do not melt the steel. Clamp the bar in a vice and slip a pipe over the end so you can bend it without burning your hands. Heat the INSIDE of the bend over a distance of about 2 inches along the bar and gently apply pressure. Don't try to bend too fast or you will distort or crack the metal. As the bend is started, move the heat to reduce the amount of bending close to the vise jaw. With some practice you can get a smooth-radius bend this way. Use as large a bend radius as possible for best results. If you crack the bar on the outside of the bend, have an expert welder weld it up with the proper rod.

Check the bar for flatness by laying it on a flat floor. If it is twisted, reheat and correct as required by bending slightly.

There are various types of end attachments for an anti-roll bar. If you want it adjustable for stiffness, a clamp-on fitting can be manufactured. In this case no end fitting on the bar itself is required.

If you want the bar to bolt onto a production car, copy the stock bar ends. Most have the ends flattened and a hole drilled through the bar for a driving link. If your bar is large enough you may not have to flatten it. If you do require a flat end you can heat the bar almost to the melting point and forge the end flat on an anvil with a big hammer. Repeated heating will be necessary. If the steel gets too hard to drill, heat it yellow and let it cool slowly.

After the ends are complete and the bar trimmed to final dimensions, take it to a heat-treating shop. Have it heat treated to a tensile strength of 160,000 to 180,000 PSI. This will be strong enough for racing or street use.

If in doubt about the bar strength, rig up a test after it is heat treated. Figure

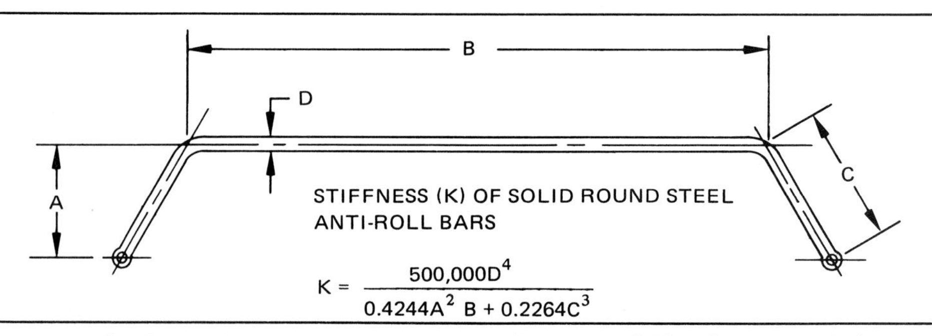

$$K = \frac{500{,}000D^4}{0.4244A^2\,B + 0.2264C^3}$$

Figure 42/Stiffness of an anti-roll bar is measured at one end. This formula includes twisting of the bar plus bending of the driving arm. It is only valid for round bars as shown. This formula does not include flexing of rubber mounting parts, which can considerably lower the stiffness.

This home-built anti-roll bar was bent too sharply. The metal is kinked in the bends. Notice the adjustable clamp-on driving-link bracket which avoids having to machine the ends of the anti-roll bar to attach the driving link.

out how far the bar will twist if one wheel is at full bump and the opposite wheel is at full droop. Then clamp one end of the bar to a heavy table or a vise and twist it that far with a long piece of pipe over the other end. If the bar is still straight after that, it should be strong enough for any use on your car. If it twists permanently in the test, take it back to the heat treater and ask for more strength.

LOCATING DEVICES

Locating devices are used to prevent excessive deflection or movement of the suspension. These are usually used to provide more precise or better location than the stock design provides. There are many types of locating devices such as traction bars, Watts linkages, radius rods, and Panhard rods. All provide better wheel location than the stock set-up.

Locating devices are commonly used at the rear of a car with solid-axle suspension. This type suspension is cheap to build by using longitudinal leaf springs to locate as well as spring the axle.

Lateral location is important when cornering. The side load on the tires is transmitted from the axle to the frame through the leaf springs. The leaf springs bend sideways to carry this load and the rubber bushings in the ends are compliant. This method of location is not very stiff, and the axle can deflect sideways relative to the frame. Thus the rear suspension has a softness which the driver feels as a vague or rubbery feeling in a corner.

The car will respond better if this sponginess is eliminated. Also sideways deflection of the axle requires more clearance for the tires to keep them from rubbing on wheel wells or springs. If you are using wide tires, this can sometimes be a real problem.

Two popular devices used to locate the rear axle laterally are a Watts linkage and a Panhard rod. Both of these do essentially the same thing. They provide a linkage which connects the axle with the frame and gives positive lateral location. The locating device is much more rigid than springs, and is able to carry almost all the lateral load. The locating device must be free to move in other directions so it does not interfere with the intended motions of the axle.

A Panhard rod is commonly used, as it is simple and cheap. A Panhard rod runs across the car, usually the entire width of the chassis. One end is pivoted on the axle and the other end is pivoted on the frame. It should be horizontal with the car in its normal riding position. A Panhard rod does not provide vertical up and down motion at the end attached to the axle because it arcs about the end pivoted on the frame and there is a slight sideways motion. If the bar is long, if it is horizontal at normal ride height, and if the axle vertical motion is small, then the sideways motion is small and not too important. This is the case for most racing cars.

On a road car with soft suspension and a fairly large amount of wheel travel, a Panhard rod may not be the best answer. The lateral motion allowed by the rod gives a rear-axle steering effect when the wheels rise and fall over bumps. This tends to make the car wander, particularly when going over undulating roads at high speeds. Small bumps will generally not cause any problem because they cause only a small amount of wheel travel.

If the Panhard rod is not horizontal you are in for trouble. Then the rod causes a large lateral movement of the axle when it rises and falls. This will be felt by the driver, and can also cause structural failure. The Panhard rod is fighting the leaf springs and large forces are developed in both com-

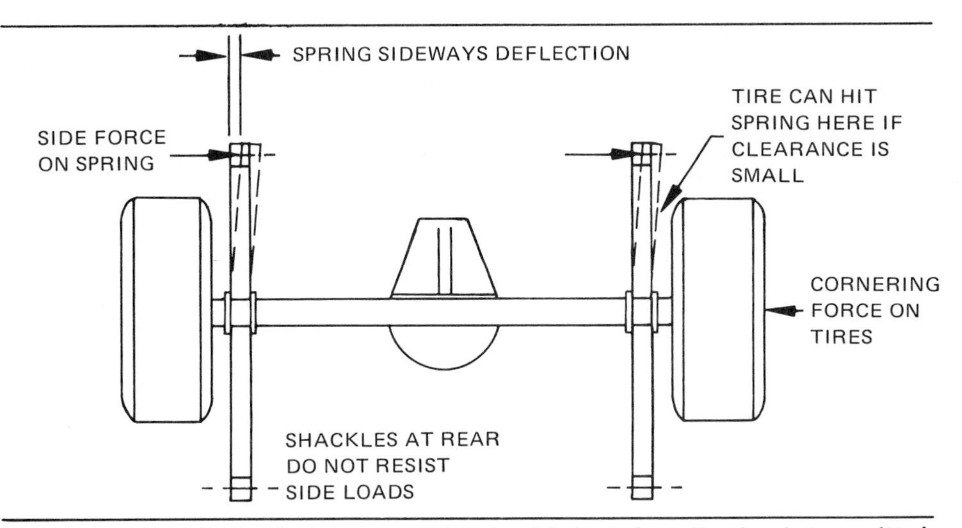

On a conventional leaf-spring rear suspension the side force from the tires is transmitted to the frame through the front of the leaf springs as shown. The springs are not very stiff and they deflect sideways under the loads. This causes the rear of the chassis to feel unstable to the driver. More precise lateral axle location results in better control, plus it gives more clearance for wide tires. Without an extra locating device the tires may hit the springs during hard cornering.

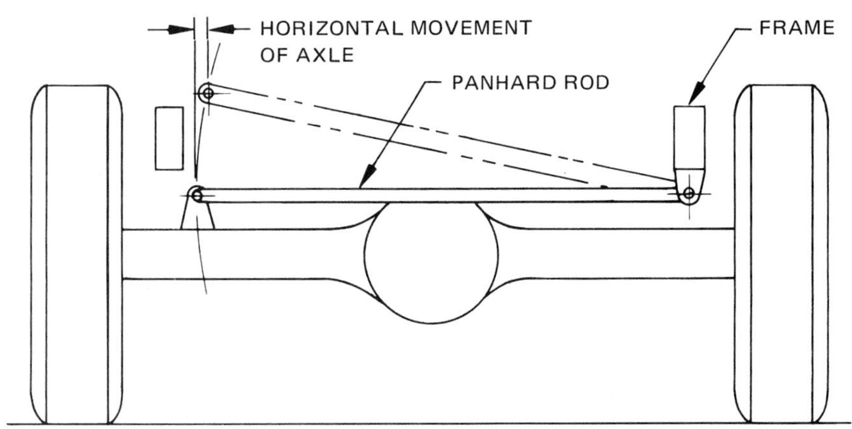

As a Panhard rod moves vertically, it also swings in an arc so the axle moves a small distance horizontally. A long Panhard rod will have a smaller horizontal movement than a short one. If the Panhard rod is not mounted in a horizontal position with the car at normal ride height, horizontal movement is larger.

This is a very unusual Panhard rod. There is a large ring in the center of the bar to allow easy gear changes in the rear end. This adds considerable weight and reduces stiffness of the rod. It represents an extreme compromise for quick gear changing.

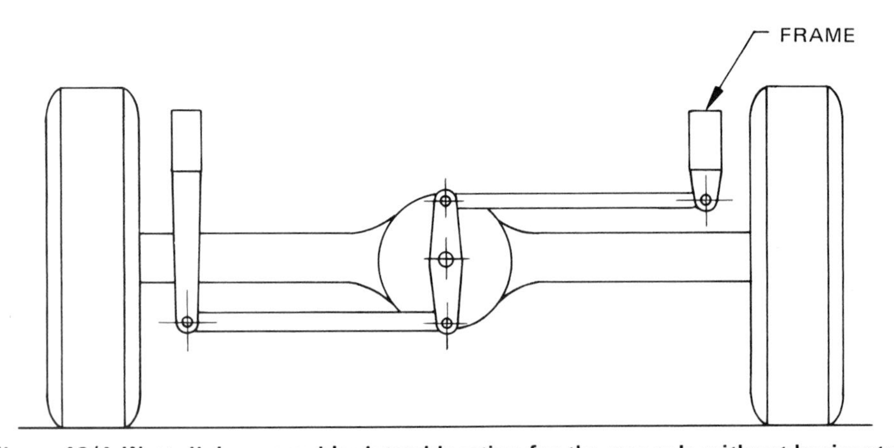

Figure 43/A Watts linkage provides lateral location for the rear axle without horizontal movement. The vertical link is pivoted to the axle housing in the center. Notice the long bracket on the left-hand frame rail to position the lower link so it is horizontal. This bracket must be reinforced to provide a rigid support.

ponents. Just make sure the Panhard rod is horizontal and you won't have this problem. If you are building one yourself this may mean building a bracket on the axle and the frame to put the ends of the Panhard rod at the proper height. The length of the rod should be as long as possible to minimize lateral motion.

A lateral locating device which does not have this problem is a Watts linkage. This is more complex and takes up more space than a Panhard rod, so it is not always the best solution. However, the geometry is better, and it should be considered if at all possible.

A Watts linkage consists of two parallel lateral links plus a vertical link connecting them. The axle is pivoted in the center of the vertical link as shown in Figure 43. If the horizontal links are parallel and equal length they will have the same amount of lateral motion at the ends attached to the vertical link. Because each lateral motion is in a different direction, this rotates the vertical link through some angle, but the center of the vertical link is not affected. This center point travels in a true vertical path, and that is where the axle is attached.

A Watts linkage can consist of tubing for the lateral links, using spherical rod ends for pivots. The vertical link is a beam, and should be designed to carry bending loads. The center pivot is usually a rigid bearing on a stud or post so it can rotate but not pivot laterally. If a spherical bearing were used here, the linkage would collapse.

To build a Watts linkage for a production car usually requires some big brackets. The links must be horizontal or the linkage won't give the proper axle motion. Some Watts linkages have been turned on their side to take up less space, and this may be the answer for your car. The attachment between the Watts linkage and the axle is a tough problem, and it must be designed strong enough to carry the side loads in cornering.

If the horizontal links on the Watts linkage are too short the linkage may not have enough vertical travel. A Watts linkage goes off perfect geometry at the extremes of its travel. Before building a Watts linkage make sure it has enough

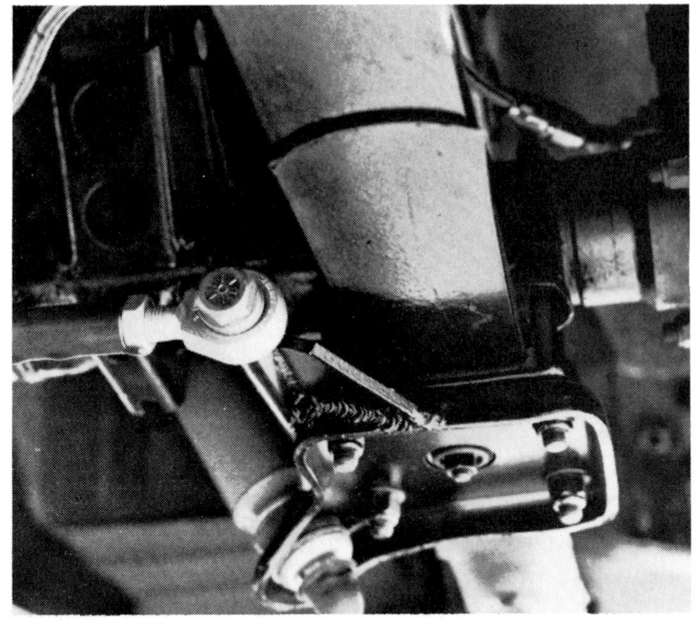

Panhard rod installed by Bill Swan is attached to frame with long bracket welded to bottom of frame rail. Length is needed to get rod horizontal at normal ride height. Axle attachment at other side of car is by a welded bracket on the spring support plate. Panhard rod is tubing threaded to accept spherical rod ends. Note use of jam nuts on rod end threads.

vertical travel to allow the full bump-to-droop motion of your suspension. To check this, use a compass and a big sheet of paper to make a scale drawing of the motion. Use the largest scale possible for accuracy. If you run out of travel make the horizontal links longer or shorten your suspension travel with suitable stops. Be sure the suspension won't hit the stops in normal driving.

You may decide to build your own Panhard rod or Watts linkage. If so you will be faced with some welding, both on the frame and axle. If you are not a good welder it is best to get some expert help. A weld failure can be dangerous if it lets go in the middle of a fast turn.

Some cars come stock with a Panhard rod, but these are not very useful for an all-out car. A stock Panhard rod is usually equipped with rubber bushings at the ends, and these are too flexible to do a lot of good for axle location. If you are trying to adapt a stock Panhard rod, replace the rubber bushings with stiffer ones or with hard bushings. Methods for doing this are mentioned in the next section.

The horizontal links of a Watts linkage or Panhard rod are easily built using heavy-wall tubing and spherical rod ends. You can tap the inside of the tubing and thread the rod end into it. Secure the rod end with a jam nut to keep the threads from causing slop in the linkage. The rod must be strong enough to take the cornering loads from the tires plus an occasional bump against a curb. Stress analysis of the rod is rather complex because buckling strength enters into the picture. If in doubt make it stout enough to support the entire weight of the car in compression.

The brackets on frame and axle are problems in building your own Watts linkage or Panhard rod. The height of the pivot point on the rear axle determines the roll center height, and normally you want this as low as possible to minimize weight transfer on the rear tires. Thus the bracket on the frame is usually quite long. Make sure this bracket is strong enough to carry the cornering load from the horizontal link. It is beyond the scope of this book to design brackets, but just keep in mind that the rod should be lined up with the center of the bracket structure so it won't twist. The best bracket designs use double shear, with one mounting plate on each side of the link.

A Panhard rod or Watts linkage will not make a big difference in the handling of a production car for street driving. These are usually reserved for use on the racetrack, where every bit of extra cornering power is important. This is not to say that you shouldn't try this on a road car. It will improve the handling some, perhaps reduce oversteer or provide better high-speed stability. On a limited budget, tires, shocks, anti-roll bars, and wide wheels will give you more improvement per dollar spent. It is for this reason that there are very few lateral location devices on the market as bolt-on kits. You will usually have to get one made by a small specialty shop or build it yourself.

There is another locating device that is very popular. This is the so-called traction bar first developed for drag racing. Traction bars will help you in a cornering situation too, but only in acceleration out of the corner. The traction bar prevents the axle from twisting about its own center during hard acceleration. This effect is shown in Figure 44.

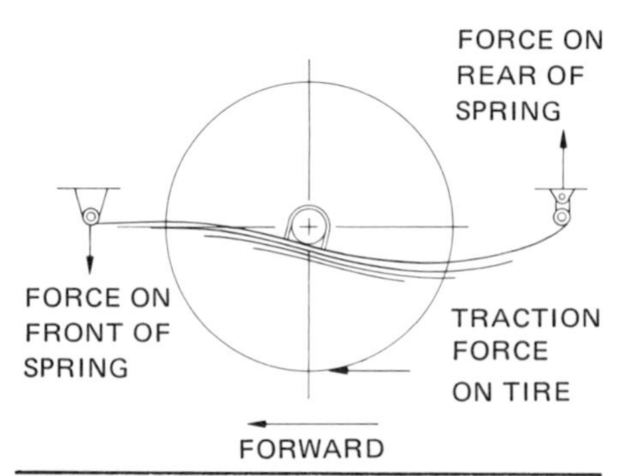

Figure 44/This shows a leaf-spring rear suspension during hard acceleration. The torque reaction on the axle twists the spring into the S-shape above. Traction bars, lift bars, and spring stiffeners are designed to locate the rear axle and keep it from twisting the springs. Traction is improved and axle hop during acceleration is eliminated by these locating devices.

On a solid rear axle with leaf springs, the axle is twisted about its own center by a torque equal to that delivered to the rear tires. This torque is transferred into the rear leaf springs, which are not able to resist it without going into an S-shape. If the axle is allowed to twist, the result is an uncontrolled vibration of the axle on the springs. This can be felt as wheel chatter when accelerating, and almost all traction is lost. In addition, this violent vibration of the axle can cause breakage of the suspension parts, and the large change in angle of the drive shaft U-joints can cause sudden embarrassing failure.

Traction bars are rigidly pivoted to the lower part of the axle near the springs—one on each side—and are pivoted ahead of the axle on the frame. When the axle tries to rotate about its center, this motion is resisted by the rigid traction bar, keeping the wheels properly located on the pavement. The geometry of the traction bar must be compatible with the vertical motion of the axle or binding will occur when the axle moves vertically or when the car rolls in a corner. Some traction bars are made strictly for drag racing and will not work on the street in a cornering situation. Look at the design before buying. Also some traction bars are extremely heavy and should not be used on a car where unsprung weight is important. Unsprung weight is less important on a smooth drag strip, but even in this type of racing there may be some small bumps to worry about.

There are a number of firms that manufacture traction bar kits, some which bolt on and some which have to be welded. Usually the welded types are lighter and simpler, but require the services of an expert welder for a safe installation. There are other similar devices produced which instead of having a separate bar, use a stiffening leaf attached to the forward part of the spring. These devices again are mostly to help in acceleration, and some even hurt the cornering by adding extra roll stiffness to the rear suspension.

There are many locating devices used on cars with coil-spring rear suspension. These are designed-in as a vital part of the suspension, because the coils offer little help in locating the axles. On most sedans these locating links use rubber pivots, which are as soft as possible to reduce road noise and ride harshness. However, these soft pivots offer rather poor location for the axle allowing unwanted deflections in cornering and acceleration. To the driver this feels like the rear end is doing its own thing and wandering from side to side under the action of side loads. Also some high-speed stability could be gained if the axle were located more precisely.

To do this several firms offer replacement parts for racing use or better road handling. TRW manufactures a line of *trac-bar* heavy-duty rear lower control arms. These are made from heavy-duty steel and have stiffer rubber pivots than

This Sunbeam Tiger has much-needed traction bars (arrow) added below the leaf springs. This car is very light and powerful, and cannot use full power without these bars to control spring windup. This car is also equipped with Koni shocks, wide rims, and racing tires.

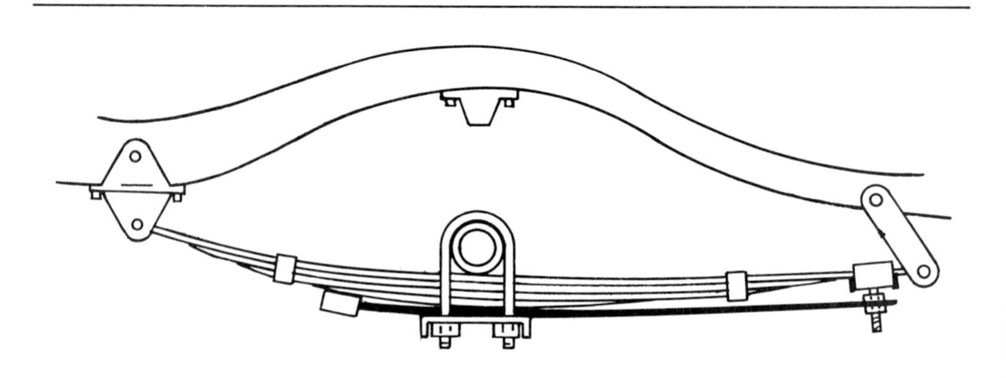

This accessory spring by Hellwig Products Co. resists rear spring wind-up during acceleration. The device is a helper spring which raises the ride height and stiffens the spring, but its design also helps reduce wheel hop in hard acceleration.

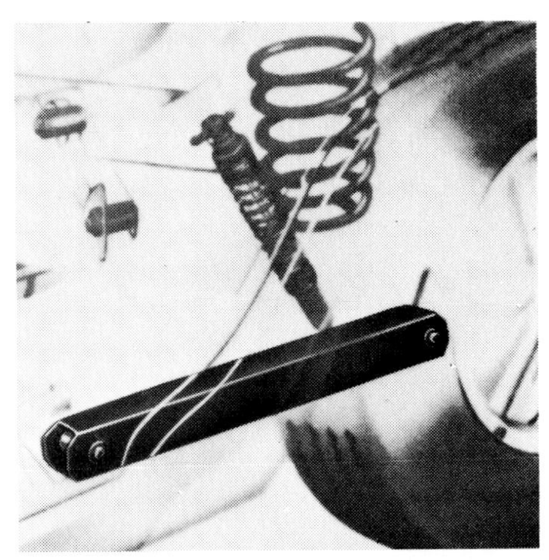

On the rear suspension of cars with coil springs the solid axle is located by links using rubber pivots. TRW manufactures a line of heavy duty links to increase the stiffness of these links for improved axle location. These TRW *trac-bar* links provide better handling on street or track.

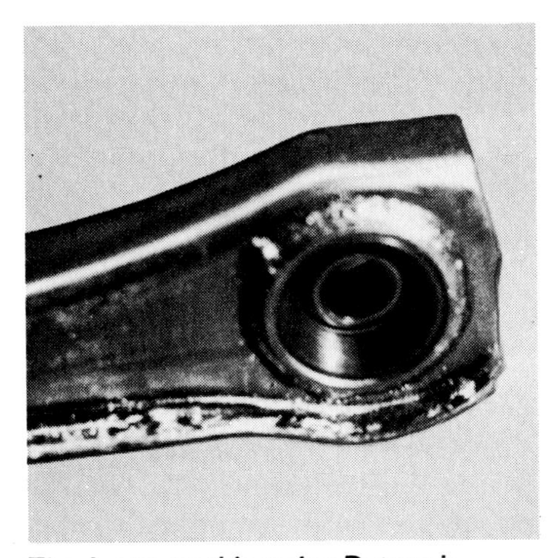

The A-arm on this racing Datsun is equipped with a pivot kit by Mac Tilton of El Segundo, California. The rubber bushing has been replaced by a spherical bearing for racing use. This is not too practical for the street, but is necessary for holding the suspension in the proper geometry in a hard racing turn. Spherical bearings are used where the link has to swivel as well as rotate. Solid metal or plastic bushings are used where there is rotation only.

the stock items. A car equipped with these arms will respond to steering better and have more traction on bumpy surfaces or when accelerating hard. High-speed stability may be improved a noticeable amount also. These are good for a road or dual-purpose car.

For all-out racing, rubber bushings are usually replaced with spherical bearings or metal bushings. This gives a rather harsh ride and transmits all the noise and vibration right through to the chassis. Also these pivot points are subject to wear and may have to be lubricated frequently. The cost is also much higher than rubber bushings.

Another solution for stiffening rubber bushings is the use of washers as shown in Figure 45. These trap the rubber so it can't move around as much, thus stiffening the bushing against unwanted deflection. However, they still isolate road noise as the stock rubber is designed to do. When installing these bushing stiffeners, trim the rubber down the thickness of the washer, say 1/8" for the design shown in Figure 45. The washer should be as tight a fit as possible to minimize flexing of the rubber.

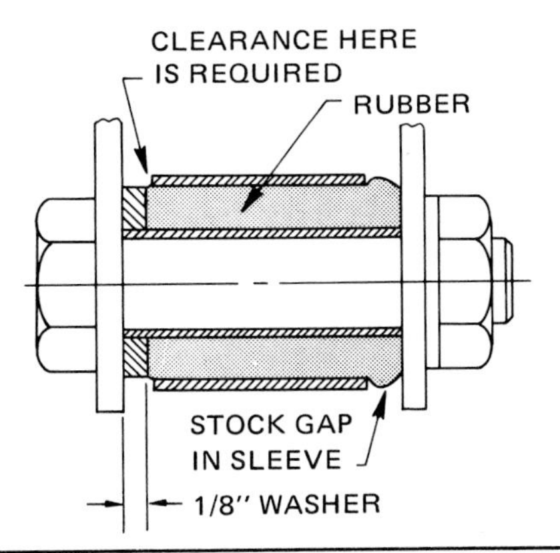

Figure 45/The right side of this bushing shows the stock gap in the sleeve surrounding the rubber. The left side shows the modification to stiffen the bushing. The rubber here is trimmed off by 1/8" and a washer of that thickness is inserted. To complete the modification the right side of the bushing should have a washer too. Be sure there is still clearance between the outside edge of the washer and the sleeve.

Other locating devices can be used to provide both lateral and torsional location of a rear axle. One of these consists of an A-arm attached to the center of the rear axle. The apex of the A-arm is a spherical bearing on the axle housing, and the two pivots of the A-arm are located on a line across the chassis. The A-arm is horizontal with the car at ride height, and the pivot points are located in a forward position so as to not fight the motion of the leaf springs. In a cornering condition, the side load is transmitted from the axle to the frame through the A-arm, so it acts as a lateral-location device. If the car rolls, the axle pivots about the spherical bearing at the tip of the A-arm, so the pivot becomes the roll center. In an acceleration condition, the A-arm acts the same as a traction bar, locating the axle against twisting.

Notice I talked about the locating device determining the roll center . It is entirely possible to re-locate the rear roll center with a lateral-locating device. The roll center with leaf springs is high, and there are some definite advantages to a

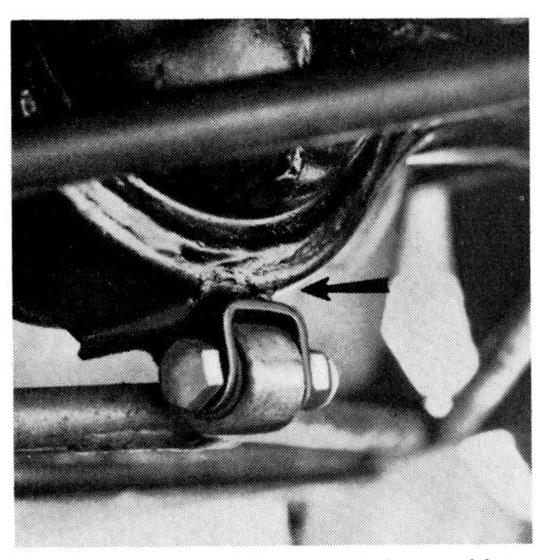

The rear axle on this Lotus is located laterally by the apex of an A-arm attached to the bottom of the axle housing. The other A-arm pivots are on the frame behind the seats. Notice how the A-arm bracket is simply welded to the bottom of a stock axle housing.

lower rear roll center. For one thing, it reduces the weight transfer at the rear wheels, thus reducing the amount of lifting of a rear wheel in a turn. On a nose-heavy car with power, this can be a real problem. Of course, lowering the rear roll center will increase the roll angle of the car, so this modification should always be done along with another modification to increase roll stiffness. The Panhard rod, the Watts linkage, or the A-arm locator can be used to change the rear roll-center location. The effect on roll-center height is shown in Figure 7.

If you have a car with conventional solid-axle rear suspension and leaf springs, the roll center is approximately at axle height. To lower the rear roll center using an axle lateral-locating device, the locating device must be a great deal stiffer laterally than the leaf springs. and the maximum body roll angle must be small.

If this is not the case the car will roll about the low roll center and put side forces into the leaf springs. The spring side forces will resist roll and add unexpected roll stiffness to the rear suspension, just as if you installed a rear anti-roll bar. This effect will increase rear weight transfer and defeat the original purpose of lowering the rear roll center. You will get some overall gain with improved rear-axle location, but the car may still tend to lift a rear wheel in a turn. To avoid this problem use a stiff axle-locating device, keep the leaf spring pivots flexible laterally, and reduce the total body roll with a stiffer anti-roll bar on the front suspension.

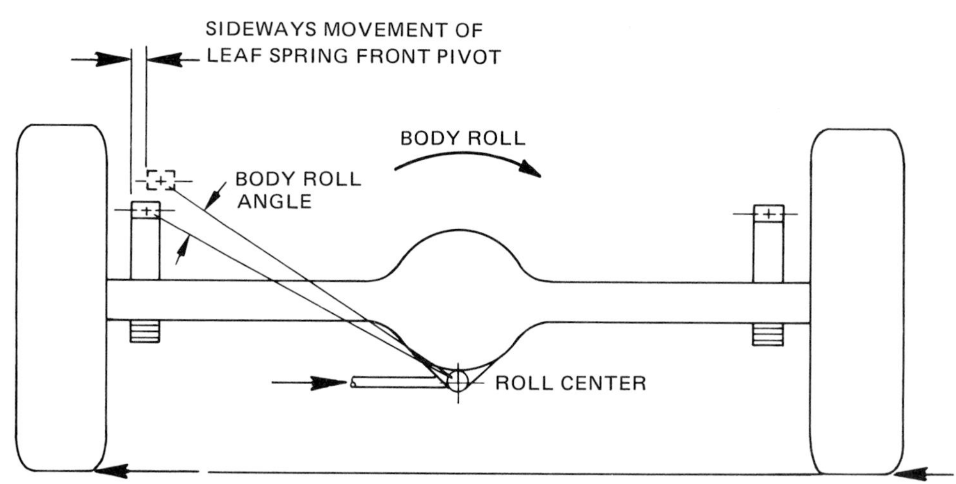

This drawing shows a solid axle with leaf spring suspension, and a lowered roll center due to a lateral locating device. The body rolls about the stiff pivot point under the differential and flexes the front leaf spring pivots sideways as shown. Because the leaf springs resist this flexing they add increased roll stiffness to the rear suspension. Thus a lowered rear roll center may not reduce the rear weight transfer as much as expected.

On this racing sports car an attempt was made to use the roll bar structure to stiffen the frame in the cockpit section. This design is better than nothing, but it would be a lot more effective if the forward diagonal brace (arrow) ran all the way to the firewall. By having this brace angled downward so sharply it connects only to the fore and aft brace across the door.

STIFFENING THE FRAME

If you are attempting to get the ultimate traction and cornering power out of your car it is important that the frame be stiff in torsion. Thus weight transfer at one end of the frame can be resisted by the suspension at the opposite end without excessive deflection of the frame structure.

What is excessive? This depends on the weight of the car and its intended use. For road use, the frame should be stiff for other reasons. A stiff frame will make the car feel more solid and will not be as apt to rattle and squeak when going over bumps. Older sports cars with rail frames were very bad in this respect. Some had such a flexible frame that you could see the dashboard shake sideways on bumps and actually feel the frame twisting under you if you drove diagonally into a driveway entrance.

To modify a frame for increased torsional stiffness is a big job that requires work on the basic frame structure. It is next to impossible for a road car, because the added frame structure usually intrudes into the passenger space. About the only thing you can do on a road car is some minor amount of stiffening. On a racing car a great deal can be done and a stiff frame helps all types of racing cars.

The usual means of stiffening a frame is to use the roll bar or roll cage for this purpose. This stiff sturdy structure can be attached to the car in such a way that it actually braces the frame structure. This is most effective on cars with open cockpits such as sports cars. In addition to using the roll bars, certain other modifications can be made. On cars where the rules allow it, doors can be welded shut all around the edges. On a sedan, this makes a closed *monocoque* box structure out of the body and really improves things. Also on unit construction cars, tack-welded body seams can be continuously welded, which provides a small increase in the joint stiffness.

On some cars the body is separated from the frame with rubber mounts. This is done to reduce noise and vibration from the road. If these mounts are made rigid, the body will stiffen the frame. Rubber mounts can be replaced with metal to remove the flexing between body and frame. On this type of car, additional mechanical connections between body and frame will also help increase stiffness.

To stiffen the frame for drag racing Stahl and Associates developed these metal bushings to replace rubber subframe bushings. The rubber bushings normally connect the body to the frame with a flexible joint to isolate road noise. Metal bushings tie the body and frame together so they act as a rigid unit for proper handling in drag racing. Stahl subframe bushings are available for Camaro, Nova and Firebird cars. (Photo courtesy Stahl & Associates)

In an open sports car with a rail frame, most of the frame twisting occurs in the cockpit area. The usual way to stiffen the frame is to use the roll bar. The typical roll bar used on a sports car consists of a hoop behind the driver plus diagonal braces running to the back of the frame. This type of roll bar does almost nothing to stiffen the frame in a twisting direction. However, if additional braces are run from high on the roll bar hoop forward to the dashboard or firewall, the frame is then greatly increased in stiffness. The idea is to triangulate the frame with the roll bar structure bridging over the flexible cockpit area. Two sets of braces are necessary running both forward and aft. These braces must be tied rigidly into the frame structure, not just tacked on. The roll bar will carry very large loads and the joints must be absolutely rigid.

Another approach is the use of a roll cage. This consists of two hoops, one at the front and one at the rear of the cockpit. These are connected with fore and aft members. To use a roll cage to stiffen the frame you should try to use triangles, not rectangles, in the design of the tube structure. A diagonal tube running across the door opening on each side is very effective. The forward roll-cage hoop should be tied rigidly to the body as high up as possible at the corners of the dashboard. It should also be connected to the main frame rails at the bottom.

The frame stiffening usually used in a sedan is the roll cage. Here many extra diagonal members can be used to increase stiffness. It should have braces running forward, through the firewall area, all the way to the front suspension crossmember. These braces should also be attached at the firewall or dashboard. The same sort of structure should go to the rear of the frame. A properly designed roll cage for a stock car is virtually the same sort of structure as a space frame. This adds safety for the driver as well as massive stiffening on these heavy cars.

When using a roll cage to stiffen the frame, pay particular attention to the attachments to the body or frame of the car. The body is usually made of light sheet metal, and it will not work to tack a heavy roll-cage tube to the thin body metal. You should always try to make the tubes attach at the strongest points of the body where there are corners or where several pieces of sheet metal are joined together. Good attachment points are where the firewall meets the dashboard, where the door sills end, and at

SCCA driver Al Sharpe has a well-designed roll cage on his Corvette. The roll cage aft hoop is braced back to the rear of the frame with diagonal braces. A diagonal across the passenger's seat adds further stiffness to the cockpit area.

This type of roll cage is used in NASCAR stock cars. In addition to the full roll cage structure there are extra diagonals through the center of the car, all meeting in a cluster of tubes behind the driver's seat. This roll cage extends all the way from the front suspension to the rear end of the frame rails. It is virtually a complete space frame of tubing attached to the stock rail-frame structure.

the corners of the floor. Back up the thin body metal with extra *doublers*—an extra thickness of metal wherever possible and spread the joint loads over as large an area as you can.

Frame stiffening often is not practical on a road car. Racing cars sometimes suffer from a lack or frame torsional stiffness, and some modifications can be helpful. One way to increase the stiffness is to tie all aluminum panels and firewalls rigidly to the frame structure. If these are fastened with only a few fasteners, use a fastener every inch or so. If you can tie a rectangular panel rigidly around all the edges it can act the same as a monocoque-frame skin panel and helps the overall frame stiffness. If you have to add some ballast to meet a weight requirement, you may wish to do it by adding stiffness to the frame. On a space frame, an easy way to do this is to put two diagonals across every rectangular section of the frame. If there is already one making two triangles out of the rectangle, just add two short tubes forming an X across the rectangle. This will greatly increase the frame stiffness with a small amount of added weight. Another area to improve stiffness is around the cockpit of a single-seat race car frame. Just run two tubes side by side along the top of the cockpit and brace the four corners of the cockpit opening. This will do a lot of good in the usually weak cockpit area.

How stiff the frame should be is a difficult problem, but extra stiffness never hurts. If you *need* extra weight, put it into the frame. It is much more useful than adding lead weights to meet the minimum weight in your racing class.

CHANGING THE CG LOCATION

As described in previous chapters the CG location has a large effect on the way your car handles. For street use you can make small changes in CG location that will be noticeable. For racing, these changes can be greater and are even more important if you want to win. The CG can be moved in any of three directions—up and down, side to side, or fore and aft.

Before starting to move the CG you should measure where it is. Then you can make an intelligent choice of how far to move it and in what direction. Also knowing the CG location is useful if you want to calculate loads on the suspension.

To find the CG location you will need at least two platform scales to weigh the car under the tires. On a very light car, bathroom scales will work, but only if you take care not to jam the mechanism. Use of scales in discussed in Chapter 3. Place the scales under the tires and take readings. Weigh the car as driven, with driver and half a tank of fuel. If you want the car set up with a greater load than this, add people or weight as required to bring it to operating weight.

This racing car was built for a class with a rather high minimum weight. The frame structure has extra diagonals to increase stiffness. The added weight is acceptable because otherwise ballast would have been required.

The *longitudinal CG location* is its position in the fore-and-aft direction. The *lateral CG location* is its position side-to-side. These locations are determinded from the wheel weights as shown in Figure 46. This gives the CG location for the entire vehicle. If you want the CG location only for the sprung weight as shown in Figure 14, then subtract the unsprung weight from each corner. For example, if the right-front wheel scale reads 650 pounds and the unsprung weight of the right-front suspension assembly is 50 pounds, then 600 pounds is the chassis weight used to calculate CG of the sprung weight.

The *vertical CG location* is the CG height, usually measured above the ground. The CG height can be measured using scales, but some extra calculations are required. You should know the CG height if you want to do any weight transfer or roll-stiffness calculations as described in Chapter 2.

The method for measuring the CG height is shown in Figure 47. First make sure the floor is level. Check the fuel tank—it should be full, empty or perhaps half full. You should know how full it is because that influences results. If you make one measurement with tank empty and another with tank full, you will know the range of CG heights. Block the suspension to prevent movement during the test. This can be done by replacing the shocks with rigid links. Determine longitudinal CG location as shown in the top part of the figure. Calculate F, the distance from the front wheel centerline to the CG.

Then elevate the rear wheels with blocks about 2 feet high. The taller the blocks the more accurate the calculations and measurements, but you will have problems jacking the car up extremely high. Place scales under the blocks and under the front tires, and drop plumb bobs from the wheel centers as shown. Calculate G from the scale readings. The two rear scale readings are added together and divided by the total of all four scales. This number multiplied by measured length L gives the value of G. L is measured horizontally between the plumb bobs.

Now for some trigonometry. Calculate θ, the angle the car is tilted. You can find the tangent of the angle θ by dividing B

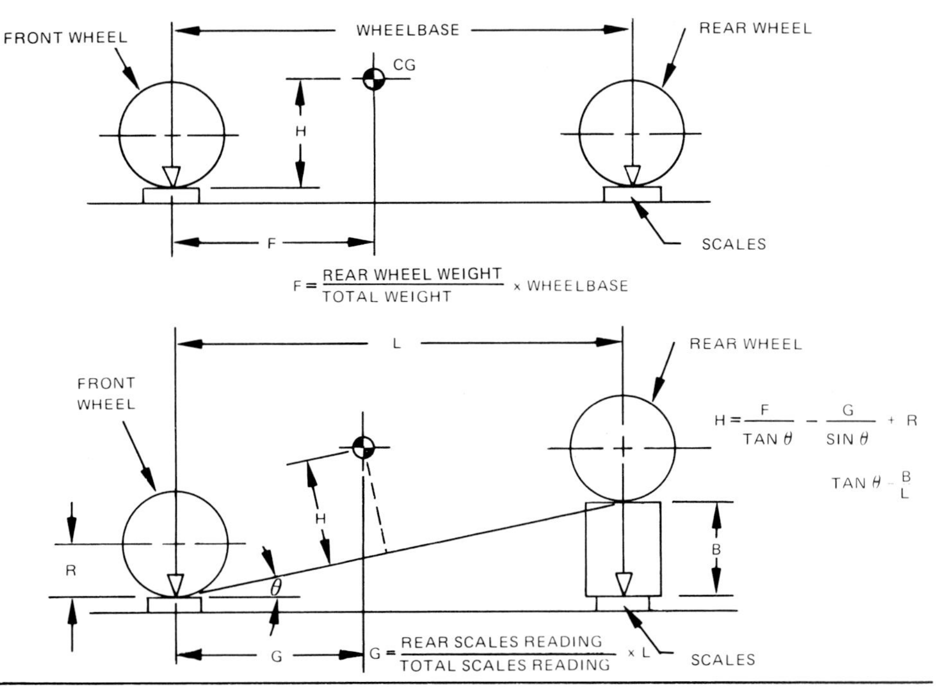

Figure 47/With these measurements you can calculate the CG height "H", determining weight transfer during acceleration, braking, or cornering.

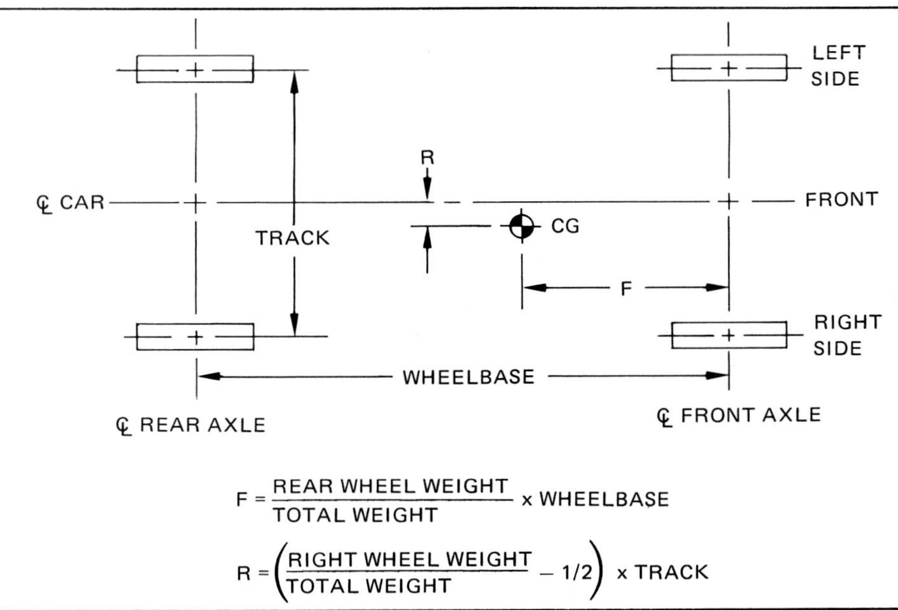

Figure 46/This is a view looking down on the car. To locate CG in the horizontal plane, place scales under all four tires and record the weights. In the equations shown, Rear Wheel Weight is the sum of left rear and right rear wheel weights. Total Weight is the sum of all four wheel weights. Right Wheel Weight is the sum of the two right-side wheel weights. Calculate longitudinal and lateral CG dimensions as shown. If the track is not the same at front and rear, you will get approximately the correct result by averaging the tracks. If the equation for "R" yields a negative answer, the CG is that amount to the left of the chassis centerline.

by L. B is the height of the blocks under the rear wheels. Now use your electronic calculator, trig tables or a slide rule to find the sine of θ.

Measure the radius of the front tire R. Now you have all values needed to solve for CG height H in the formula given. It really isn't all that difficult unless you have trouble with the trigonometry. If so, you can find H by making an accurate drawing of the lengths and angles measured. You will end up with something very similar to Figure 47.

To do it graphically, notice the perpendicular through the CG in the top drawing of Figure 47. Transfer this perpendicular to the bottom drawing as shown by the dotted line. Then construct the perpendicular shown in the bottom drawing which also passes through the CG. The CG is at the intersection of these two lines. Measure the height by scaling your drawing.

To find the CG height of the *sprung weight only,* use the following formula:

$$H_s = \frac{(W \times H) - (U_f \times R_f) - (U_r \times R_r)}{W - (U_f + U_r)}$$

where,

- W = total weight as measured by all four scales
- H = total car CG height as calculated from Figure 47
- U_f = unsprung weight of both front wheel assemblies
- U_r = unsprung weight of both rear wheel assemblies
- R_f = radius of front tire
- R_r = radius of rear tire
- H_s = CG height of sprung weight only, above ground level

This formula makes the simplifying assumption that the CG of each unsprung wheel assembly is located at the center of the wheel. This gives reasonably accurate answers for most suspension systems, as almost all the heavy unsprung items are centered around the wheel centerline.

Now that you know the CG location you can decide where to move it. For either street driving or racing you may wish to change the vertical CG location. For cornering you want the CG as low as possible. This holds true for street driving or most types of racing. The exception to this is drag racing, where a raised CG will increase weight transfer off the line. On a front-engine car this is quite important for winning drag races. This tends to spoil the handling in other types of driving.

The most effective way to change the CG height is to change the ride height. Chapter 3 goes into detail on the methods of adjusting ride height and the limits to lowering the car. Study this carefully before starting to modify anything. In addition to suspension adjustments there are many bolt-on kits on the market to either raise or lower the ride height.

In addition to changing ride height you can raise or lower various components of the car to change the CG height. For everything but drag racing you want every item as low as possible. Always start with the heaviest things first, as they make more difference to the CG height. Consider lowering the engine, the driver, and the fuel tank. The battery is a heavy item that can usually be moved easily. If you are racing in a minimum-weight class and have to add ballast, put it low in the car to lower the CG as much as possible. For drag racing, reverse all this and place movable items high in the car.

For maximum cornering power you want the CG as low as possible. This stock car has the ride height reduced to the minimum possible value for racing on paved ovals. Lowered CG means weight transfer in the corner is reduced.

For drag racing with a front-engine sedan you need maximum weight transfer during acceleration. To do this, raise the CG by increasing the ride height. Here the rear ride height has been increased by means of long shackles. These raised cars corner very poorly, but they do accelerate well off the line.

The lateral CG location has to be a compromise unless you are racing only. If only the driver is in the car his weight tends to move the CG to his side of the car. If two people are aboard this effect is cancelled out if they weigh the same. For street driving you will have to choose which condition you set the car up for. It will not be perfect for other loading conditions. For street use, you can shift certain items laterally in the car such as the battery, spare tire, and fuel tank. Only a small change in CG location is possible without ruining the car for street use.

For racing you can do more radical modifications. For road racing you would like to have the CG dead center in the car for equal handling turning both right and left. On a sports car the driver's weight will usually put the CG off center. Offsetting or tilting the engine is sometimes used to help correct this, as well as moving certain items attached to the chassis. If you have to add ballast to meet the minimum class weight, put it on the side opposite the driver as far away from center as possible. Sometimes in road racing a compromise is made because there are usually more right turns than left turns. A car with right-hand drive may be used to deliberately shift the driver's CG to the right to help in the right-turns. Of course this tends to hurt cornering in left turns.

In oval-track racing there are only left turns, so it helps to have the CG offset to the left. The limit to this is when the car becomes unstable in hard acceleration or braking due to unequal traction forces on the two sides of the car. Oval-track cars usually must live with this problem to gain cornering speed. They are almost always set up with the CG heavily offset to the left. Cars for tight tracks can use a greater CG offset than cars for high-speed ovals. At high speed, stability becomes more important.

For a drag-racing car with solid-axle rear suspension, the CG should be offset slightly to the right. This counteracts the driveshaft torque reaction that tends to unload the right rear tire and add vertical load to the left rear tire. Because drag racing with passenger cars is done with only the driver, the CG usually ends up offset to the left. The need to make the car as light as possible usually prevents having the CG offset to the right. Driveshaft torque reaction can also be compensated by weight jacking as described in Chapter 3.

Ron Moore selected a right-hand drive Mini for road racing to have an advantage in right-hand turns. Here at Riverside Raceway, where he does most of his racing, there are more right turns than left. At this point in a hard right turn he is able to apply power, where a left-hand drive car would have to wait an instant later before accelerating out of the turn. This works against him in left-hand turns.

On drag-racing cars without rear suspension or with independent rear suspension there is no driveshaft torque reaction to worry about. The differential on these cars is mounted solidly to the frame and the driveshaft torque is taken on the differential mounts without affecting the loads on the rear springs. On these cars the CG should be centered laterally to give equal traction forces on both rear tires. The car will not run straight unless the right and left traction forces are equal.

The longitudinal CG location is important for all types of cars. For street driving its location determines steer characteristics and sometimes high-speed stability. For racing it greatly affects the overall performance of the car in acceleration, braking, and cornering.

Most street-driven cars understeer in stock form. This can be reduced by moving the CG farther aft. An extreme example of this is a pickup truck, which has an extremely forward CG location without a load in the bed. A pickup is one of the worst-handling vehicles made, due mostly to this extremely unbalanced longitudinal CG location.

In steady-state cornering a pickup understeers heavily, but if power is applied in a turn the lightly loaded rear tires tend to lose traction due to a combination of side load and forward thrust. Thus the understeer changes instantly to extreme oversteer. A similar thing happens if you hit the brakes in a turn in a pickup, usually resulting in a spinout or worse. The solution for these vehicles is to do everything possible to move the CG towards the rear.

Mount the spare tire on the rear, buy a big rear bumper, and put gas tanks as far back as possible. Even this won't solve the inherent problem, but it will help. Only with a partial load in the rear is a pickup reasonably balanced in a turn.

For racing, having the CG toward the rear wheels is a must. For drag racing this is particularly important. The Funny Car category was invented with this in mind. On these cars both the front wheels and the rear wheels are moved forward relative to the body. This effectively moves the CG back in relation to the wheels. Also engine and driver location are moved to the rear.

A tail-heavy weight distribution is helpful in other types of racing too. It helps acceleration out of the corners, particularly if larger rear tires are used. A tail heavy car does not always oversteer, as explained in earlier chapters. If a powerful racing car were nose-heavy, full power could not be applied as early in a corner without losing rear-wheel traction. Also braking would suffer, because the forward weight transfer adds to the already nose-heavy condition. A car has maximum braking traction with equal loads on all four tires.

So far I have shown how to find the CG of a car and given some general guidelines about its location in cars of various types. Exact methods for changing the CG are not possible to describe in detail because it will vary with every car design.

To move the engine, you must fabricate motor mounts and install them appropriately on the chassis at the new location. This may require a longer or shorter driveshaft. If the driveshaft takes a new angle, you must consider the effect on the U-joints. Moving anything supported by brackets involves the same general procedure: find a new location where it will fit, make new brackets to support it.

Before cutting and welding, it may be a good idea to do your modification on paper to see how much the CG is affected and where the new location is. Once the CG is known by the experimental method described earlier, it is fairly easy to calculate the effect of moving any part on the car.

I say fairly easy because it depends on how much you remember of high-school physics. Here's a quick review.

The CG of a car is a fiction which assumes all parts of the car become weightless except the CG which is a tiny point where all the weight is magically concentrated.

In a real car, the effect of a part with real weight is determined both by how much it weighs and where it is located on the chassis. If you move the battery forward, it will put more weight on the front wheels; less on the rear wheels; move the CG forward; but not change the total weight of the car.

The owner of this slalom buggy felt a need for more weight on the front tires so he installed this ballast as a front bumper. Usually a handling problem can be cured by more efficient means, but if the CG has to be moved this is one way to do it.

Both weight and location or distance from a reference point are important in determining the effect of a part. A handy point of view is to consider each part as though it is trying to rotate the vehicle around some reference axis such as the rear axle. From that point of view, the radiator is trying to move the front end of the car downwards, rotating the car about the rear axle.

The torque caused by a part is equal to the weight of the part multiplied by the distance from the reference axis— in this case the rear axle.

In physics, this kind of torque is called a *moment* but it's basically a torque. Because the car is not actually rotating around its rear axle, the moment produced by one part must be exactly opposed and balanced by some other moment. If we add up all of the moments which are trying to make the front end of the car move downwards into the ground while rotating around the rear axle, the opposing torque or moment is the force of the ground against the bottoms of the front tires.

When all of the weight is considered to be concentrated at the CG, then the total moment trying to rotate the front of the car downwards is equal to the total weight of the car multiplied by the distance from the CG to the rear axle.

But in a real car, that total moment is the sum of all the moments of the individual parts. Move one part and you affect the moment of that part and the total moment at the CG.

The easiest way to show how to figure the effect of moving one part is by example. The accompanying sketch shows a 2,000-pound car with the CG located 60 inches ahead of the rear axle. The radiator filled with water weighs 100 pounds and is located 100 inches ahead of the rear axle. What would be the effect of re-locating the radiator 10-inches farther forward?

First calculate the total moment of the existing car, using the rear axle as the reference axis.

Moment = 2,000 pounds x 60 inches = 120,000 lb.-in.

Now imagine you take the radiator completely out of the car. That would reduce the total moment by whatever moment the radiator itself was contributing. Calculate the moment of the radiator at its old location:

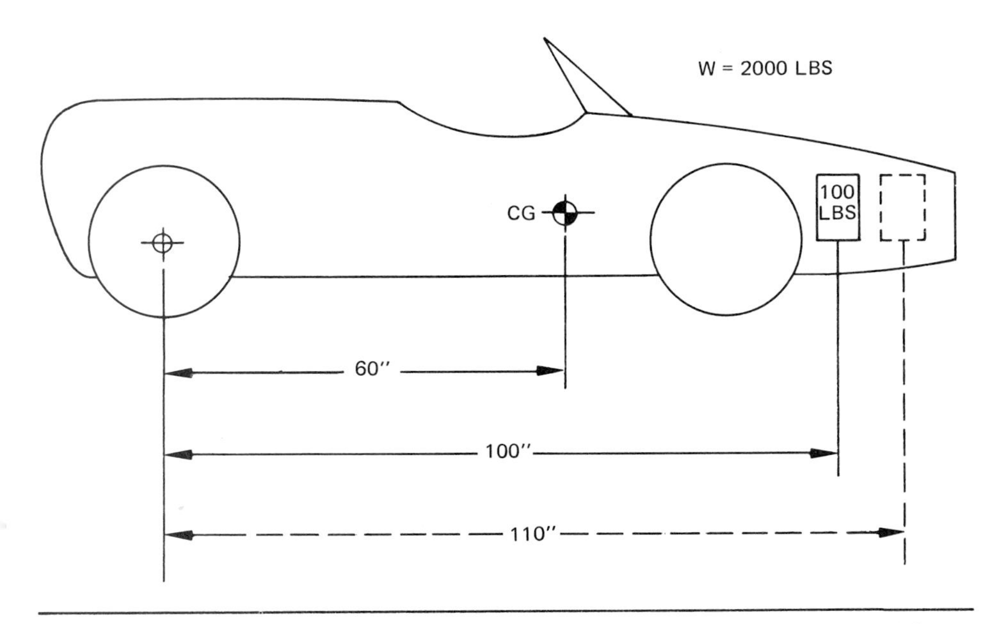

By the methods described in the text, you can calculate the effect of moving weight on the chassis before you do it.

Moment of radiator at old location =
100 x 100 = 10,000 lb.-in.

Subtract that from the total moment with radiator installed.

Moment without radiator =
120,000 - 10,000 = 110,000 lb.-in.

That will move the CG back toward the rear axle, but you don't care because you aren't going to run the car without a radiator.

Now, figure the new moment of the radiator moved 10 inches forward.

New moment of radiator =
100 x 110 - 11,000 in.-lb.

Add that back into the total moment of the car.

New moment of car =
110,000 + 11,000 = 121,000 lb.-in.

For that new moment, the weight of the car didn't change, but the distance of the CG from the rear axle did change. Find the new distance by dividing moment by weight.

CG distance from rear axle =
121,000/2,000 - 60.5 inches

Moving the radiator forward by 10 inches only moved the CG forward by half an inch.

It doesn't matter which axle you choose as the reference axis. It is handy to use the rear axle when you plan to move something forward and the front axle when you plan to move something backward on the chassis—the arithmetic is simpler.

You might have to work for two days to move that radiator forward and the result would probably not be noticeable in handling of the car—as far as steady-state cornering is concerned. The change *might* be felt in transient response when entering a corner.

If there are several parts you can move or if you are considering moving the engine, calculations like this can tell you the effect before you do it. In general, unless you can move the CG by a few inches, it may not be worth doing.

The best fore-and-aft location of the CG is not easy to decide. You definitely want more weight on the rear wheels than the front, but just how far to go is the big question. If a racing car is nose-heavy it will have traction problems when accelerating out of slow corners. Cars with high power-to-weight ratios such as racing cars cannot tolerate nose-heavy weight distribution. But if the car is *too* tail-heavy it can have other problems. In acceleration at low speeds the front tires may lift off the road. During cornering there will be a tendency to lift the inside front tire off the road. A car with good acceleration out of a corner requires a high percentage of the lateral weight transfer to be taken by the front tires, so the rear tires will have some traction available for acceleration. A car that is very light on the nose is forced to take most of the lateral weight transfer at the rear, being limited by lifting of the inside front tire.

CHANGING THE POLAR MOMENT OF INERTIA

The polar moment of inertia is very difficult to change. On a production car for street use there is very little that you can do to change it. Moving a few items in location is about all that can be done within practical limits. However, on a racing car, more can be done with considerable effort.

Polar moment of inertia takes some understanding as discussed in Chapter 2. It is a computed number that involves both the weight of the car and the distribution of its weight. Thus two cars with the same total weight can have different polar moments of inertia. A high polar moment of inertia means that the car has a high resistance to changing direction of travel. If it has a high polar moment of inertia it will go into a spin very slowly, but once the spin starts it will take longer to stop. As a disadvantage, a high polar moment means that the car will not respond quickly to the steering, and thus it will tend to understeer when entering a tight turn. It will be more stable at high speeds, however, because disturbing forces will not easily change its direction.

Note that the polar moment of inertia (PMOI) involves both weight and distance. The distances in the calculations are squared, so that makes them much more important than the weight. However, because there is not very much you can do with distance of various items in the car from the CG, it is not easy to make big changes in the PMOI. The means of changing it involve the wheelbase, the track, the locations of the fuel tanks, and the locations of some heavy items such as the battery and the radiator. Large production sports cars and sedans have a high polar moment of inertia because of their weight, size, and engine location, and thus are more sluggish in handling characteristics. An all-out racing car is usually much lighter, and so its PMOI is automatically less than a big production car. Larger racing cars, such as A-sports racing cars have a high PMOI relative to their smaller brothers such as D-sports racers. The larger weight and size automatically makes the bigger car have a larger PMOI. Thus if you design a D-sports racer you may be in trouble if you copy a big car. A high-polar-moment design may be needed on a tiny light car, just to make it more driveable. A large heavy car on the other hand may require all the tricks possible to make the PMOI lower.

The following tricks are used to *reduce* the polar moment of inertia:

Short wheelbase. (Also increases weight transfer in accelerating and braking.)
Narrow track. (Also increases weight transfer in cornering.)
Fuel near the center of the car. (May require a longer wheelbase.)
Driver more upright in the seat. (Raises the CG and sometimes the air drag.)
Central-mounted radiators. (May require bigger radiators.)
All systems located near the center of the car. (Space problems.)
Small amount of overhang front and rear. (Aerodynamic drag increase.)

Obviously the opposite tricks can be used to increase the polar moment of inertia. Because of all the various trade-offs in the design you may not wish to use all the tricks available, and thus the polar moment of inertia will end up at some value which you cannot change very much.

What is the ideal value for polar moment of inertia? This is another question where drivers disagree. If you prefer a slower responding and more stable car you should keep the value high. If you have quick reflexes and want ultra-quick response to the steering, then reduce the PMOI. It is my opinion that small racing cars such as those under 1000 pounds and under 150 horsepower can use a high polar moment of inertia to advantage. Cars larger than this may be better handling with a lower polar moment of inertia design. You are the judge when modifying your own car.

BRAKE SYSTEM MODIFICATIONS

For both street driving and racing an efficient braking system is essential. For good handling your brakes should be consistent, use only a moderate amount of pedal force, and be free of fade in the hardest use. Methods of brake system modification are essentially the same for street or track.

Before starting brake-system modifications give the system a complete check. Take all the wheels off and look closely at the wheel cylinders. Check for signs of leaking fluid. If there is any doubt, give the cylinders an overhaul. New seals are cheap compared to the damage caused by bad brakes. Also look at the master cylinder and overhaul or replace it if any leaks are evident. If the brake linings are worn or contaminated with oil or brake fluid replace them with new ones. Fix leaking oil seals.

It is possible to greatly improve braking consistency by making the disks and drums run true. Chattering or uneven braking is often due to rough, scored, or uneven braking surfaces. For disk brakes the disks should be ground to remove all grooves. Make sure the disks are the same thickness on both sides of the car, so heat will be equally distributed. When you install disks make sure they don't suffer from excess lateral runout. Put a dial indicator against the side of the disk and rotate the wheel. The disk should not wobble sideways more than 0.005" at the edge. If it does, shim under the disk where it mounts or have it re-ground to true it up. A tiny amount of runout is desired to keep the pad from dragging against the disk all the time. About 0.002" runout is ideal.

Check lateral runout of brake disks with a dial indicator as shown. Excess disk runout will cause added pedal travel, because the pads are pushed farther away from their full contact position. Ideally the lateral runout should be about 0.002".

Steel-braided brake hoses are used to make a firmer pedal and to prevent hose damage. These hoses are available for virtually any car, and are good insurance on either road or track. They are a must for hard use such as off-road racing.

Drum brakes should be turned to make them smooth and round. Make sure the diameters are equalized. This is very important because different drum diameters make the car swerve to the side when you hit the brakes hard. You can test the brakes for roundness on the car if you remove the backing plates. Then install the drums and use a dial indicator on the inside of the drum. If they are not round and concentric, shim or replace as required.

Brake drums and disks are heavy and affect the balance of the wheels. Make sure the dynamic balance of the brake rotating parts is checked if you are trying to get perfect wheel balance. If you balance the wheels on the car this takes care of the problem, but then the wheels cannot be swapped without changing the overall balance. It is best to have the brakes balanced separately if you can afford the extra money.

After installing rebuilt brakes complete the system check by inspecting the brake lines and hoses. If there is any doubt about the condition of the hoses replace them. Pay particular attention to the possibility of a brake hose rubbing on the chassis. The results of a blown hose are disastrous. Leaking brake line fittings should be fixed. If the lines are damaged or corroded replace them. Steel brake lines are available in various lengths with flared ends. It is bad practice to attempt to make your own flares. Brakes require a special reinforced flare, which the standard cheap flaring tool will not do. If you need to bend brake lines, use a proper bending tool. A crimped line should never be used.

When the brakes are used hard the hydraulic pressure causes brake lines to expand. Most of the expansion occurs in the rubber brake hoses at the wheels. The expansion creates a softer pedal which requires more stroke. To increase pedal firmness you can install steel-braided flex hoses in place of the rubber hoses. These are high-pressure, high-temperature aircraft-type hydraulic hoses, lined with teflon and reinforced with stainless braid These hoses are available from Earl's Supply Co., Lawndale, California, and other suppliers. Earl's offers kits for almost any car, including those with metric fittings. These metal-braided lines are more resistant to damage from rocks and rubbing on chassis components. They are a must for racing and good insurance for street use.

Braking can be improved by bleeding the fluid and replacing it with new high-performance brake fluid. When brake fluid is old it deteriorates from heat and by picking up moisture from the air. Old fluid has a reduced boiling point. Boiling brake fluid causes a mushy pedal, which can happen suddenly in hard use. Boiling fluid is the major cause of brake problems with disk brakes, and can be a problem with drums too.

If the fluid is old, drain the system completely and fill with new fluid from a previously unopened can. Don't store brake fluid unless it is tightly sealed to keep out moisture. Use high-performance brake fluid with a boiling point of at least 550°F. Check with your car manufacturer's recommendations for brands of fluid. Some fluids can cause damage to the rubber parts. If you cannot find out what is recommended, use American fluid in an American car, English fluid in an English

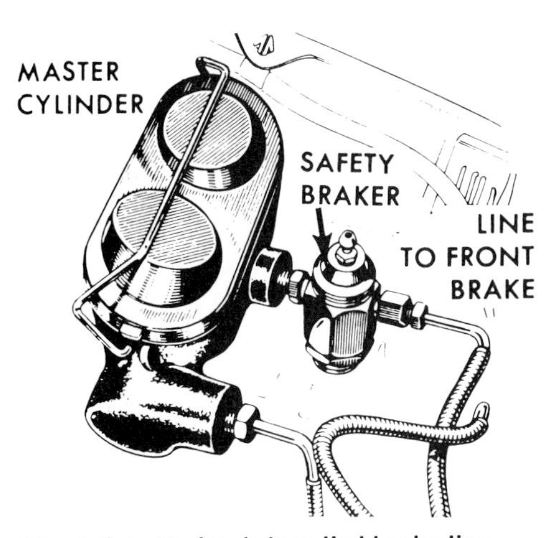

The Safety Braker is installed in the line to the front brakes as shown. This works on cars with any type of braking system, power-assisted or not. On cars with a single master cylinder the Safety Braker installs in the single line from the master cylinder. For the ultimate set-up with a dual braking system, use a Safety Braker in the line to the rear brakes in addition to the one serving the front brakes. The device is more helpful in the front system than the rear, but every little bit helps. This device reduces the tendency for one wheel to lock up prematurely in a hard stop.

car and so forth. Never mix brands of fluid.

A good fluid to use is silicone-based fluid. Silicone fluid has the advantage of not absorbing moisture, thus maintaining its boiling point of over 750°F. Also, it is compatible with all brake-system seals. When converting to silicone fluid, you have to flush the entire brake system with alcohol to purge the old fluid and moisture. To do a 100% job, the cylinders should be rebuilt at this time to avoid the chance of having old fluid trapped in the system.

If you are racing, bleed the brakes after each race and replace the fluid each racing season. This removes any contaminated fluid from the brake system. If you are driving on the street, bleed the brakes once a year for top performance.

If your brakes are too sensitive and prone to lock there is a bolt-on gadget that may help. This is the *Safety Braker* manufactured by G & O Manufacturing Co. of San Rafael, CA. This is a little chamber in the brake system which

BASIC HYDRAULICS

There are only about three basic principles needed to understand hydraulic brakes and similar systems.

In a closed brake system, the hydraulic pressure is the same everywhere. Pressure is measured in units such as pounds per square inch (psi) which means that there is some force measured in pounds exerted outward on every square inch of the inner surface of the brake system—the master cylinder, the wheel cylinders, and all of the connecting lines and hoses which are filled with fluid.

When you apply force to the master cylinder by using the brake pedal, this force tends to move a piston in that cylinder and causes the fluid in the brake system to have a higher pressure. When pressure at one point increases, pressure everywhere increases to the same value. A simple explanation for this is the fact that fluid will always flow from a high-pressure area to a low-pressure area until the pressures are equal.

The pressure in the system is caused by the force which you apply through the brake pedal. To pick easy numbers, suppose your right foot, acting through the leverage of the brake-pedal mechanism, applies a force of 100 pounds to the piston in the master cylinder and the area of the master-cylinder piston is one square inch. The force you apply is distributed over the entire area of the piston, so you are applying 100 pounds per square inch. On the opposite side of the piston, the hydraulic fluid receives a *pressure* of 100 psi due to your right foot on the brake pedal. Every part of the hydraulic system has a pressure of 100 psi.

The wheel cylinders act "backwards" from the master cylinder. At the wheels, these cylinders convert hydraulic pressure back into mechanical force which is used to move the brake lining against the drum or disk. The amount of mechanical force which can be derived from hydraulic pressure is determined by the area of the wheel cylinder piston. If these pistons are also one square inch in area, then each is being forced outward by 100 pounds of hydraulic force. Each will exert a mechanical force of 100 pounds.

The mechanical force produced at each wheel cylinder can be changed by changing the area of the piston. If the piston area becomes 2 inches, the force will be 200 pounds, and so forth.

It appears from what has been said so far that any amount of mechanical force can be derived from a hydraulic pressure of 100 psi just by increasing the size of the piston in the wheel cylinder. You could get 1,000 pounds by using a piston of 10 square inches, without having to push a bit harder with your foot. That is true but a practical problem results from the fact that the pistons in the wheel cylinders *move.*

Because the wheel cylinder pistons move to apply the brakes, the volume of fluid contained in each wheel cylinder increases when the brakes are applied. Therefore the master cylinder not only pressurizes the hydraulic fluid, it must also act *as a pump.* The master cylinder must pump out as much fluid as the wheel cylinders require when the wheel cylinder pistons move outward to apply the brakes. This affects *pedal stroke.*

To pump more fluid from a master cylinder of fixed diameter, the pedal stroke has to be longer. There is a limit to how much you can extend your leg and toe to operate the brake pedal and apply a reasonable amount of force. There is a related limit of pedal travel which is the location of the floor board or a mechanical stop which does the same job. When the mechanical limit of pedal travel is reached, or your anatomical limit, that's all the stroke there is. It is good practice not to use all available stroke in applying the brakes so as to leave a little in reserve.

It is obvious that more movement at the wheel cylinders also causes more stroke at the pedal, which is why brake systems are set up with small travel of the brake shoes or pads between brakes fully off and fully on.

Anyway it turns out that force multiplication hydraulically by using larger pistons in the wheel cylinders has a practical limit. More force multiplication requires more pedal travel. Which is exactly the same as it used to be in the days when brakes were totally mechanical, operated by a system with leverage. In a hydraulic brake system, the ratio of piston areas is equivalent to the ratio of lever arms in a mechanical system.

reduces the pressure pulses that occur when brakes tend to lock. It makes the braking effort more consistent and easier for the driver to control. The *Safety Braker* bolts into a brake line, and on most cars it makes a noticeable improvement in driver control during braking.

There are other modifications that can be used to change the brake pedal force. For maximum control during hard braking the driver should have to push hard, but not so hard that it takes maximum effort to lock the wheels. Pedal force that is too low, such as on some cars with power brakes, is as dangerous as pedal force that is too high. On cars with too low a pedal force, rapid application of the brakes causes wheel locking and a loss of control. Maximum braking occurs just short of locking the wheels, and the driver must be able to maintain this by controlled force on the pedal.

If your pedal force is too high, the car can be converted to power brakes. Most cars these days offer power brakes as an option. You can perhaps buy all the required parts from a junk yard, overhaul them, and install them on your car. On some cars this is an easy bolt-on conversion. If your model never had power

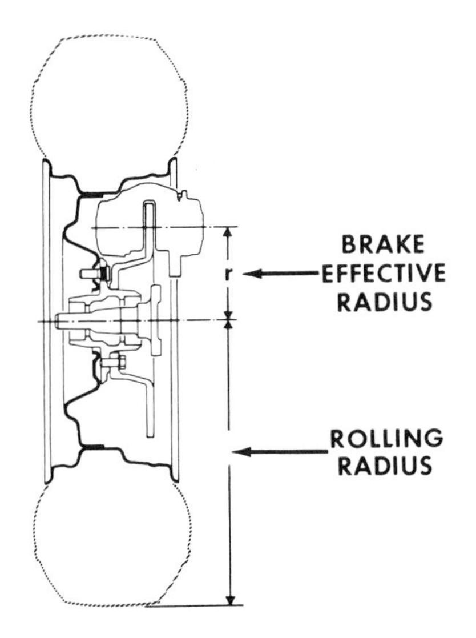

The brake effective radius is the distance from the center of the axle to the center of the brake lining contact area. In the disk brake shown, the effective radius is to the center of lining pad. On a drum brake the effective radius is the actual inside radius of the drum. Drum and disk brakes cannot be compared simply on the basis of effective radius because the drum brake usually has a self-servo action causing it to have more braking force. The tire rolling radius also affects braking.

brakes it might be possible to adapt one off another car. Here you will have to be very careful on picking the right system to get a pedal force that is what you require. It pays to get the help of a mechanic or an engineer to help calculate the pressures and forces involved. Each car is different, and the specific calculations require a complete knowledge of how the power assist unit works that you are thinking of using. A trip to the library might be of help in finding specs on a system from another car.

If you have a light car, say under 2500 pounds, you probably won't need power brakes. The pedal force can be changed by putting a different size master cylinder on the car. Calculations on master cylinders are easy. You must know the diameter of the old master cylinder bore and the new one.

$$\text{New pedal Force} = \text{Old pedal force} \times \frac{(\text{New master cylinder bore diameter})^2}{(\text{Old master cylinder bore diameter})^2}$$

Let's take an example. Assume your car requires 100 pounds of pedal force to lock the wheels, and you want to reduce it for easier braking. The old master cylinder has a bore diameter of 1.00". Let's assume you want to install a new master cylinder with a 0.75" diameter bore.

$$\text{New pedal force} = 100 \times \frac{(0.75)^2}{(1.00)^2} = 56 \text{ pounds}$$

As you make the master cylinder smaller, the pedal force gets smaller for a given amount of braking force at the wheels.

When you install a smaller master cylinder the pedal stroke will increase. Check this out before you do a modification like this. If the pedal stroke becomes so long your brake pedal nearly hits the floor, you cannot use a smaller master cylinder to reduce the pedal force. Pedal stroke can be calculated as follows:

$$\text{New pedal stroke} = \text{Old pedal stroke} \times \frac{(\text{Old master cylinder bore diameter})^2}{(\text{New master cylinder bore diameter})^2}$$

Brake diameters also affect pedal force. Using a bigger diameter brake drum or disk will reduce pedal force. The critical dimension is the distance from the center of the axle to the center of the brake lining. This is the *brake effective radius.* On a drum brake this is merely the inside radius of the brake drum. On a disk brake it is the distance from the center of the axle to the center of the brake lining pad. Unlike changing master-cylinder diameter, brake effective radius does not change the pedal stroke. *Putting bigger brakes on a car is the most effective way of reducing pedal force.* In addition, the bigger brakes will be more resistant to fade, because they don't get as hot. Bigger brakes have more metal in them, and the more metal, the lower the brake temperature.

If you put bigger brakes on your car, here is the calculation of the new pedal force. This assumes that the hydraulic cylinders are the same size, and only the effective radius of the brake changes.

$$\text{New pedal force} = \text{Old pedal force} \times \frac{\text{Old brake effective radius}}{\text{New brake effective radius}}$$

The rolling radius of the tire has the opposite effect from the brake effective radius. If you increase the tire rolling radius you increase the brake pedal force. This can be a noticeable effect if you put much taller tires on a car, say for drag racing. If you put taller tires on the rear only, it will reduce the braking force on the rear brakes only, and the front will remain the same. This upsets brake balance as well as increasing the pedal force.

To calculate the new pedal force if all four tires are changed equally in rolling radius, use the following formula:

$$\text{New pedal force} = \text{Old pedal force} \times \frac{\text{New tire rolling radius}}{\text{Old tire rolling radius}}$$

You can reduce pedal force with softer brake linings. However, this may not be a good idea if you drive hard. For fast driving and hard braking you need hard linings. A soft lining on drum brakes tends to fade more under severe use. On either disks or drums, a soft lining will wear out faster.

For severe use there are metallic linings, such as the *Velvetouch* brand marketed by Lakewood Industries of Cleveland, Ohio. *Velvetouch* linings are outstanding for drum brakes, because they are virtually fade free. Conventional organic linings suffer from fade caused by gas emitted from the hot linings which acts as a lubricant between lining and drum. If this fade is really severe the brakes feel like there is oil on the linings. The pedal force goes up so high that it is impossible to lock the wheels. The *Velvetouch* metallic linings virtually eliminate this problem. These linings are also available for disk brakes.

Fade is caused by excess heat. If brakes can be made to run cooler fade will disappear. Getting more cooling air to the brake is a simple and easy way to reduce fade. On drum brakes the backing plates can be drilled with cooling holes. Scoops or ducts can be added to vent the

Front brake scoops are built into the chin spoiler on this Mustang Cobra. This set-up does not have ducts running to the brakes, but enough cooling air reaches the brakes for street use. Ducts would be easy to install, and would direct more air where it is needed. A spoiler without brake-cooling provisions will hurt brake performance as air flow under the car is restricted. (Tom Monroe photo.)

This racing Porsche has front brake ducts with flex hoses leading from the front spoiler to the brake calipers. The duct blowing air to the inside of the wheel, and the rounded body sucking air out of the fender opening combine to provide air flow through the holes in the wheel. These ducts are obviously too low for street use, and are not necessary on this car for ordinary street driving.

interior of the brake drum. However, these cooling holes let water in, so braking in the rain is likely to be scary until the brakes get good and hot. Vented backing plates may not be suitable for street use in the rain.

On disk brakes the cooling air should be directed at the caliper. Fade is usually caused by fluid boiling in the caliper. The duct should exit as close to the caliper as possible. The ultimate solution is to fabricate a duct and attach it to the spindle, then attach the flex hose to the duct. A spoiler on the front of the car may contain brake-vent holes, and you can connect these to heater ducts to bring cooling air to the calipers.

Cooling air for the brakes increases if it has an exit out the fender openings. On most cars this can be helped by ventilating the wheels. If your car has solid-disk wheels, switching to spoke-type wheels can help cool the brakes, but they should be aided by cooling ducts bringing fresh air into the inside of the brake. If your car has a rounded nose and front fender shape, the air flowing around the sides of the fenders creates a suction on the front fender openings. This alone may be strong enough to suck cool air through the wheels to exit air from the brakes. You can tell what is happening by observing the car driving in the rain. If you see spray flying out of the front fender openings, the body is doing a good job of extracting air.

Braking can be aided by drilling holes in the braking surfaces of disks or drums. On drum brakes this drilling lets hot gas out that normally would be causing fade. Approximately 3/16-inch holes 2 inches apart are suitable, drilled through the brake drum surface. Chamfer the holes to keep the lining from snagging on the edges.

On disks, drill across the disk through the braking surfaces. Use 3/8-inch holes spaced about 2 inches apart in a regular pattern. The holes must be chamfered to prevent damage to the lining. Cross-drilling disks gives a more consistent braking action and keeps the linings from glazing due to heat. After drilling either disks or drums, the rotating part should be balanced.

The brakes on a big-engine Corvette really need lots of cooling air for road racing. These brakes are cooled by ducts in the front bodywork, plus each wheel has a cooling fan attached to extract air through the holes in the wheel. These are Corvair engine-cooling fans, and they are very light. The added unsprung weight of the fan is hardly noticeable on a big heavy car such as this.

If your car has drum brakes, perhaps you will want to try converting to disk brakes. Disks are more consistent in action and virtually free from fade. They are affected less by water and dirt. Disk brake linings are easy to inspect for wear, and a snap to change.

Most drum brakes have a self-servo action, which reduces pedal force. The friction of the lining against the drum provides a torque about the pivot point of the shoe, which presses the lining harder against the drum. This has a power-assist effect, which allows the use of non-power brakes on big heavy cars. However, if fade occurs on this type of brake the effect of fade is multiplied by the servo action, requiring a *greater* increase in pedal force than if there was no servo action.

Disk brakes are not servo-assisted. If you push twice as hard on the pedal you get twice as much stopping force at the tires. Thus with disks the car is easier to control in a quick stop.

The lack of servo action on disk brakes means that the pedal force will be higher than drum brakes with the same effective radius. Most disk brakes have larger hydraulic cylinders at the brake to increase the force of the lining against the braking surface, but this is limited by pedal stroke. Also there still may not be adequate braking force available at a reasonable pedal force, so power boost may be required. Most disk brakes on big cars require power boost.

Because there are so many things that can affect pedal force when converting from drums to disks, it is necessary to do some careful planning. To eliminate the complex factors such as self-servo action of the drums, and power assist, use the following method:

1. Pick a disk-brake system from a car of approximately the same weight as yours. Plan on using the power-assist unit that comes on the car with the disk brakes if it has such a unit. Also plan on using the master cylinder that comes on the disk brake system.

2. Make the assumption that the disk-brake-equipped car is well designed and has a correct amount of pedal force. If you can, drive the car with the disk brakes and prove to yourself that the pedal force is OK. If the pedal force is too high or too low, plan to compensate on your car.

3. Assume that the disk brakes require a pedal force of 100 pounds on the car they came from. This is only an assumption that makes calculating easy. Then calculate the new pedal force that results from all the differences from the original system that occur on your car. The resulting pedal force will be compared to the assumed 100 pounds to see how much higher or lower it will be. This will tell you the effect on pedal force of converting your car to disk brakes.

4. Calculate the new pedal force by correcting for the pedal ratio in your car. The *pedal ratio* is the mechanical advantage or leverage of the brake pedal. It is the distance between the pedal pad and the pedal pivot, divided by the distance between the master cylinder pushrod and the pedal pivot. The new pedal force is given by:

$$\text{New pedal force} = \text{Old pedal force} \times \frac{\text{Old pedal ratio}}{\text{New pedal ratio}}$$

Notice the holes in the brake disk on this race car. Cross drilling disks aids braking in several ways. Besides making the disk slightly lighter it reduces any buildup of hot gas between the lining and the disk. Also there is a reduced tendency to glaze the brake pads. These are slight improvements but every bit helps in racing.

For your calculation based on the assumption of 100 pounds of pedal force, the formula becomes:

$$\text{New pedal force} = 100 \times \frac{\text{Pedal ratio from disk brake equipped car}}{\text{Pedal ratio from your car}}$$

This new pedal force is to be used in the next calculation.

5. Calculate the effect of the tire rolling radius. If your tires are the same rolling radius as the tires on the disk brake equipped car, there is no change in pedal force. However, if you have different size tires use the following formula:

$$\text{New pedal force} = \text{Pedal force calculated in step 4} \times \frac{\text{Your tire radius}}{\text{Tire radius on disk brake equipped car}}$$

This new pedal force is used in step 6.

6. Calculate the effect of the weight of the car. If your car is exactly the same weight as the disk-brake-equipped car there will be no change in pedal force. Otherwise use the following formula:

$$\text{New pedal force} = \text{pedal force calculated in step 5} \times \frac{\text{Your car weight}}{\text{Weight of disk brake equipped car}}$$

If the weight distribution of the car equipped with disk brakes is greatly different than your car the pedal force will be the same as calculated in this step, but the brake balance between front and rear will be wrong on your car. You will have to plan on making a balance adjustment to your car if you want to use this disk-brake system. I discuss how to modify the brake balance later in this chapter.

7. If you are using a complete disk-brake system on your car including master cylinder and power-assist unit, if any, the job of calculating the new pedal force is over. Compare the pedal-force calculated in step 6 to the original 100 pounds you assumed. If the new pedal force is between 80 and 120 pounds the disk-brake system will probably work well. If it is outside these limits you can plan on quite a difference in pedal force from the amount the car with the original disk-brake system had. If this is satisfactory you may want to use the disk-brake system anyway, but be aware of what is likely to happen after you install the disk brakes.

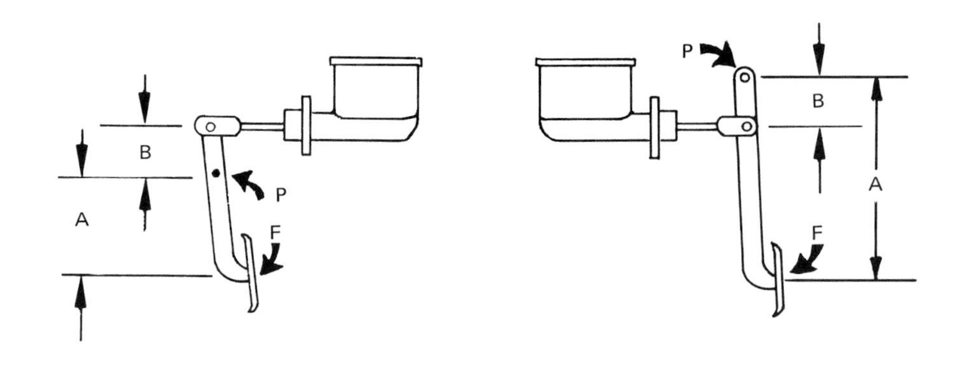

Pedal ratio is given by the following formula: Pedal Ratio = $\frac{A}{B}$

In the drawings, the pedal pivot is at point "P." The force "F" is applied to the pedal pad by the driver, and it is multiplied by the pedal ratio as a larger force on the master pushrod. Thus if the driver pushes with 50 pounds on the pedal and the pedal ratio is 4, force on the master cylinder is 200 pounds.

If the original car with disk brakes had drums on the rear you may not wish to use them. If your existing rear drums are the same effective brake radius as the ones on the rear of the disk-brake-equipped car, chances are the system will work as calculated above. However, if the size of rear drums is greatly different you can plan on brake-balance problems and some change in pedal force. If the drums on your car are smaller than the ones used on the disk-brake-equipped car, and you use the same master cylinder the disk brakes had, you will have less braking on the rear than you want.

A system with disks and drums usually uses a pressure-limiting valve for the rear drum brakes. If your drum brakes are the same size as the ones from the disk brake equipped car, then you may wish to use the pressure-limiting valve off the disk-brake system. However, you may choose instead to use the adjustable pressure-limiting valve described later in this chapter. Read on, and make your decision afterwards.

The mechanical problems of bolting on a disk-brake system are numerous. On some cars it is possible to get a disk-brake system that was optional on another model and simply bolt on the proper connections. This is the easiest method. There are kits available to convert some cars to disk brakes.

It is also possible to buy a caliper and a disk and machine special parts to mount them on your car. Along with the disk it is best to get the hub too. The bearings may have to be spaced or changed to another size to adapt it to your spindle. Be careful when modifying bearings not to make a drastic change in bearing spacing or size. Keep in mind the loads which the bearings were designed to carry. If you are increasing the bearing loads due to a heavier car, closer bearing spacing, or switching to smaller bearings, be very careful. If in doubt, consult an expert mechanic or an engineer.

Mounting the caliper is usually the biggest problem. A bracket will have to be fabricated. Look at some caliper mounts in a junk yard, and you will get an idea what will work. Use a husky bracket, at least as strong as what the disk-brake system originally used. If you use a flimsy caliper bracket you may have a disaster. If in doubt, get expert help on this critical component.

This Corvette pressure limiting valve is adjustable. It's a quick and simple way to add adjustable brake balance to a car with a dual braking system. The adjustment is by turning the large threaded shaft on the left side of the photo to change the maximum pressure setting of the valve.

Keep in mind the wheel clearance when installing disk brakes. The caliper often protrudes outward toward the wheel, and may cause problems. You may wish to switch wheels at the same time you convert, because wheel offset, clearance, or even the bolt pattern may change. Take careful measurements of the caliper position before you start work, and make sure your wheels will fit.

Disk brakes are often used on the front wheels, with drums on the rear. This gives the advantage of disk brakes on the front where the most braking force is applied. If you use a disk-and-drum system, there may be a problem. If the rear drum brakes are self-servo type, the braking force increases out of proportion to the pedal force. On the front disks there is a proportional relationship between pedal force and braking force. Thus the rear wheels will tend to lock up in a hard stop.

The solution to brake-balance problems is an adjustable pressure-limiting valve. There is one used on 65 to 68 Corvettes, Chevrolet part number 3878944. It should be installed in the rear brake line, in a position where it can be adjusted easily. This valve has a large jam nut on a threaded rod. To adjust it, you loosen the jam nut and turn the threaded rod. Turning the rod inward increases the braking force on the rear wheels, and turning the rod outward reduces the braking force on the rear wheels. This valve has the overall effect of reducing the rear braking force, so if the rear wheels never lock up even with the rod screwed all the way in, then you must increase the braking force on the rear brakes some other way. Perhaps you can install a smaller master cylinder for the rear brakes only, or use softer brake linings at the rear and harder at the front.

The Corvette pressure-limiting valve can also be used to adjust brake balance on cars which do not have a modified brake system. Just install the valve in one of the lines, and it will reduce the braking effort at that end of the car.

If you do not have a dual braking system, the valve will not work. You must first convert to a dual master cylinder for this or any other brake-balancing method to work. In addition a dual master cylinder is highly desired for safety, and is now required on all new cars.

To convert to a dual master cylinder takes some planning. The brake-pedal force has to be shared between the front and rear systems, so the *total area* of the master cylinders must be the same as the single original cylinder. If your original system used a 1-inch diameter master cylinder and you put on two 1-inch cylinders, you get only half the braking force at the wheels. Thus the pedal force is twice as high with the twin 1-inch cylinders as it was originally with the single 1-inch cylinder.

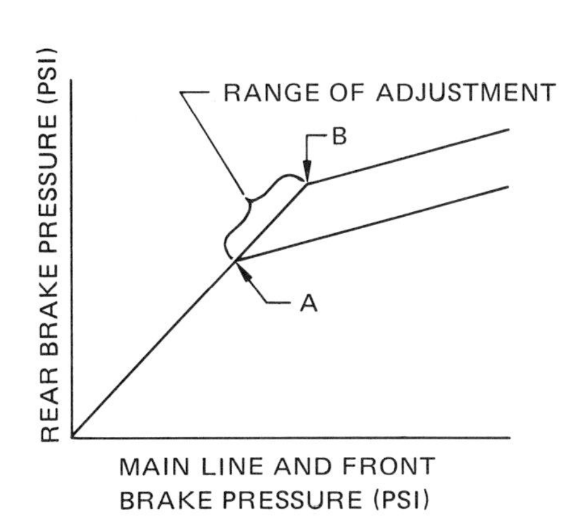

Points *A & B* describe the minimum and maximum limits of the brake proportioning valve split points. Front and rear pressures are equal until pressure reaches the split point.

The area of a master cylinder can be computed as follows:

Cylinder Area = 0.7854 x (Cylinder diameter)^2

New master cylinder area must equal old master cylinder area for equal pedal force. On some cars it is possible to bolt on a dual cylinder in place of a single one. Many cars converted to dual system when it was required by law, and some conversions are easy. On other cars, space will prevent an easy conversion. On these cars you may wish to buy a complete new pedal assembly with dual cylinders.

Neal Products of San Diego, California supplies a well-made pedal assembly with dual master cylinders and an adjustable balance bar. This is the best set-up for adjusting front-to-rear brake balance. The pressure-limiting valves mentioned earlier reduce the pressure at one end of the car to get a balanced ratio of braking effort. With a balance bar, one end of the system is reduced and the other end of the system is increased. Thus the total pedal force is not changed with a balance adjustment. Balance bars are used in racing cars,

Here is the Neal brake pedal, complete with adjustable balance bar and dual master cylinders. This unit is used primarily for specially built cars such as racing cars, dune buggies, and street rods. (Photo courtesy Neal Products)

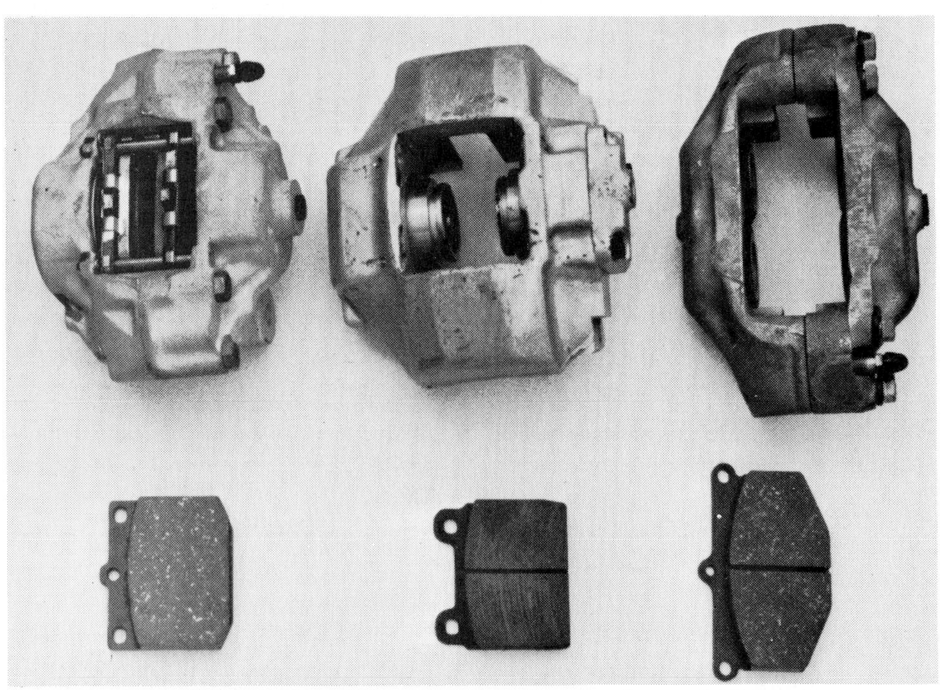

Here are three different calipers that were tried on Balboa Datsun's 510 Sedan racer. The one in the center is the lightest, because it is aluminum. Best results were obtained with the iron caliper on the left. An iron caliper sometimes gives better braking than an aluminum one because iron is stiffer than aluminum. You should test calipers to be absolutely sure which is best.

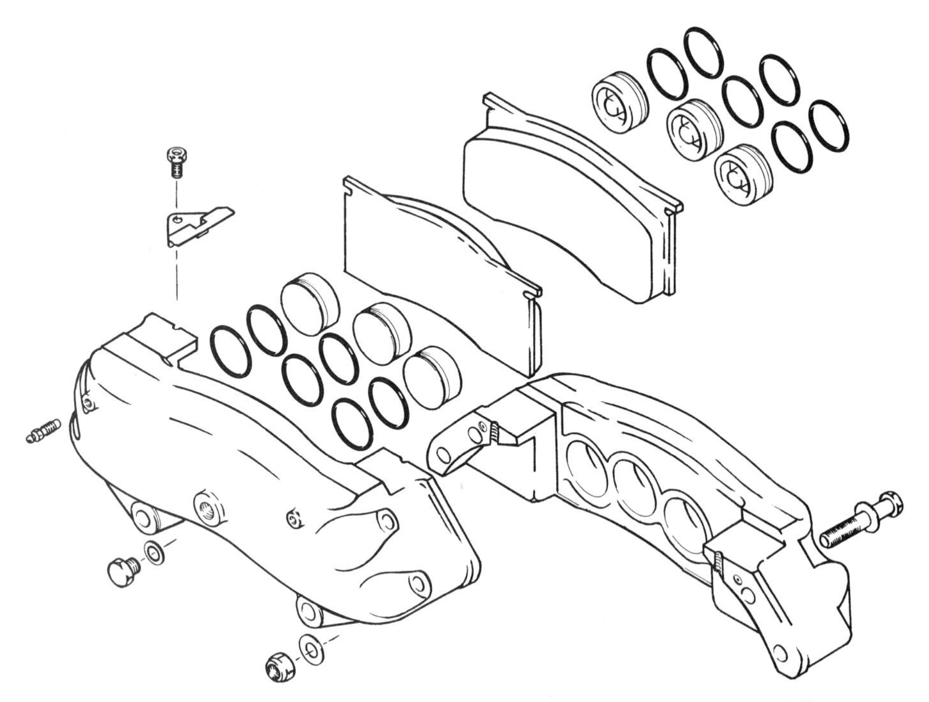

This drawing shows the Edco triple-piston caliper. The three pistons are stiffer structurally than one big piston, plus it allows a more compact caliper design. Racing calipers such as this one have pistons on both sides of the disk. Cheaper calipers used on production sedans have pistons on only one side and the caliper moves sideways when the brakes are applied. (Drawing courtesy of Edco Specialty Products)

where perfect braking is essential. Methods of adjusting and testing brake balance are discussed in Chapter 3.

If you have disk brakes, a modification that reduces unsprung weight is the installation of light-alloy calipers. This is expensive for street use, but for racing every bit of unsprung weight savings is essential. Light-alloy calipers are made of aluminum or magnesium. They save about half the weight of the original iron caliper. Light-alloy calipers are used on all racing cars and some street sports cars such as Porsche.

Light-alloy calipers are manufactured by Airheart and Edco among others. They offer catalogues with engineering information including dimensions and specifications of the calipers. The manufacturers can also help in adapting one of their calipers to an existing car. In addition to the dimensions and mounting problems, the new calipers should have approximately the same hydraulic-cylinder area as the original calipers. If the cylinder area is less, pedal force will be increased. Changing the cylinder size at the caliper works exactly opposite to changing the master-cylinder size. For reduced pedal force, calipers with multiple pistons are used.

AERODYNAMIC DEVICES

There is so much talk about automotive aerodynamics these days that the high-performance enthusiast has got to be interested in what he can do for his own car. Due to the relatively slow speeds on the street, little can be done to street handling with aerodynamics. There are some small changes that will help, usually in high-speed stability. You will not be able to make any noticeable change in cornering power at legal road speeds. On a racer a great deal can be done.

For street use, the aerodynamic devices that are on the market consist of spoilers for the nose and tail and also wings. The difference between the two is that a wing has smooth airflow on both its upper and lower surfaces, while a spoiler has one of its edges attached to the body of the car so it gets smooth airflow only on the forward surface. As the names imply, spoilers kill lift by *spoiling* the airflow. Wings generate lift, or *downforce* in race-car applications. Some of these devices are offered by car manufacturers as options, particularly on American sedans. They will probably be of no value in ordinary driving.

Of all the street-type aerodynamic devices, the nose spoiler is most useful. It always gets undisturbed air hitting it, so you know it is actually applying some force to the body of the car. It will tend to keep the nose of the car from lifting at high speeds by blocking some of the airflow normally passing under the car. Also on a front-engine car it can help engine cooling by causing a low pressure under the nose of the car, which aids airflow through the radiator. It will also reduce aerodynamic drag as lift is reduced.

The spoiler fitted to this Mustang's deck lid is designed to reduce lift and also reduce drag slightly. A lot of aerodynamic downforce could be generated with a huge spoiler, however drag would increase dramatically. (Tom Monroe photo.)

A wing this size on a heavy car will have no noticeable effect at normal highway speeds. These are sold mostly for looks as accessory items on road cars. For racing it has some small benefit at very high speeds.

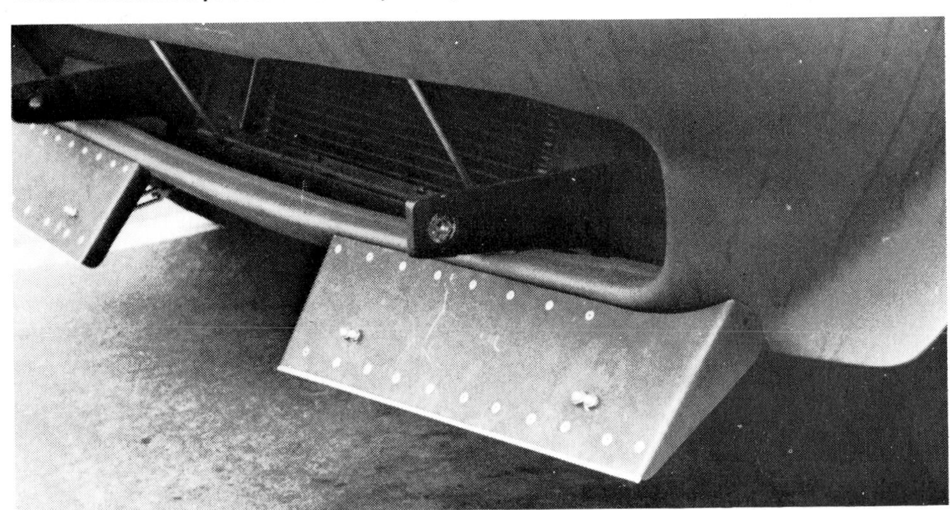

There are a number of bolt-on accessory spoilers on the market, but this owner chose to build his own from aluminum. This is not only a neat functional spoiler, but also it is adjustable from underneath. The adjustment will have little effect on this heavy car.

The effect of a front spoiler on handling is to gain a small amount of traction on the front wheels at high speeds. Usually this will result in less understeer, but the effect on handling is different on different cars. The reason is, various cars lift different amounts at high speeds. Most tend to oversteer at high speeds, but some understeer and some are neutral. The addition of a nose spoiler may help the handling or may hurt it, depending on the car's aerodynamics. Also the amount of air flow actually hitting the spoiler will vary from one car to another. All you can do is ask someone who has already tried the spoiler on the same make of car, or else make your own experiment.

A rear spoiler is designed to reduce lift on the rear end of the car by deflecting air upward. It is a desirable addition because reducing rear-end lift is an aid to stability at high speeds. On some cars, particularly very powerful cars, an efficient rear spoiler can actually cause downforce on the rear of the car thus improving rear tire adhesion with increasing speed. The thrust from the rear tires is large at high speed just to overcome drag on the body. This reduces the amount of additional rear tire adhesion available for cornering. Lift on the rear of the body makes this worse, and extreme oversteer can result.

Unfortunately it is next to impossible for a rear spoiler to help at normal highway speeds on a typical sedan or sports car. The spoiler is usually mounted low on the body behind the rear window or the windshield of an open car. The air at this point is totally separated from the body surface and is extremely turbulent. A spoiler requires reasonably smooth air flow to work effectively. Many spoilers on production cars are for looks only. If you want a rear spoiler to work on the road it must be quite large, and even then it won't do much for you at legal speeds.

Some road cars have been built with a small wing as an option. This device is about the same size as the typical rear spoiler, and it is equally useless at legal highway speeds. Tests have shown that some of these street wings actually increase the rear-end lift at high speeds. Like the tail spoiler, a rear wing is usually in very turbulent air. Even if properly designed it won't work in that turbulent air flow. Street wings are mostly for looks.

There are certain other aerodynamic modifications you can make for street use. If you lower the front ride height, the car will have less tendency to lift the nose at high speed. If ride height is adjustable, you might try this as an experiment. Don't raise the rear end in an effort to help

BMW Sedan shows all-out aerodynamic modifications for road racing. Notice the forward-projecting chin spoiler almost touching the ground and the rear wing placed high where it will get "clean" air flowing over it. The slat located at the rear of the roof helps smooth the normally turbulent air behind the rear window so the rear wing will be even more effective. This car was driven by David Hobbs in IMSA road racing, not at the Bonneville Salt Flats where the downforce/drag included by these devices would be a hinderence, not a help. (Tom Monroe photo.)

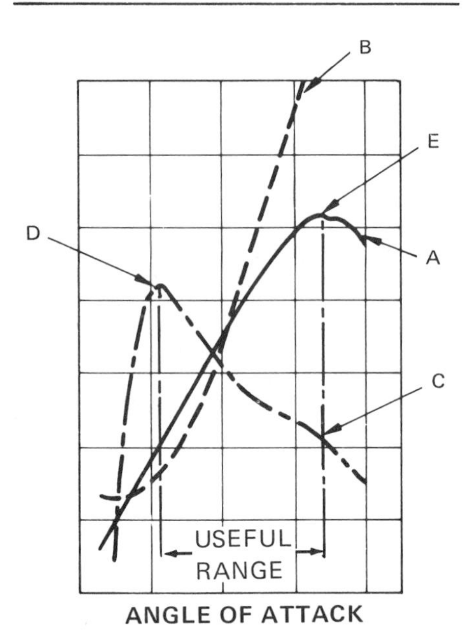

Figure 48/This graph shows how wing downforce "A" and drag "B" vary with angle of attack. Notice there is also a curve "C" showing how downforce-to-drag ratio varies. This curve is a measure of wing efficiency. Most racing cars operate their wings somewhere between maximum wing efficiency "D" and maximum downforce "E". Do not ever go outside this range—it will hurt overall performance and handling.

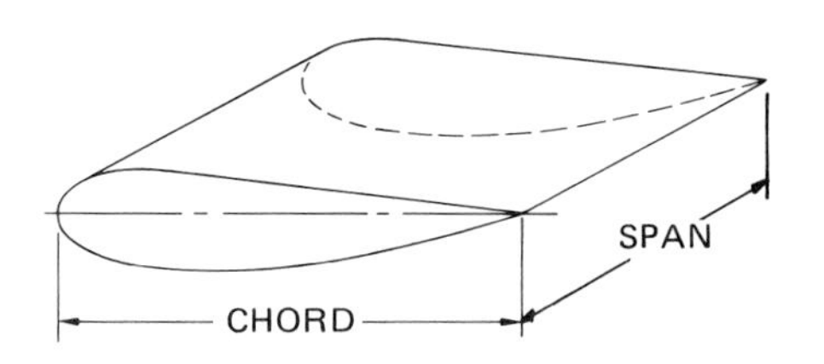

$$\text{ASPECT RATIO} = \frac{\text{SPAN}}{\text{CHORD}}$$

WING AREA = SPAN x CHORD

Here are some of the terms used to describe size of a wing. The larger the aspect ratio the greater the wing efficiency. The larger the wing area the greater the downforce and drag. Wing efficiency is the ratio of downforce to drag.

the aerodynamics. The increased CG height will cancel out any gain in traction you get from aerodynamics at legal road speeds.

Aerodynamics really are important in racing, due to the much higher speeds. The forces due to air flow are zero with the car at rest, and increase with the square of the car's speed. Thus if you double the speed you get four times the aerodynamic force. Aerodynamic forces on a racing stock car at 180 MPH are 9 times as great as the forces on the stock sedan traveling at 60 MPH on the road. This is why a small spoiler on a stock car is useful, whereas the same device on the street would have no noticeable effect.

For racing, the most useful aerodynamic device is a wing. It is the most efficient means of producing downforce, if it is allowed by the racing rules. The *efficiency* of a wing is the ratio of its downforce to drag. A wing has various amounts of downforce depending on the *angle of attack*—tilt of the wing—as shown in Figure 48. Methods for adjusting downforce are discussed in Chapter 3. Drag also increases with angle of attack, and each wing has a certain angle of attack where the downforce-to-drag ratio is a maximum. This is the most efficient angle of attack for that wing. A wing has some drag even at zero downforce, just because it is an object in the airstream. Also if the angle of attack is too large, the wing stalls and the downforce goes virtually to zero. There is a huge amount of drag on a stalled wing, so don't get carried away with the angle of attack.

The size of the wing determines the amount of downforce it produces. The width is called the *span,* and the length is *chord.* If you divide span by chord you get the *aspect ratio,* a measure of the wing's efficiency. A high aspect ratio gives the most efficient wing, as you can see by observing the design of a glider. Span multiplied by chord is the *area* of the wing, and this area determines the total downforce. The larger the area the larger the downforce. Thus the wing should always have the maximum possible span, as limited by the racing rules, and the appropriate chord to give the wing area required.

To compute the amount of downforce on the wing you should have some data from wind tunnel or full-scale testing. The area of the wing, aspect ratio, cross sectional *airfoil* shape, car speed, and wing angle of attack are all important in determining the actual downforce and drag.

The manufacturer of a wing should be able to give you the figures. If you are designing your own wing, start with a book on aircraft wing design for shapes and force information. Wing design is a subject in itself, and another book could be written on it.

Once the wings are decided on for the front and rear of the car, the forces on the tires can be computed. Wings should be balanced so as not to overload either end of the car at high speeds. The final adjustment of wing angles can be done at the track to fine-tune the chassis, as described in Chapter 3. For design purposes keep the ratio of the wing areas such that the front and rear downforces on the tires are roughly in proportion to the static weight on the tires.

Don't forget that the location of the wing on the body is important in determining the force applied to the tires. A wing ahead of the front wheels has a leverage on the front tires and actually reduces the force on the rear tires. Thus this must be balanced by a suitable rear wing. Mount the rear wing as high as possible to pick up clean flowing air.

The air hitting the rear wing is usually turbulent due to all the objects in front of it. Thus for maximum rear-wing efficiency, all items such as the driver's head and engine air intakes should be streamlined with care. A long tapering fairing behind the objects will avoid much of the turbulence that spoils the downforce of a rear wing.

D-Sports racer with an add-on wing. This wing can be adjusted to vary the amount of downforce on the front tires. The wing's induced drag could be reduced by increasing its aspect ratio by reducing its chord and increasing its span so the wing would mount between the inside edges of the fenders.

This wing is mounted as high and as far back as the rules permit so it will catch as much clean-flowing air as possible. Notice the end plates which prevent air from spilling over the ends of the wing, effectively increasing the wing's span.

Because there is high pressure on top of a wing and a low pressure under it, air tends to spill over the tips of the wing. This *tip loss* reduces wing efficiency and should be minimized. End fences are used for this purpose and can be combined into wing mounts or front tire fairings. Fences should be used on race-car wings because of their limited span. They are not often used on airplanes because airplane wings are not limited by race-car rules.

The wing adjustments should be designed so the angle of attack can be varied for chassis tuning. The designer of the wing should know the minimum drag position, the position of highest wing efficiency, and the maximum downforce position. Make sure your adjustment device covers this entire range of angles. Do not allow positions outside of this range, because they will be very bad for performance. If the angle of attack is too low the wing could have lift rather than downforce. If the angle is too high the wing is stalled, with very

Saudia Williams Formula-One car has a streamlined fairing behind the driver's head to reduce drag, but more importantly, to provide clean air flow to the rear wing. Body sides are as low as possible to provide clean air flow under the wing. The body nose is narrow so the front wings will have more span and the radiator/ground-effects venturi opening is wider. (Photo courtesy of the Goodyear Tire and Rubber Company.)

The rear wing on this McLaren is mounted on supports which also serve as end fences. In this way the wing supporting the structure adds little or no extra drag to the design. Only the slender angle-of-attack adjusting links are in the air stream under the wing.

The rear wing on this Porsche racing car is mounted on steel tubes bolted to the back of the gearbox. This looks rather frail, but it works on this car. Usually it is best to go extra stiff on wing mounting brackets to avoid destructive vibration.

high drag and a large reduction in downforce.

The forces on a wing can be quite large, particularly if the wing has a large area and the speeds are high. On the fastest racing cars the wing forces exceed 1000 pounds. Mount the wing so the forces are carried directly into the frame structure of the car. Try to avoid mounting the wing on the body, as bodies have a tendency to fail under aerodynamic loads. To avoid cutting holes in the body, the wing can be mounted on flat plates on the upper surface of the body. These plates then bolt or pin through the body to rigid plates under the body, which are braced down into the frame structure. The bolts or pins are removed to open the body, and the wing remains attached to the top of the body.

In building wing mounts don't ignore the drag loads—they can be large too. Also vibration can cause wing mounts to fail. A wing mounted on flexible brackets can have a vibration caused by the aerodynamic forces. This is known as *flutter,* and it can quickly tear a wing structure to pieces. A stiff mount is the prevention.

Adding a nose spoiler for racing is a common modification. This is often done in conjunction with a rear wing or tail spoiler. In addition, most cars are set with the front ride height lower than the rear ride height. All these items contribute to reducing lift or increasing downforce.

A nose spoiler can be easily made of sheet metal. It should come as close to the road surface as possible. Usually the shape is less important than the distance between the spoiler and the road. On some cars, the addition of a nose spoiler will increase the top speed by reducing the drag caused by the air passing under the car. On some cars the nose spoiler increases drag. In almost every case, the nose spoiler helps high-speed stability and handling, so they are almost universal in racing.

As an alternate to front wings on open-wheel racers the full-width nose has been tried. These are often added to an existing car when a rear wing is added. A full-width nose has a similar effect to a nose spoiler on a sedan or sports car, plus

The full-width nose has been used on some open-wheel cars to streamline the front suspension and to eliminate interference drag. Dimensions are often restricted by the rules, so the designer must study the various possibilities for streamlining. A well-designed front fairing will provide a significant amount of downforce to balance a rear wing.

A front-tire fairing reduces interference drag between tire and body. In addition this fairing acts as an end fence for the wing, improving its efficiency. This is a good method of drag reduction on open-wheel cars.

it has a streamlining effect on the front tires and suspension. The drag caused by exposed front tires is a major part of the drag of an open-wheel car, so this approach has a double benefit.

Another add-on device on open-wheel racers is fairings in front of the front tires. These have a drag-reducing effect, and can also serve as end fences for nose wings. The fairings deflect the air up and outwards to prevent it from entering the space between the front tire and the body. Without the fairing the air tends to jam up in the opening causing a large amount of *interference drag.* Widening the front track reduces this interference drag, but not as efficiently as a proper fairing.

MAKING THINGS ADJUSTABLE

Chassis adjustments can make your car handle better, but if they are not easily adjustable you cannot take advantage of chassis tuning. Making non-adjustable parts into easily adjustable ones are modifications that can be useful for both street and racing.

Adjustable bump steer really helps. As previously mentioned bump steer on the front suspension of most cars is adjusted by changing the height of one end of the steering tie rods. This applies to cars with conventional double A-arm independent front suspension or MacPherson strut front suspension. The tie rod can be raised or lowered at the outward end by replacing the tie rod ball joint with a large spherical rod end. This requires a bolt through the steering-arm hole rather than the ball joint tapered stud. The steering-arm hole is opened up slightly to accept a bolt shank and the spherical rod end is bolted to the steering arm with shim washers between the arm and the ball. These shims are changed in thickness to adjust bump steer.

This modification is not well suited to street use unless you are prepared to go to some trouble. The rod end is not built to hold as much lubricant as the tie rod ball joint, so grease jobs should be more frequent. Also the rod end may have limited angular movement, so bump and droop travel of the wheel may have to be limited to avoid binding the rod end and subsequent breakage.

An easier method of making adjustable bump steer can be used on cars with rack-and-pinion steering. If the rack brackets are horizontal, shims can be used under the rack to adjust bump steer. If the rack has to be lowered, special thinner rack-support blocks can be built to allow the rack to be lowered. Shims can be used for fine adjustment.

Many cars require a method for easily adjusting camber. There are a number of methods for doing this depending on the exact design of the car. Some devices that have been used are shown in the accompanying photos. Some method of adjusting camber is essential for proper chassis tuning on any car.

Most cars do not have an easy method of adjusting ride height. A number of rather difficult methods have been used, such as shimming the springs or changing the arch of leaf springs. All of these are very time consuming, so some quick method of adjustment is highly desirable. For leaf-spring cars, Air Lift helper springs or air-adjustable shocks are a good method of quickly adjusting ride height. Another method is rear shackles with several holes in them for quick but large adjustments.

For coil springs, ride-height adjustment can be made with a threaded mount for the spring. Stock cars use this type of device on the front springs to aid in setting up quickly for oval tracks. This consists of a large bolt threaded through a nut welded on the frame directly above the spring. The bolt is attached to a spring support plate which presses against the top of the coil spring. Turning the bolt down raises the ride height. These devices can be purchased from stock-car chassis suppliers and adapted to your car.

On cars with coil springs mounted directly on the shocks, such as on many MacPherson strut suspensions, the answer is a threaded spring mount on the shock. An example is shown in the accompanying photo of a racing Datsun. This could come in handy if you are a weekend racer. The ride height can be lowered for competition and brought back to practical street clearance after the race.

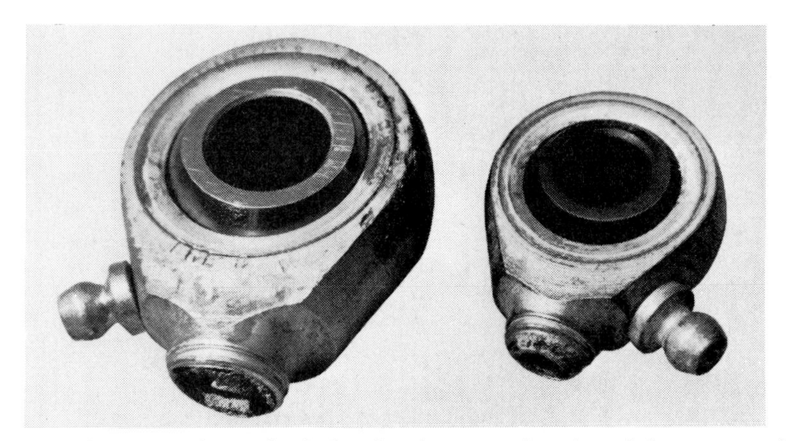

Here's what happens when spherical rod ends on steering tie rods bottom out before the suspension hits the stops. The broken rod end can be very dangerous because you lose steering. When using rod ends to make things adjustable, be sure to check for binding at both extremes of suspension travel, including all positions of steering movement.

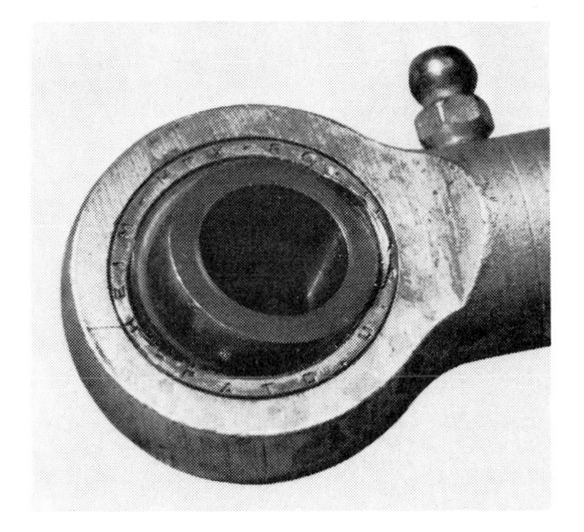

This rod end shows signs of bottoming out against the bolt through the ball. Notice the dents in the edge of the bearing race. This was taken off the steering linkage of an old racing car. The problem was cured by reducing the suspension travel with rubber bumpers placed over the shock absorber shaft.

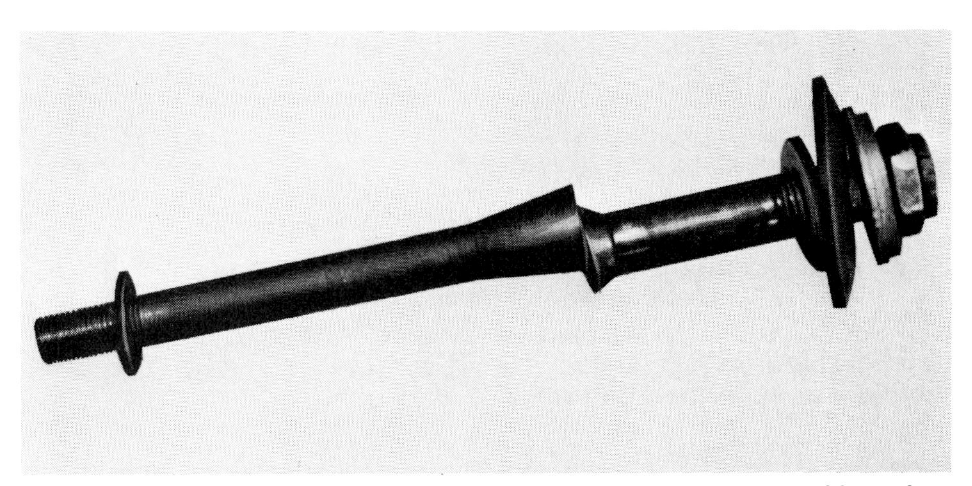

The lower front A-arm bolt is replaced with this special eccentric bolt to enable camber to be adjusted easily. Loosening and rotating the bolt changes camber, but it also changes the vertical location of the A-arm pivot slightly. This vertical change is usually not important compared to the change due to camber adjustment.

STEERING MODIFICATIONS

On most passenger cars the standard steering is too slow and vague for good control. Making the steering faster is an aid to good handling, as is stiffening the steering linkage.

There are several ways to make the steering faster. Fast steering implies that there will be fewer turns of the wheel from lock to lock. Faster steering will also create higher steering effort for the driver. On some cars this will be a problem in parking. Therefore manual steering systems on heavy cars are usually quite slow and cannot be made much faster without requiring a lot of muscle to turn. Power steering is required equipment on these cars if fast steering is needed.

On many American cars different steering gears can be installed in the same steering box. A friendly parts person can show you what is available. Putting in lower numerical ratio steering gears is the best way to make the steering faster. All aspects of steering geometry are unchanged by this modification.

If different steering gears are desired for rack-and-pinion steering systems you must change the entire rack. One example of this modification is the special rack assemblies manufactured by Mitchell Engineering for racing cars and street use. Their rack for a Datsun provides stiffer mounts as well as a faster steering ratio. It is an excellent way to improve the reaction of the car to steering.

For some cars there are bolt-on steering arms or idler arms to give faster steering. This changes the geometry of the steering linkage, so bump steer should always be checked after installing such a device. People have made shorter steering arms or longer idler arms by cutting and welding, but this method has its problems. You should be sure to use a quality welder, preferably with heliarc. The welds must be checked for cracks by Magnaflux. After welding the parts should be stress-relieved and heat treated to their original strength.

The top mounting for this MacPherson strut has been modified with slotted holes to allow camber adjustment. The bolts are loosened, the strut moved in the slots, a measurement is taken, and the bolt clamped tight. A similar modification can be used for any car with MacPherson strut front or rear suspension.

Here is a modification for adjustable rear camber and toe-in on a Datsun 510. The brackets on the rear crossmember are slotted to allow the A-arm to be moved. One bracket controls the camber and the other controls the toe. The adjustment is prevented from slipping by welding a serrated plate to the bracket and using a serrated washer under the bolt. The serrations close together allow very fine increments of adjustment.

Another way to improve steering response is to install a smaller steering wheel. This requires less motion of your hands when turning, and gives faster steering. Many good steering wheels are on the market for reasonable prices. When you buy a steering wheel be sure you are satisfied with the size and feel of the grip. This makes quite a difference on long trips.

This MacPherson strut has been modified to include an externally threaded sleeve on the outside of the strut. The spring mounting pad is welded to the tube which has internal threads. Rotating the collar with a spanner wrench changes the ride height. This is a very expensive modification due to all the machined parts required.

This is a custom-built rack and pinion steering assembly built by Mitchell Engineering. The rack is very sturdy with adjustable end fittings. The housing is welded and machined from aluminum tubing. These racks are stronger and stiffer than many stock units and are available in faster steering ratios. This one has a center steering design for a racing car, but the pinion can be placed to the side for two seaters.

Oval-track racing is all left turns, usually banked, on a dirt or paved surface. The cars are front-engine designs with solid axles in the sprint and midget classes. A midget is shown here in the typical oversteering cornering attitude. Other oval-track cars include stock cars and the rear engine championship cars. (Don Larsen photo)

Cars for use on the highway or for racing on a paved road course are normally designed or modified for balanced handling features which allow them to turn well in both directions, accelerate, stop, and perhaps use aerodynamic aids to high-speed handling.

There are drivers and driving activities which emphasize one aspect of handling more than others and specialized vehicles are built to cater to such special interests. Usually some aspects of handling are improved by sacrificing performance in other ways.

If entered in a road race, a drag-racing car would likely lead the pack into the first turn but not out of it. Reverse the situation and the road car drivers will have the opportunity to inspect the backside of a lot of smoke while trying to compete on the drag strip.

The ideas and methods described in preceding chapters are used by chassis tuners for all kinds of cars, but in different degrees for different applications.

This chapter describes a few specialized vehicles and the particular handling characteristics desired for each type.

OVAL-TRACK RACING

Oval-track racing is done with stock cars, jalopies, foreign sedans, midgets, sprint cars, open-wheel race cars with all size engines, and other specialized racers. The tracks range in size from 1/4 mile or less to the giant speed ovals such as Daytona or Indianapolis Speedway. All these types of cars and races have one thing in common. The corners are all in one direction, almost always to the left. If the track has both kinds of turns the car should be set up similar to a road-racing car.

To favor left turns requires certain modifications to suspension which help when turning left and hurt when turning right. Obviously having the CG of the car off-center toward the left will help in left turns. Another technique used is *weight jacking.*

This is setting the diagonal weight distribution so the right front wheel is heavy, the left front wheel is light, and the opposite weight shift occurs on the rear wheels. In a left turn the rear weight transfer counteracts the weight jacking and the rear tires end up with nearly equal forces on them. The front tires, however, end up with weight transfer adding to the weight jacking. The result is sometimes a lifting of the left front wheel.

Wheel lifting does the car no good at all, and should be avoided by using less weight jacking or more roll stiffness in the rear suspension.

The effect of weight jacking is the same as a lot of roll stiffness in the front suspension, say with a large front anti-roll bar. However, the advantage of weight jacking is that it does not stiffen the suspension when one wheel hits a bump. It does nothing to change the spring rate of the suspension. The disadvantage of weight jacking is that the rear tires have unequal loads on the straights, and thus can develop unstable tendencies. It is common to see a powerful race car on a dirt oval track trying to go sideways on the straight. Traction is greater on the more heavily loaded left-rear wheel, and with the commonly used locked rear end, the major part of the forward thrust is on the left tire. This tends to steer the car to the right on the straight, requiring a great deal of attention from the driver. The solution is less weight jacking. It is common for little or no weight jacking to be used on high-speed ovals, where high-speed stability is vitally important.

Because oval tracks are usually lined with solid walls, a car that understeers in the turns is considered deadly. Weight jacking is usually used as an instant adjustment to cure understeer. Some cars are even set up to allow the driver to adjust weight jacking from the cockpit. It is a quick adjustment on most oval track cars. You commonly see ride height adjusters or weight jackers mounted externally on the suspension so pit crew or driver can quickly make changes. Some of these devices are shown in the accompanying drawings.

If the race car understeers in left turns more weight is added to the right rear tire. If the car oversteers the weight on the right rear tire is reduced. This weight jacking procedure only works if the left front tire stays on the track in the turns. If it lifts, the car is a three-wheeled vehicle, and weight jacking will merely change how high the wheel lifts.

The aerodynamics of an oval-track car can be unsymmetrical too. The left side of the car is sometimes given more downforce to counteract weight transfer in a turn. This obviously involves a trade-off

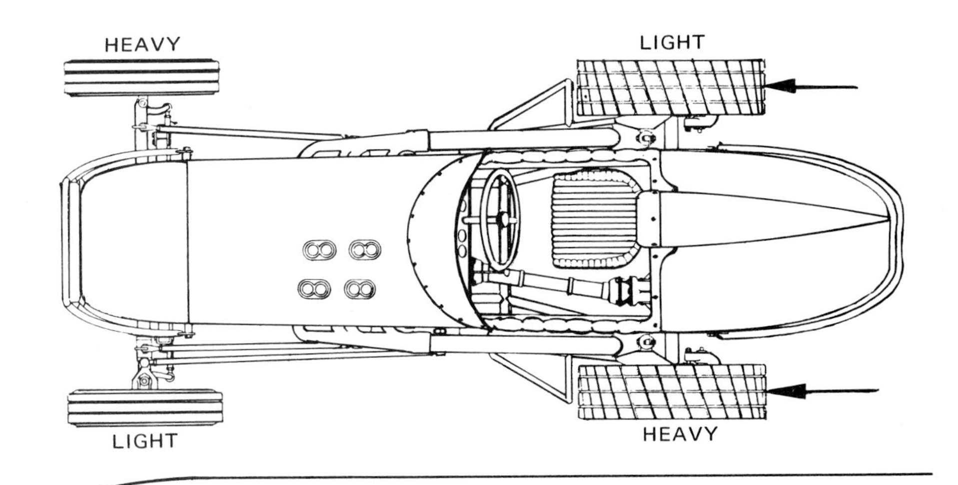

When a dirt-track car is set up properly for left turns it wants to turn right on the straight when under power. The inside-rear wheel is heavy and the outside-rear wheel light due to weight jacking. The greater thrust on the left tire tends to turn the car to the right. In a turn this unstable tendency disappears, giving maximum traction out of the turns. This car also has an offset engine, which moves the CG of the car to the left of center. "Stagger" of the rear tires is sometimes used to help in turning left. With stagger, the right-rear tire is slightly larger in diameter than the left-rear tire. This tends to eliminate tire scrub in left turns when running a locked rear end. (Drawing Courtesy CAE)

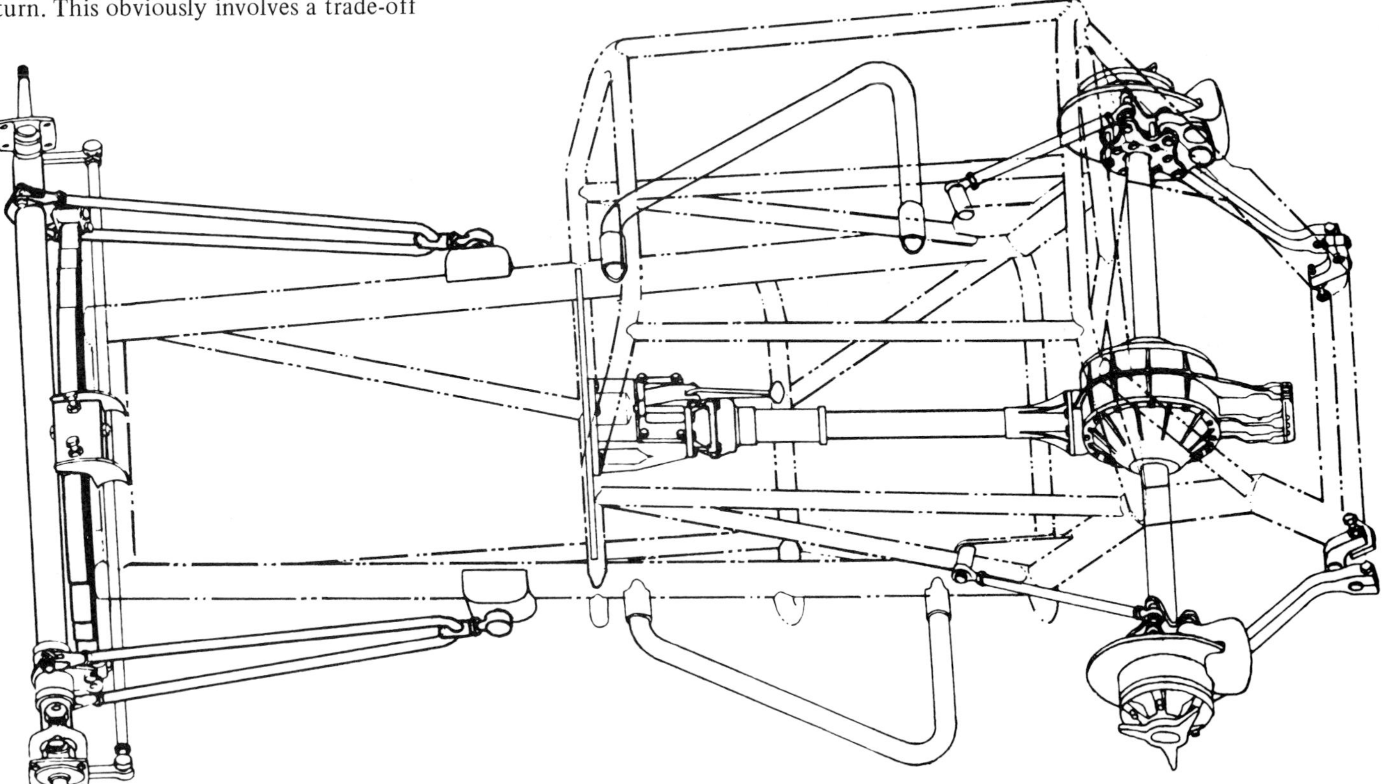

This drawing shows the CAE sprint Super Modified chassis. The weight jackers are the two bolts on the front leaf-spring mounting bracket and the torsion bar adjusting bolts at the extreme rear corners of the frame. This car has the traditional solid axle suspension front and rear, suitable for both dirt track and paved ovals. (Drawing courtesy CAE)

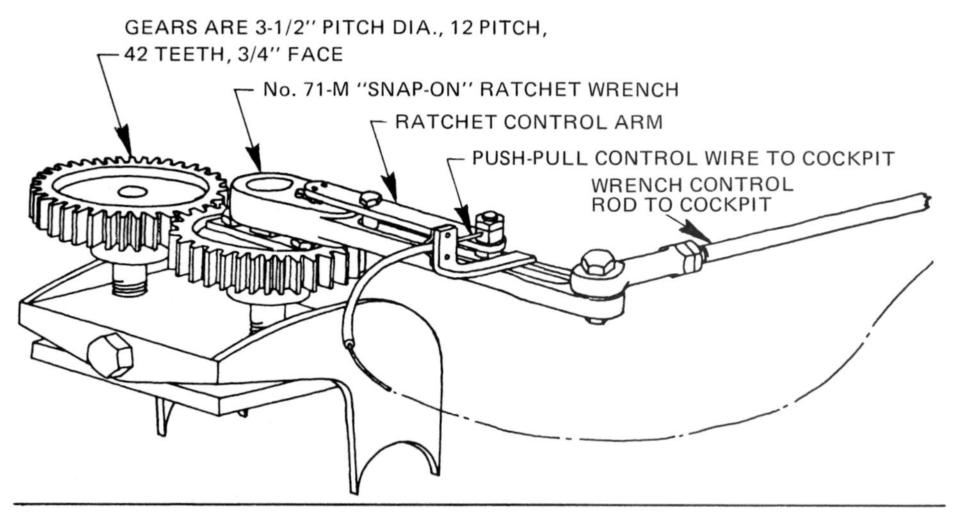

Here is CAE's cockpit-control weight jacker. The driver can operate the ratchet wrench by moving the control rod and thus tilt the leaf-spring mounting plate under the gears. The control wire is attached to the direction control on the ratchet wrench so the motion can be reversed. This method is simple and effective. (Drawing courtesy CAE)

Another way to adjust the torsion bars from the driver's seat. The anchor for the torsion bar is a bolt attached to a ratchet handle. The handle can be operated by a linkage from the cockpit. The other control allows the driver to reverse the direction of the ratchet motion. This device is used to change the weight distribution of a sprint car during a race, as track conditions change.

The rubber washer on the shaft on this shock is a suspension-travel indicator. The washer is pushed down against the shock body in the pits and its final position shows the amount of suspension travel after a run. These shocks are on a NASCAR stock car which runs on some steeply banked tracks.

Grand National, NASCAR and other circle-track cars steer left only, therefore additional caster is adjusted into the right-front suspension to assist the driver in turning left. This helps reduce fatigue during long-distance races. For example, the left-front caster may be 0° and the right front adjusted to 3°+ caster.

with high-speed stability, so care must be used with such modifications. Generally, the unsymmetrical settings that help in the turns result in less stability on the straights. These tricks are more effective on shorter tracks.

Many oval tracks have steeply banked corners. A banked corner tends to compress the suspension, similar to the effect of downforce on the car. Oval-track cars are set up with rather high static ride height to get the correct ride height while in the banked turns. The springs on an oval-track car are also quite stiff to allow less total movement of the suspension in banked corners.

One trick which has been tried on oval-track cars is to use the jacking effect due to roll-center height to raise the ride height in a corner. This tends to compensate for the reduced ride height caused by the track banking. This only works on cars with independent suspension and has to be designed into the suspension geometry from the beginning. Cars with solid axles do not experience any change in ride height due to roll-center height—but they also don't change camber with ride height.

The amount of suspension travel is very critical when running on a banked track. A simple device is used to tell how much suspension travel is left on the banking. This consists of a small rubber bushing fitted on the shaft of the shock absorbers. The pit crew pushes the bushing down against the body of the shock with the car at rest and their final position is noted after the car is run on the track. The rubber bushing is pushed up the shaft by the shock movement, but friction prevents it from falling back down against the shock body. Thus it indicates the maximum shock travel.

Stiffer springs are used on steeply-banked tracks to limit the suspension travel. Soft springs are used wherever possible on flatter tracks for better handling on bumps. If the car has independent suspension, anti-roll bars are then adjusted to limit body roll. Body roll can also be determined from the suspension-travel indicators by noting the difference between the left and right-side indicators. Note that this technique only works if the turns are all in the same direction.

Oval-track cars, particularly those used on dirt tracks, are among the last users of solid-axle suspension on both ends of the car. Although the suspension seems outdated it does have some unique advantages. It is simple and rugged, a very important characteristic on a car in frequent contact with other cars or the unyielding crash wall. Another, perhaps more important advantage, is that as long as the four tires are on the track the camber remains zero under all conditions of ride height. Thus oval-track cars subjected to high aerodynamic loads or in steeply banked turns or with greatly varying fuel weights always have the ideal tire geometry. This may outweigh some of the disadvantages of solid axles.

Also, the high roll centers inherent with solid axles make the car quite resistant to body roll. Anti-roll bars are usually not used to limit roll on cars with solid-axle suspension and weight jacking is used to proportion the weight transfer effectively between the front and rear suspensions. Also with solid front-axle suspension, there is little worry about bump steer. With the steering systems used there are no problems in obtaining perfect bump-steer characteristics.

The main disadvantage of a solid axle suspension is the effect on one tire when the other hits a bump. This can upset the balance of the car in a bumpy corner, and many oval track cars suffer from this. Also the rear axle with its unsprung ring and pinion suffers from high unsprung weight, losing traction on bumps. A solid axle gives outstanding performance compared to any independent suspension if the track is smooth, but on rough tracks it is questionable. It is safe to say that the solid-axle suspensions used on current oval-track cars are very highly developed, and it would take a really outstanding piece of engineering to replace them with another suspension type.

DRAG RACING

Drag-racing cars are designed for one purpose—to accelerate as rapidly as possible over a quarter mile. To achieve this, the suspensions are among the most specialized of any racing-car class. Most drag-racing cars have suspensions that bear little relationship to a street driven automobile.

This Midget shows the typical solid front-axle suspension used on oval track racers. These cars run on both dirt and pavement, and merely change tire and suspension settings for the different surfaces. The solid front axle may seem obsolete, but it is highly developed and works quite well on these specialized racing cars. This car uses cross torsion bar springs plus coil springs on the shocks. Notice adjusting knob on lower portion of shock.

The tires on a typical drag-racing car are narrow and light on the front and huge on the rear. The rear tires are light too, but put a very large contact patch on the ground. This car uses special magnesium wheels on the front, suitable for drag racing only.

There are two types of drag-racing cars—those with rear suspension and those without. Racing rules are written so those cars without rear suspension compete in the same classes. The tracks are so smooth in drag racing that the need for suspension is minimal, and there are advantages to be gained by mounting the driving wheels solidly to the frame. What suspension the car needs is provided by the huge soft tires.

The most important part of the drag race is the initial jump off the line. Tires play a vital part of this, and have been developed through the years just for this purpose. The front tires of drag racing cars are as narrow and light as possible for minimum resistance to acceleration. Large-diameter front tires are common in spite of the weight penalty to get an advantage at the timing lights which measure the start of a race. A larger-diameter front tire can be moving with a slight head start compared to a smaller tire without breaking the starting lights prematurely. This slight advantage is vital in a drag race, where success and failure are measured in thousandths of a second.

The rear tires on a drag-racing car are huge slicks, giving the largest possible contact patch on the ground. They are run at low pressures so the contact patch is long as well as wide. These soft tires grow in diameter as the speed increases, thus changing the overall gear ratio. The effect is like an automatic transmission which gradually upshifts as speed increases.

The rear tires have maximum grip when heated to operating temperature. The *burnout* is used to warm the tires before the run. Bleach or other special liquid traction compound is poured on the pavement in front of the rear tires. The car is accelerated hard through the liquid, causing intense wheelspin. The tires spinning on the dry pavement heat quickly and soften the rubber. In addition the rubber melted off the tires adheres to the pavement giving an ideal rubber-to-rubber contact with the track surface. At the starting line of a drag strip your shoes literally stick to the pavement after a race.

The rear tires are operated at such a low pressure that the sidewalls wrinkle

This front suspension assembly is manufactured by Strange Engineering for use on Funny Cars. It is similar to a road-racing independent suspension except for the brakes. These brakes are very small and light for a car of this speed, since most of the stopping is done with the chute. Notice the aluminum double-adjustable Koni racing shock.

Pit crew member pours liquid on the pavement in front of the rear tires prior to a burnout. Wheelspin in the burnout heats the tires so they get maximum traction during the race. The tires must remain hot to be effective.

This sedan is typical of a drag-race car that is also driven on the street. It is raised up for improved weight transfer, and drag slicks are installed for the race. Notice the large contact patch of this low-pressure tire. The car is accelerating without wheelspin, and the torque wrinkles the sidewalls of the tire.

during hard acceleration. There isn't enough pressure inside the tire to stabilize the thin sidewalls. The tires do a good job of transmitting torque from the wheel to the tread if the tire beads are mechanically fastened to the wheel rim. Sheet-metal screws through wheel and tire bead are the traditional way of keeping the tires from slipping on the rim. These are simple but not very strong for the loads involved. Stronger fasteners for tire retention should be developed.

For maximum acceleration, rear tires must not break traction during the run. Traction increases can be obtained with wider tires, but the increased weight and rolling drag tends to cancel out much of the gain. There are definite limits to how wide you can make the rear tires.

Maximum acceleration with a given set of tires is obtained with 100% of the car's weight on the rear driving wheels. Because the rules in each class control weight distribution of the car, additional weight on the driving wheels must be obtained by weight transfer during acceleration. All sorts of tricks are used to increase weight transfer on cars where the CG is near the center of the wheelbase.

As an aid to weight transfer some cars are equipped with special shocks. They are very soft on extension and stiff on compression, say a 90/10 damping ratio. The effect is to allow the nose of the car to rise up on acceleration, but resist its coming down rapidly. The effect on the rear prevents squatting. The result is a higher CG and more weight transfer. Some foolish drivers use no front shocks at all to try to get the same effect. It not only doesn't work as well, but also it is very dangerous at high speed.

Some classes allow moving the engine in the frame, and this is done to the extremes allowed. Some cars have the engine raised as well as moved back, often combined with a short wheelbase. The result is great weight transfer, but also a lack of stability at high speeds. This is particularly obvious in the AA/Altered class, where a wild ride seems to be normal. Wings are sometimes used as an aid to high-speed stability.

This modified Anglia is equipped with big drag slicks and is raised for better weight transfer. It also has a very heavy rear push bar to move its CG back slightly. This trick is not legal in the stock class.

A bit of nostalgia here, but only the "office" is from the good ole days. The engine is moved as far back and as high as the rules allow. Front-engine cars have to resort to these tricks to get weight transfer on the rear tires for maximum traction.

This drag racer has a set of bolt-on lift bars to locate the rear axle. These prevent wheel hop and tend to raise the CG slightly during acceleration. Some lift bars with forward pivots are bad for cornering as they tend to twist the rear axle housing when the car leans. The result is the same as a super-stiff rear anti-roll bar. The lift bar shown here uses a rubber bumper (arrow) to contact the frame, making it suitable for street driving if there are few bumps in the road.

In classes where sufficient weight is allowed on the rear tires by the rules, some sense of stability returns. Pro stock cars and Funny Cars use aerodynamics with spoilers and body rake to get some downforce. Aerodynamic lift is a problem in the Funny Car classes because of the combination of light weight, very high speed, and a large body area to cause high total forces. Early experiments with these cars resulted in some flights the Wright brothers would have been proud of. After the problem was realized, builders went to extreme rake angles and spoilers to give a net downforce at high speeds.

Virtually all drag-racing cars with rear suspension use a solid axle from a production sedan. Axle location is very critical, so various devices are used to aid traction. Wheel hop, rear-suspension squat, and lateral weight transfer due to driveshaft torque are all to be avoided.

Favorite devices for rear-axle location are *lift bars.* These are similar to traction bars but attached rigidly to the rear axle housing. A traction bar is pivoted at both front and rear. The difference is shown in Figure 49. Lift bars not only prevent wheel hop, but they also provide an upward force on the body to combat rear-suspension squat. They also push down on the tires, increasing traction. Lift bars are great for drag racing, but some are very poor for cornering. The ones using pivots at the forward end try to twist the rear-axle housing when the car leans, creating a huge amount of roll stiffness. A car with these lift bars is virtually locked in roll, and transmits almost 100% of the lateral weight transfer through the rear suspension. This results in rear-wheel lifting and oversteer.

To allow lift bars to be used on the street, some are designed with rubber pads on the front end to replace the forward pivots. The pads rise and hit the underside of the frame when the axle tries to twist, thus pushing up on the frame and down on the rear axle. These lift bars do not twist the axle during cornering because the rubber pad on the inside wheel is not touching the frame in a turn. However, these lift bars do restrict suspension movement over bumps, so they stiffen the ride. Some are adjustable so the suspension travel can be increased for street driving.

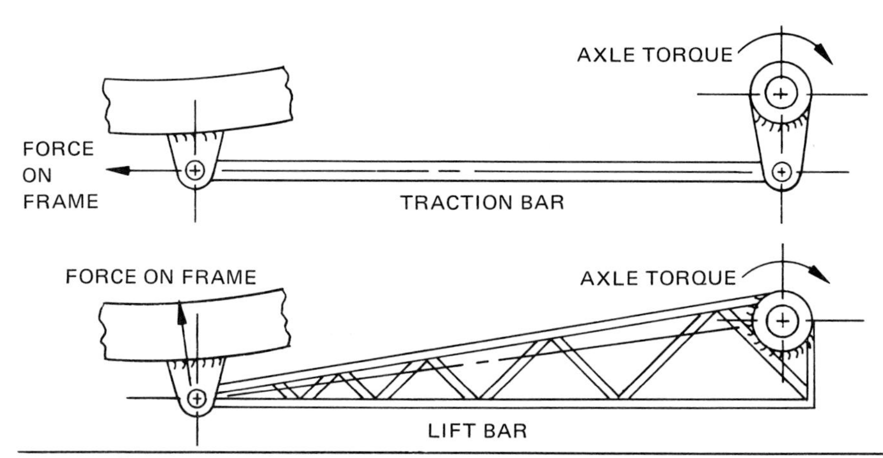

Figure 49/Traction bars and lift bars look similar but they apply totally different forces to the frame during acceleration. With the traction bar the force acts down the center of the bar and pushes forward on the frame. Lift bars create an upward force on the frame, its direction determined by the size and angle of the bar. The lift bar force acts at the pivot in a direction perpendicular to a line between the pivot and the center of the axle. The shorter the lift bar the greater the lifting force.

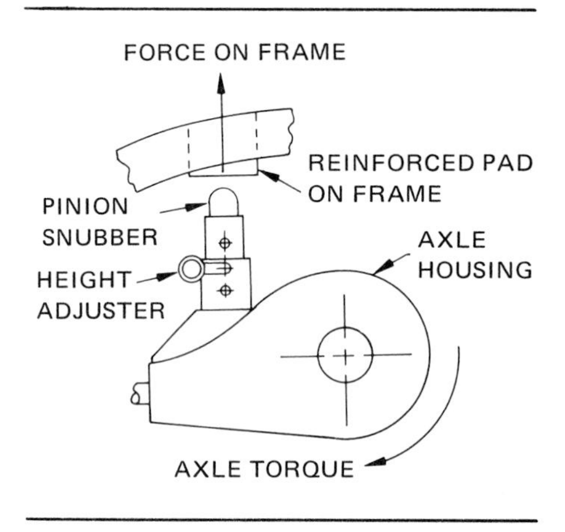

A pinion snubber mounts on the top of the differential housing and hits the frame when the axle tries to twist. This provides the same effect as lift bars. The height adjuster allows the rubber pad to be lowered for street driving, but severe bumps are still likely to cause bottoming. A pinion snubber will not affect cornering.

A *pinion snubber* is similar to lift bars. This is an extension mounted on the front of the differential housing where the pinion bearings are housed. It has a rubber pad next to the frame, and when the rear axle twists on its springs during acceleration the rubber hits the frame. The result is similar to lift bars—an upward force tending to lift the body and a downward force pressing the tires harder against the pavement. A pinion snubber is not an axle-locating device, so other methods such as traction bars must be used. A pinion snubber is not good for street use because it severely limits suspension bump travel, but some have an adjustment allowing driving on smooth roads.

As mentioned previously drive-shaft torque causes a shift of weight from the right rear tire to the left rear tire during acceleration. To get equal thrust from both tires, weight jacking is used to combat driveshaft-torque reaction. For drag racing the right rear tire is set up with more weight than the left rear—just the opposite from weight jacking on an oval-track car.

Weight jacking can be done with the springs as described earlier. Another handy method is to use air shocks or air-filled helper springs. These can be adjusted from

Here are some products for drag racing manufactured by Lakewood Industries. Lower left is a lift bar with a pivot at the forward end. The lift bar at the upper right uses a rubber pad rather than a pivot, so it will be more suitable for cornering. Upper left is a traction bar, this particular one used on a Corvette. Lower right shows a pinion snubber to fit a Chrysler Corporation axle housing. (Photos courtesy Lakewood Industries)

This drag racing rear suspension uses extra coil springs and air shocks for ride height adjustment. The air shocks can also be used to combat drive-shaft torque reaction by putting more pressure in the right-hand shock. This pre-loads the right tire, similar to weight jacking used for oval-track racing, but in the opposite way.

outside the car, and are easily changed if required. For converting back to street driving the air pressure is made equal after the race. Air shocks or air springs also can be used to raise the CG height for additional weight transfer.

In the early days of drag racing it was discovered that driveshaft-torque reaction was eliminated if the car had no rear suspension. Thus the early dragsters had the rear axle mounted solidly to the frame, and this idea is still in use today. Besides no torque reaction, the lack of rear suspension saves weight and eliminates the need for axle-locating devices.

Early cars placed the driver behind the rear axle to get as much rearward weight as possible. These cars are very long and low, but still have enough weight transfer to lift the front tires off the road. Thus the dragster achieves the ultimate condition of 100% of the car's weight on the driving wheels.

Dragsters have recently changed to rear-engine designs. These cars have several advantages over the older front-engine cars. Besides being slightly more streamlined, rear-engine cars have safety advantages. The driver of a front-engine dragster sits near the rear axle and close behind the engine. Any sort of engine or axle failure can cause serious injury to the driver from burns or flying parts. Also the rear-engine design gives the driver better visibility during his wild ride down the quarter mile.

Dragsters are the most powerful racing cars in the world. They easily exceed accelerations of 1.0 g's, once thought to be the limit to automobile acceleration. Because the class is basically unlimited, advances in tires or chassis design will continue to push speeds higher and higher.

Besides the quarter-mile races on paving, dragsters are also used on a sand surface. These sand drags are shorter and the speeds are considerably reduced, but the cars are similar. The rear tires are much larger than asphalt slicks, and are equipped with paddles to dig into the sand like the paddle wheels of an old steamship. A great amount of forward thrust can be transmitted with these big paddle tires. Sand dragsters have similar suspension problems to the pavement racers, but speeds and stability problems are reduced.

This supercharged rear engine dragster burns the tires to warm them up before the run. The rear tires grow considerably in diameter as they spin at high RPM. During the run the clutch slips rather than the tire, giving maximum grip. Near the end of the run the rear tires expand. This has the same effect as shifting to a higher gear.

This off-road racer has two shocks per wheel. The shocks have fins added to dissipate the heat generated when travelling over rough terrain at high speeds. Even with all this, there is a spare shock mounted to the frame rail (arrow). Suspensions on these cars take a real pounding.

This sand dragster is making a perfect start with the front tires just off the ground. The car uses special paddle-tread tires at the rear which dig into the sand and give excellent traction. (Photo by Shirley Depasse)

OFF-ROAD RACING

The sport of off-road racing is fast growing in popularity, and preparation of the suspension is becoming a highly advanced science. In this type of competition the suspension must be designed primarily to accept large bumps. Thus the ride height must be large, the suspension travel long, and the springs, shocks and other components must be extremely rugged. In addition, unsprung weight is really critical, as this directly affects the loads in suspension components. An excellent book on the details of preparing a VW-based car for off-road racing is H. P. Books *Baja-Prepping VW Sedans & Dune Buggies* by Bob Waar.

A great many cars use the VW suspension because of its rugged simplicity and large wheel travel. However, some cars now use specially designed suspensions offering more advantages. The large camber change of the VW rear suspension is not particularly good because the rear wheel can be tucked under the car when the car hits the ground after traveling in the air between bumps. The ideal suspension for this type of racing is some type of independent linkage that keeps the wheels nearly vertical at all times.

Shocks for off-road racing get particular abuse, and failures are quite common. Bilstein shocks with their gas-chamber feature are quite popular for off-road racing. These shocks resist fade by reducing oil foaming and the gas chamber provides some extra cushioning at the end of the stroke. You often see double shocks on each rear wheel on an off-road racing car due to the large loads and heat input to the shocks.

The wheels must be very strong yet light and able to take the occasional overstressing without ultimate failure. Here cast wheels are not the best unless they are well shielded by the bulging tire sidewalls. The monocoque aluminum wheels which I designed, now marketed by Cragar as the *Super Trick,* are outstanding in off-road racing due to their ability to bend into a pretzel without breaking. They are also as light as magnesium racing wheels and adaptable to the popular VW bolt pattern. On heavier cars steel wheels are almost the only way to go due to their ability to absorb impacts at the rim and still remain in one bent piece.

Aerodynamics and cornering power of an off-road racer is usually ignored in favor of rugged simple construction and performance over bumps. The frames are usually tubular steel, with rare attempts at other types of construction. The roll cage is usually an integral part of the frame structure to save weight and to provide maximum protection. Wrecks are not at all rare in this rugged competition.

Off-road tires are specialized in nature, usually large diameter and of very tough construction. Traction is usually not too important except on soft sand or dirt. Tire strength with reasonably light weight is the most important consideration. Off-road racers are plagued with flats, and they use inner liners or tubes with tubeless sealed wheels for double protection. There are many experimental methods being tried to improve tire reliability.

SPEED RECORD VEHICLES

Specialized cars designed strictly for top speed can range in size from tiny to immense, but they all share the same problems and design goals. The problem is to reduce air drag and rolling drag and also maintain traction and stability at high speeds.

The suspension on speed record cars is simple and straightforward, usually with zero camber for maximum high-speed stability. Cornering is not important but obviously bump steer is deadly. There have been some cars built with no suspension at all, but the combination of rock-

2.3-Liter Pinto-powered two-seater, off-road racer. Light and rugged Centerline wheels help handling over the rough terrain at high speed by reducing the unsprung weight. The monocoque wheel has become highly popular in off-road competition due to its light weight and high-impact strength. (Tom Monroe photo.)

The rugged off-road racing tire is shown next to an inner liner. The liner fits inside the tire and has its own sealed-off air supply, at a higher pressure than the outer tire. This is protection against flats and helps keep the tire on the rim even against side impacts.

Gary Gabelich shows the speed at the Bonneville Salt Flats that gained him the world's land speed record at 622 MPH. The special Goodyear racing tires are mounted on hand-made Cragar wheels. Both tires and wheels are made to an incredible degree of perfection for such speeds. The technology that goes into this equipment is similar to that required for high-performance aircraft or space vehicles. (Photo courtesy Cragar Industries)

A speed record car has specialized tires and aerodynamics. Notice the rounded tires with minimum contact patch for low rolling resistance. These are run at rock-hard pressures to keep flexing to a minimum. The car is not designed for downforce, but for the lowest possible aerodynamic drag. Even the wheels are streamlined with flush aluminum covers.

hard tires and ultra-high speeds on a dirt or salt course is not a happy one. Most cars today have working suspensions.

The tires are special, with very high air pressures and stiff casings to combat heat. The resulting traction is usually not very good, so the cars are built to have a large percentage of weight on the driving wheels. These cars cannot use large aerodynamic downforce, because this adds too much drag. The weight of the car is sometimes a bit high to get better traction, but this must be balanced against the ability to accelerate to top speed in whatever distance is available. Even though the courses are several miles long, a heavy car just doesn't have enough room to peak out. Thus the designer is faced with some critical compromises.

The aerodynamic design of the car is critical. The effort is made to have minimum drag without any lift or instability. This means that on really fast cars, wind tunnel testing is the way to go. A 400 mile per hour car can quite easily fly or crash if the aerodynamics aren't just right.

For the enthusiast interested in safer, better handling cars, the future holds a great deal of promise. Today's racing cars tend to become tomorrow's road cars, particularly in the chassis and handling department. We can look for dramatic improvements in handling on our everyday transportation cars, as more effort is put into this area by the automotive producers. Along with improved handling will come even better riding qualities, as suspension designers continue progress.

GROUND-EFFECT CARS

In this section I discuss those highly-specialized racing cars which use skirts on the body that rub on the road. These skirts are able to seal the low pressure which is produced one way or another under the car and provides a downforce on the body.

Ground-effect machines are heavily regulated, or not permitted at all in many types of racing due to the tremendous cornering speeds, complexity and costs that have resulted. Consequently, ground effects may be banned altogether. The concept is impractical for normal road use as the skirts would be damaged on rough surfaces.

Three types of ground-effect cars have been used in recent times. All work best with skirts around the bottom periphery of the body to "seal" the low pressure under the car. The three types used to produce this low pressure are:

1. Sucker Car—A fan pumps air from under the body like a vacuum cleaner.
2. Venturi Car—Air flowing through curved ducts under the sides of the car as it moves creates the low pressure.
3. Rear-Suction Car—Low pressure generated at the back of a car as it moves through the air creates low pressure under the body.

Sucker Car—The sucker car has been tried or disguised in several ways. Usually as soon as the fan is spotted for what it really is in the disguised version, it is outlawed. Notable examples of sucker cars are Jim Hall's 2J Chaparral Can Am car and the Brabham BT46B Formula-1 car.

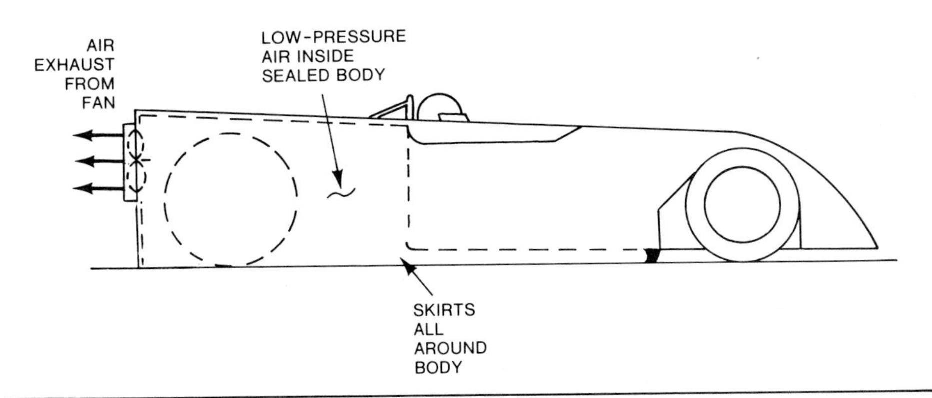

Figure 50/The sucker car used a fan or some type of air pump to reduce the pressure under the sealed body. On a drag car, the fan was replaced with suction from the engine's supercharger. The partial vacuum under the body is sealed by skirts all around the bottom edge of the body. The suction, or vacuum downforce works even with the car standing still as long as the fan is running.

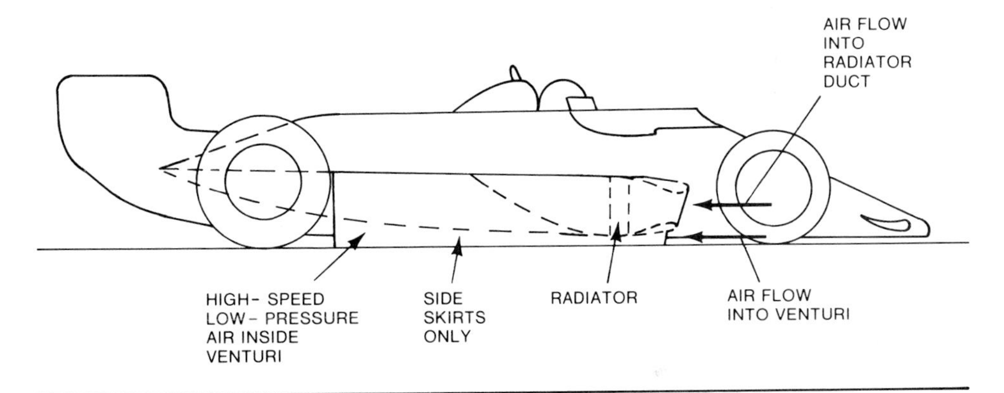

Figure 51/The venturi car relys on air flowing through a duct between the body and the pavement. Above the venturi is the radiator duct which does little to provide downforce. The venturi is larger at the rear than at the front, and is carefully shaped to prevent air-flow separation and turbulence. Only side skirts are used to seal the low pressure in the venturi.

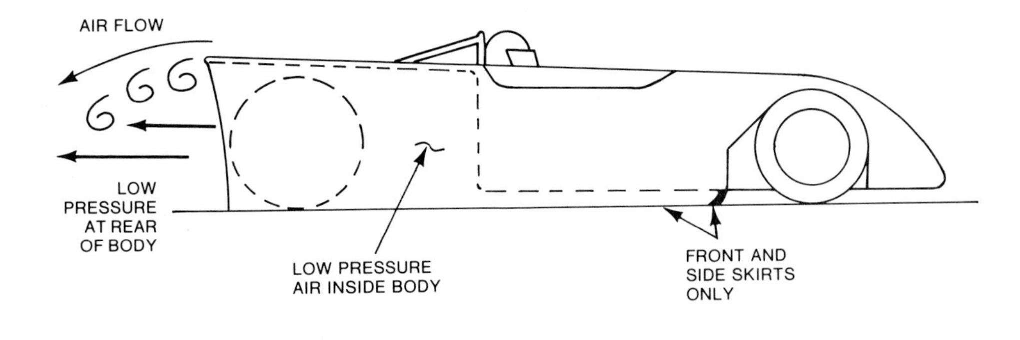

Figure 51/The rear-suction car relies on the low pressure that always exists behind a blunt object moving through the air. On this type of ground-effect car, the rear of the body is left open so the low pressure behind the car will extend forward under the body to the front skirt.

The Chaparral 2J used a small on-board engine to drive the suction fan which blew air out the back of the body. There was no attempt to disguise this design, and it was banned as a "movable aerodynamic device" after setting several track records.

The Brabham was a little sneakier. An oversized *cooling fan* driven off the back of the transaxle and blowing through a rear-mounted radiator produced the needed low pressure. Cooling fans are legal, but when this fan resulted in a lap record, the car starting from the pole position and winning its first time out, it was banned.

A sucker car with skirts that seal all around the body works at all speeds as long as the fan is blowing. If the rear of the car is open, it only works at maximum when the car is at speed. So this type of ground-effects car is a combination of types 1 and 3. See Figure 50 for an illustration of a sucker car.

Venturi car—This has been the most popular type of ground-effect car, however it is the hardest to design. It was first shown successful on the Lotus 79 Grand Prix car, and then later when other car designers and builders adapted the idea. The concept uses skirts only at the sides of the car. Consequently, air enters between the bottom of the car and the road at the front and exits straight through and out the rear. The bottom surface of the car is a smoothly curved duct, forming a venturi between the car and the road. As the air travels through this venturi—relative to the car—it speeds up, then slows down again. In speeding up, the pressure reduces, thus sucking down on the body. A venturi car is illustrated in Figure 51.

The venturi action of the ground-effect car uses the same principle as a carburetor. Consider the throat cross-section of a carburetor. It gets smaller in the venturi section. Because the throat is smaller, air speed increases, thus lowering the pressure. This low pressure sucks the fuel out of the float bowl, or in the case of a ground-effect car, sucks the car down.

Jim Hall pioneered the ground-effects car, so it's no surprise that he came up with this winning Indy 500 car. On Johnny Rutherford's Chaparral venturi car, air is ducted between the sidepod lower surfaces and the track surface to provide downforce. The side skirts seal the low-pressure underside of the car from the higher atmospheric pressure to maintain the necessary pressure differential. Front and rear wings provide additional downforce, but more importantly they provide a way to make fine aerodynamic adjustments. If you think these front wings are designed for maximum downforce, look again. They are symmetrical airfoils—the same shape on the top and bottom—for low drag. The left one is set at zero angle of attack for zero downforce whereas the right wing has some angle for additional right-front tire grip.

As more production sedans are equipped with front-wheel drive, you'll see more on the racetrack. Racers have to be aware of the differences between rear-drive and front-drive cars to work with the newer cars. Putting higher roll stiffness at the rear instead of the front, dealing with torque-steer problems and keeping the highly loaded front tires and brakes from overheating are some of the obvious problems to be faced. Drivers will also have to change style to compensate for a car which understeers under power. (Photo courtesy of the Goodyear Tire and Rubber Company.)

The problem with designing a venturi car is knowing the size and location of the resulting downforce. And once you have it, it is impossible or very difficult to change. Also, if the location moves fore-and-aft, or the forces change with movement of the car's suspension, you can have horrible problems. Some of these cars can get very unstable and even "porpoise" up and down on the suspension at high speeds. To cure this, the underbody must be reshaped, and this be very difficult. Consequently, testing models in a wind-tunnel is essential when designing a venturi car.

Rear-suction car—The rear-suction car may be the easiest to build as it can be adapted to a conventionally designed body. With this design, the skirt across the nose restricts air from flowing under the car, thus creating a low pressure behind it. The side skirts extend from the front skirt and continue to the back of the car. The rear is left open so the low pressure created behind the car will further reduce the underside pressure.

The low pressure behind a moving car is what causes a front-skirt machine to operate. When a car moves, it "punches a hole" in the air, making a high pressure in front of the car and a low pressure behind it. Because of the absence of the rear skirt, the low pressure generated at the rear extends all the way forward underneath the car to the front skirt and between the side skirts. This low pressure under the car generates the downforce. A rear-suction car is shown in Figure 52.

Skirts—The best skirts are a carbon-fiber/plastic composition, developed for use as control surfaces by the aircraft industry. Sometimes they are fitted with three or four aluminum "rub blocks" to reduce skirt wear. They are replaced frequently simply because they wear out from abrasion. The rules allowing, light springs force the skirts down against the road surface. The skirts are free to move up and down with the suspension in guides. However, if a skirt sticks in the up position, downforce is suddenly reduced. Consequently, a reliable skirt system is a must for an optimum ground-effect car, regardless of the type. On many skirt designs, an off-road excursion will tear up the skirts and ruin the handling. Thus the driver of a ground-effect machine *must* stay on the smooth pavement.

If the rules don't permit moveable skirts, skirts that flex as the car moves up and down on the suspension are used. With these "fixed" skirts, the skirt material wears away rapidly during a race, resulting in a deteriorating downforce as the skirts get "leaky."

ROAD CARS

As with progress in the racing field, improved road-car handling is led by improvements in tires. Today we can see trends towards much better tires as radials become standard equipment on more cars. With radial tires come improved traction plus better mileage too. The ride is as good as ever because the new cars are designed around these improved tires.

Also we will see lower profile tires come into all size ranges, with even lower profiles yet to be introduced. Particularly for sports cars, tires that match racing-tire profiles will one day be available to improve cornering ability further. Along with improved tire shapes we expect to see new materials to improve tire strength and reduce their total weight. Metals in increasing amounts plus new plastics will be used in the tires of the future. Also improvements in synthetic rubbers can be expected, as research in this area continues.

We should also get improvements in tires for wet and freezing weather as better tread patterns and materials are developed. Someday we can expect that hydroplaning on wet roads will be virtually eliminated by better tires, and traction on icy roads will likewise be improved. We may see inner liners or safety tubes become standard equipment so a car can run safely on a flat tire. This sort of device is already used on off-road-racing and stock-car-racing tires, and has seen limited use on family sedans. The result will be elimination of the spare tire, reducing weight and wasted space in the car.

Improved materials will surely be used in the future to reduce weight on the chassis. We will most likely see more aluminum and plastic components in our cars to reduce weight. In the future, cars will be smaller and lighter for better fuel economy, and the cars will surely handle better. Lower weight always has benefited this department.

Someday an all-plastic chassis may be mass produced to save weight. Plastic for springs and wheels is currently being worked on. In fact, a production plastic spring is now a reality, beginning with the GRP (graphite-reinforced plastic) transverse rear-suspension leaf spring of the '81 Corvette. Plastics are becoming a practical alternative to metals. Some plastics even have a strength-to-weight ratio greater than any know metal.

Aerodynamics should be an area of great improvement even on road cars. At legal speeds, air drag is still the major user of power and fuel. We can expect streamlining to become more important and smaller engines used as designers look for further ways to improve fuel economy. Stability will be a factor as always, and race-car technology will surely find its way to the street with wings, spoilers, and similar devices. We will see cars with fully streamlined undersides to reduce air drag.

As tires tend to get lower and wider, wheels will have to have wider rims.

An additional improvement in ride will result from the use of inboard brakes.

Lighter and stiffer wheels will be more popular, and will tend to follow racing designs. We can expect forged-aluminum wheels to be used on sports cars, and an increased use of cast-aluminum wheels. A stamped-steel wheel resembling monocoque racing wheels will someday be produced for street use.

Because of the future emphasis in fuel economy, tomorrow's car will be lighter. To keep the ride comfortable, reduction in unsprung weight will be necessary. There will be increased use of independent rear suspension, as well as lighter wheels, tires, and suspension components. The use of air springs or air-oil suspension will come into use to improve the ride further. Automatic-leveling systems will see increased use to control ride height of light cars with soft springs.

Front disk brakes are now found on almost all cars. Rear disk brakes will become more common as time passes. Anti-skid braking systems as currently used on aircraft are being developed for cars. They are already available on luxury models and some trucks and will eliminate the brake-balance compromise built into today's braking systems.

With the drastic drops in micro-computer prices, future technology will include their use to aid automobile handling. Imagine a system that automatically senses road conditions and compensates for the situation at the speed of light. For example, consider sway bars that change stiffness depending on the front-to-rear weight distribution of the car. Such things are possible today, and I predict you will see them on road cars of the future.

Rising fuel cost are dictating smaller cars, but interior space is still required for people. Thus efficient packaging has become a vital design criteria. Independent suspensions will become more plentiful as they'll replace solid-axle suspensions due to their more efficient use of space. Also, as car manufacturers become more space-conscious, they are turning to front-wheel-drive cars. Front-wheel drive offers a very compact power train, leaving the remainder of the car for passengers and luggage.

RACING CARS

Racing tires will continue to be improved with smaller wheel diameters expected for the future. Along with this will be increased use of inboard brakes to allow use of smaller wheels and tires. We can expect plastics to be used more in chassis and suspension components as aircraft composites become more widely available and less expensive. Metals such as titanium and beryllium will find increasing use in racing cars, with less use of the traditional mild steel. We can expect materials with higher strength-to-weight ratios to take over.

Eventually there will be a breakthough in suspension design to eliminate the compromise between camber change and suspension stiffness. I expect a practical system to be developed which will vary the suspension geometry for body roll as opposed to vertical motion. Also we may see more use of hydraulic and pneumatic devices in the suspension system to give lighter components and to provide more adjustable features. Interconnections between front and rear suspensions will gain favor and these may use a hydraulic system similar to the BLMH system. We can expect suspension adjustments to be available to the driver so he can tune the suspension during the course of a race.

There will be more progress in aerodynamics as designers become more scientific in testing. Wind-tunnel testing will become necessary for all racing cars as designs get more sophisticated. We can expect aircraft tricks such as boundary layer control to be tried and developed on the more expensive cars. Streamlining will become more important, particularly with smaller cars, and we will see some radical-looking bodies come on the scene. A great deal of work will be done in internal ducting and air flow, and we can expect drag reduction in that area in the future.

Ground-effect research will continue, particularly toward designs that do not require skirts touching the pavement. Radical changes in the internal-component layout may be required to accomplish such a ground-effect system, like placing the driver and engine on opposite sides of the car and using the center of the body as a venturi duct. We will certainly see a great deal of wind-tunnel work as aerodynamic testing is advanced to correctly simulate the road surface in the wind tunnel.

In drag racing I expect a great deal of change in aerodynamics. Speeds are already in the range of land speed-record cars. Top-fuel records exceed the top speed of every other type of racing machine. I expect to see some use of a practical ground-effect system such as using the engine's intake or exhaust system to change the pressure distribution on the body. In addition there may possibly be a change in drag-racing tire technology which will require the use of rear suspensions.

THE FUTURE FOR THE AUTOMOBILE ENTHUSIAST

In the technical area, the past is often an indicator of the future. Some innovations cast long shadows which presage popular use and such use can be forecast with reasonable certainty.

Where the development of automobiles is influenced by economic, political and social factors, any forecasting should more honestly be called *guesswork*. It's a good guess that fuel may not always be used casually for recreation or sport due to high prices, shortages, or both.

A worsening shortage of fuel will undoubtedly provoke laws and regulations governing its use and will also affect the general attitude of the public. The Sunday drive and other forms of automotive recreation may become less popular or even restricted.

It is probably obvious from this book that I am an automobile enthusiast. Performance cars have been both hobby and business for me. Even though I have no business connection with the industry at this time, my interest is unabated and the fun I get from working with cars is an important recreation.

My best guess is that people who like to build and drive racing cars, enter slaloms and rallies, or whatever branch of the sport is of interest, can expect to continue doing it for fun or profit in the foreseeable future. The costs of doing it will probably increase. If and when fuel availability problems have a major effect on the general public, automobile enthusiasts will suffer also.

On the other hand, the entire spectrum of automobiles is becoming more interesting to the chassis tuner. Even low-cost production sedans will become better cars for reasons already discussed—lighter overall weight and chassis improvements to maintain good ride and handling.

When engines become smaller for fuel economy, more people will become interested in the handling of personal vehicles and will learn to savor a well executed steering maneuver and controlled braking as much as some now enjoy an informal drag race.

The future lies ahead and we're all going to find out what it holds. I think it holds promise and interest for the person who can visualize the effects of bump steer, camber change and weight transfer in a one-g corner—and wants to do something about it.

TIRES

Armstrong Rubber Co.
500 Sargent Dr.
New Haven, CT 06507

Bandag, Inc.
Bandag Center
Muscatine, Iowa 52761

Bridgestone Tire Co. of America, Inc.
2160 W. 190th St.
Torrance, CA 90509

CEAT Representative Office, Inc.
645 Madison Ave.
New York, NY 10022

Continental Tires
1200 Wall St. West
Lyndhurst, NJ 07071

Cooper Tire & Rubber Co.
Lima and Western Aves.
Findlay, OH 45840

Dayton Tire & Rubber Co.
P.O. Box 1026
Dayton, OH 45407

Dunlop Tire & Rubber Corp.
Box 1109
Buffalo, NY 14240

Firestone Tire & Rubber Co.
1200 Firestone Pky.
Akron, OH 44317

Formula One Tires
Div. of Javelin Corp.
614 N. 1st St.
Minneapolis, MN 55401

BF Goodrich Co.
500 S. Main St.
Akron, OH 44318

Goodyear Tire & Rubber Co.
1144 E. Market St.
Akron, OH 44316

Hercules Tire & Rubber Co.
1300 Morrical Blvd.
Findlay, OH 45840

Kelly Springfield Tire Co.
P.O. Box 300
Cumberland, MD 21502

Kleber Tires
The French Tire Co., Inc.
615 Franklin Turnpike
Ridgeway, NJ 07450

M & H Tires
Jacono Industries, Inc.
2811 W. 2nd St.
Chester, PA 19013

McCreary Tire & Rubber Co.
P.O. Box 749
Indiana, PA 15701

Metzler Tires
Euro-tire
567 Route 46
Fairfield, NJ 07006

Michelin Tire Corp.
One Marcus Ave.
Lake Success, NY 11042

Mohawk Rubber Co.
1340 Home Ave.
Akron, OH 44305

Phoenix Tires
Meon, Inc.
221-18 Merrick Blvd.
Springfield Gardens, NY 11413

Pirelli Tire Corp.
110 East 59th St.
New York, NY 10022

Pos-A-Traction Industries, Inc.
17920 S. Wilmington Ave.
Compton, CA 90221

Pro-Trac Tire Co.
425 N. Robertson Blvd.
Los Angeles, CA 90048

Semperit of America, Inc.
156 Ludlow Ave.
Northvale, NJ 07647

Uniroyal, Inc.
1230 Avenue of the Americas
New York, NY 10020

Yokohama Tire Corp.
1530 Church Rd.
Montebello, CA 90640

WHEELS

American Racing Equipment
2600 Monterey St.
Torrance, CA 90501

American Revolution, Inc.
4641 N. Greenview
Chicago, IL 60640

Ansen Wheel Corp.
709 E. Walnut St.
Carson, CA 90746

Appliance Industries
23920 S. Vermont Ave.
Harbor City, CA 90710

BBS Wheels
Intermag
2201 4th St.
Berkeley, CA 94710

Borrani Wheels
H & H Specialties, Inc.
20 Reid Rd.
Chelmsford, MA 01824

BWA Wheels
Conti Enterprise Corp.
20-23 119th St.
College Point, NY 11356

Cal Chrome
936 Mahar Ave.
Wilmington, CA 90744

Campagnolo USA, Inc.
P.O. Box 37426
Houston, TX 77036

Centerline
13521 Freeway Dr.
Santa Fe Springs, CA 90670

Compomotive USA, Inc.
1699 Bryant St.
Daly City, CA 94015

Cosmic Wheels
Mini City Ltd.
876 Turk Hill Rd.
Fairport, NY 14450

Cragar Industries
19007 S. Reyes Ave.
Compton, CA 90221

Cromodora Wheels
Intermag
2201 Fourth St.
Berkeley, CA 94710

Dayton Wheel Products, Inc.
1147 S. Broadway St.
Dayton, OH 45408

Gotti Wheels
European Racing, Inc.
2660 Walnut Ave., Suite G
Tustin, CA 92680

Halibrand
ARC Industries
8157 Wing Ave
ElCajon, CA 92020

Hardy & Beck Performance, Inc.
1799 4th St.
Berkeley, CA 94710

Hayashi Racing USA
15535 Minnesota Ave.
Paramount, CA 90723

Jackman Wheels, Inc.
1035 Pioneer Way
El Cajon, CA 92020

Jongbloed Modular Wheels
1521 E. McFadden Ave., Suite G
Santa Ana, CA 92705

Keystone Automotive Products, Inc.
1333 S. Bon View Ave.
Ontario, CA 91761

LeGrand Race Cars
3491-75th St. West
Willow Springs, CA 93560

Melber Wheels
Shankle Engineering
15451 Cabrito Rd., Bldg. F
Van Nuys, CA 91406

Minilite Wheels
Linkport, Inc.
P.O. Box 1842
Matthews, NC 28105

Monocoque Wheel Co.
10658 Prospect Ave., Suite D
Santee, CA 92071

Motor Wheel Co.
1600 N. Larch St.
Lansing, MI 48914

Panasport Wheels
Formula Mag, Inc.
1801 Avenue of the Stars, Suite 344
Los Angeles, CA 90067

Precision Wheel
411 South West
Fresno, CA 93778

Rial Wheels
Euro-Tire
567 Route 46
Fairfield, NJ 07006

Rocket Industries, Inc.
9935 Beverly Blvd.
Pico Rivera, CA 90660

Ronal Wheels
European Racing
2660 Walnut Ave., Suite G
Tustin, CA 92680

Ruspa Wheels
Serra Hi Performance Products, Inc.
279-C Adams St.
Bedford Hills, NY 10507

Seral Wheels
Intermag
Box 2
Berkeley, CA 94701

Shelby Industries, Inc.
19021 S. Figueroa St.
Gardena, CA 90248

Speedway Mags
11015 Roswell Ave.
Pomona, CA 91766

Stilauto Wheels
Walt Mag
P.O. Box 415
Mount Laurel, NJ 08054

Superior Industries International
7800 Woodley Ave.
Van Nuys, CA 91409

Tru-spoke, Inc.
1800 Talbot Way
Anaheim, CA 92805

Vial Wheels
ADCE-Ronal
15692 Computer Lane
Huntington Beach, CA 92649

Universal Wheels, Inc.
17210 Marquardt
Cerritos, CA 90701

Weldwheels, Inc.
933 Mulberry St.
Kansas City, MO 64101

Western Wheel
Rockwell International
6861 Walker St.
La Palma, CA 90623

Wolfrace Wheels
Atlantic Union, Inc.
2784 Phoenix
Las Vegas, NV 89121

Zenith Wire Wheel Co.
155 Kennedy Ave.
Campbell, CA 95008

SHOCKS

Airshox, Inc.
421 W. Cheltenham Ave.
Melrose Park, PA 19126

Armstrong Hydraulics, Inc.
3480 S. Clinton Ave.
S. Plainfield, NJ 07080

Bilstein Corp. of America
11760 Sorrento Valley Rd.
San Diego, CA 92121

Carrera Racing Shocks
5412 New Peach Tree Rd.
Atlanta, GA 30341

Columbus Parts
One John Goerlich Sq.
Toledo, OH 43694

Delco Products Div.
2000 Forrer Bl.
Dayton, OH 45401

Gabriel Shock Absorbers
Div. of Maremont Corp.
200 E. Randolph
Chicago, IL 60601

Hurst Performance, Inc.
50 W. Street Rd.
Warminster, PA 18974

KYB Corporation of America
207 Eisenhower Lane S.
Lomard, Il 60148

Koni America, Inc.
111 W. Lovers Ln.
P.O. Box 40
Culpeper, VA 22701

Monroe Auto Equipment Co.
International Dr.
Monroe, MI 48161

Spax Shocks
Altered Engineering
P.O. Box 3758
Industry, CA 91744

ANTI-ROLL BARS

Addco Industries
Watertower Rd.
Lake Park, FL 33403

DM Engineering
3710 W. 167th St.
Markham, IL 60426

H & H Specialties, Inc.
20 Reid Road
Chelmsford, MA 01824

Hellwig Products Company, Inc.
16237 Avenue 296
Visalia, CA 93277

INTERPART
230 W. Rosecrans
Gardena, CA 90248

Spearco Performance Products, Inc.
2054 Broadway
Santa Monica, CA 90404

WINCO
10329 Detroit Ave.
Cleveland, OH 44102

AERODYNAMIC DEVICES

A & A Fiberglass, Inc.
1534 Nabell Ave.
Atlanta, GA 30344

Airheart Brake Company
1028 S. 3rd St.
Minneapolis, MN 55415

American Racing Equipment
540 Hawaii Ave.
Torrance, CA 90503

Cannon Industries, Inc.
9073 Washington Blvd.
Culver City, CA 90230

CP Auto Products
3869 E. Medford St.
Los Angeles, CA 90063

MAS Racing Products
2538 Hennepin Ave.
S. Minneapolis, MN 55405

McCoy Tire Service Centers
3333 McHenry
Modesto, CA 95350

Perfect Plastics, Inc.
1304 Third Ave.
New Kensington, PA 15068

Sbarbaro Racing Enterprises
1481 Laurenita Way
Alamo, CA 94507

Spearco Performance Products, Inc.
2054 Broadway
Santa Monica, CA 90404

Unique Metal Products
8745 Magnolia Ave., Unit 4
Santee, CA 92071

BRAKES

Contemporary Chassis Design
15506 Vermont St.
Paramount, CA 90723
(Disk Brake Conversions)

Earl's Supply Co.
P.O. Box 265
Lawndale, CA 90260
(Steel Braided Brake Hoses)

Edco Specialty Products
1525-15 W. MacArthur Blvd.
Costa Mesa, CA 92626

G & O Manufacturing Co.
77 Mark Drive.
San Rafael, CA 94903
(Safety Braker)

Grey-Rock Division
Raybestos-Manhattan, Inc.
205 Middle St.
Bridgeport, CT 06603

Henry's Engineering Co.
Prince Frederick, MD 20678
(Steel Braided Brake Hoses)

Kelsey-Hayes Co., Inc.
38481 Huron River Drive
Romulus, MI 48174

Lakewood Industries
4800 Briar Rd.
Cleveland, OH 44135
(Velvetouch Linings)

Lamb Components
150 W. College Way
La Verne, CA 91750

Lee Manufacturing Co.
Division of Rolero
P.O. Box 7187
Cleveland, OH 44128

Minnesota Automotive, Inc.
1911 Lee Blvd.
North Mankato, MI 56001

Morak Brakes, Inc.
9902 Ave. D
Brooklyn, NY 11236

Neal Products
5231 Cushman Place
San Diego, CA 92110
(Pedals, Balance Bars)

Omega Engineering, Inc.
Box 4047, Springdale Station
Stamford, CT 06907
(Brake Temperature Indicators)

Troutman
3198L Airport Loop Dr.
Costa Mesa, CA 92626

Titton Engineering, Inc.
McMurray Rd. & Easy St.
P.O. Box 1787
Buellton, CA 93427

SUSPENSION COMPONENTS

ARC-Halilbrand
8157 Wing Ave.
El Cajon, CA 92020

Dick Guldstrand Enterprises
11914 W. Jefferson Blvd.
Culver City, CA 90230

Eelco Manufacturing & Supply Co.
P.O. Box 4095
Inglewood, CA 90309

Frankland Racing Equipment
P.O. Box 278
Ruskin, FL 33570

Genuine Suspension
1645 Superior St.
Costa Mesa, CA 92627

Hellwig Products, Inc.
16237 Avenue 296
Visalia, CA 93277

Huffaker Engineering, Inc.
22 Mark Dr.
San Rafael, CA 94903

JR Products Corp.
1222 Delevan Dr.
San Diego, CA 92102

Le Baron Racing Equipment
135 Hopper Ave.
Waldwick, NJ 07463

Le Grand Race Cars
1104 Aurora
Sylmar, CA 91342

MAS Racing Products
2538 Hennepin Ave.
S. Minneapolis, MI 55405

Mickey Marollo Race Cars & Parts
RD ε2
Creek Rd.
Pinecity, NY 14871

Mr. Gasket Co.
4566 Spring Rd.
Cleveland, OH 44131

Mueller Fabricators
10872 Stanford Ave.
Lynwood, CA 90262

Northwest Race Cars
23445 S.W. 82nd Ave.
Tualatin, OR 97062

PSI Industries, Inc.
9103 E. Garvey Ave.
Rosemead, CA 91770

SGR Industries
7041 Dogwood
Mentor, OH 44060

Speed Sport Specialties
403-13th St.
San Diego, CA 92101

Stahl & Associates
337-341 West King St.
York, PA 17404

Strange Engineering Co.
739 W. Howard St.
Evanston, IL 60202

Summers Brothers
530 So. Mountain Ave.
Ontario, CA 91761

Total Performance Inc.
328 State St.
New Haven, CT 91761

Tilton Engineering, Inc.
114 Center St.
El Segundo, CA 90245

Wing Engineering Development Co.
3038 Scott Blvd.
Santa Clara, CA 95050

TRACTION BARS

Ansen Automotive Engineering
13720 S. Western Ave.
Gardena, CA 90249

Britting Engineering
10885 Kalama River Ave., Suite C
Fountain Valley, CA 92708

Don Hardy Race Cars
202 W. Missouri St.
Floydada, TX 70235

Lakewood Industries
4800 Briar Rd.
Cleveland, OH 44135

SPECIAL TOOLS

Consumer Electronics
P.O. Box 43055
Cincinnati, OH 45243
(Tire Pyrometer)

Performance Marketing
3199 A Airport Loop Dr.
Costa Mesa, CA 92626

LRE
Rt. 44
Millerton, NY 12546
(Dunlop Alignment Gages)

Mitchell Engineering
P.O. Box 2335
Leucadia, CA 92024
(Camber Gauge)

Omega Engineering, Inc.
Box 4047, Springdale Station
Stamford, CT 06907
(Pryometer, Temperature Measuring Devices)

TRW Replacement Division
8001 East Pleasant Valley Rd.
Cleveland, OH 44131
(Coil Spring Compressor)

Wacho Products Co.
2171 Cleveland Ave.
Columbus, OH 43211
(Spring Checker, Toe-in Gauge, Tire Heater, Weight Checker)

NMW Manufacturing Co.
355 E. 54th St.
Elmwood Park, NJ 07407

Traction Master Co.
2917 W. Olympic Blvd.
Los Angeles, CA 90006

METRIC CUSTOMARY-UNIT EQUIVALENTS

Multiply:		by:		to get:	Multiply: by:			to get:
LINEAR								
inches	X	25.4	=	millimeters(mm)	X	0.03937	=	inches
feet	X	0.3048	=	meters (m)	X	3.281	=	feet
yards	X	0.9144	=	meters (m)	X	1.0936	=	yards
AREA								
inches^2	X	645.16	=	millimeters^2(mm^2)	X	0.00155	=	inches^2
feet^2	X	0.0929	=	meters^2(m^2)	X	10.764	=	feet^2
VOLUME								
quarts	X	0.94635	=	liters (l)	X	1.0567	=	quarts
gallons	X	3.7854	=	liters (l)	X	0.2642	=	gallons
feet^3	X	28.317	=	liters (l)	X	0.03531	=	feet^3
feet^3	X	0.02832	=	meters^3(m^3)	X	35.315	=	feet^3
fluid oz	X	29.57	=	milliliters (ml)	X	0.03381	=	fluid oz
MASS								
ounces (av)	X	28.35	=	grams (g)	X	0.03527	=	ounces (av)
pounds (av)	X	0.4536	=	kilograms (kg)	X	2.2046	=	pounds (av)
FORCE								
ounces—f(av)	X	0.278	=	newtons (N)	X	3.597	=	ounces—f(av)
pounds—f(av)	X	4.448	=	newtons (N)	X	0.2248	=	pounds—f(av)
TEMPERATURE								
Degrees Celsius (C) = 0.556 (F - 32)					Degree Fahrenheit (F) = (1.8C) + 32			
ENERGY OR WORK (Watt-second=joule=newton-meter)								
foot-pounds	X	1.3558	=	joules (J)	X	0.7376	=	foot-pounds
Btu	X	1055	=	joules (J)	X	0.000948	=	Btu
PRESSURE OR STRESS								
pounds/sq in.	X	6.895	=	kilopascals (kPa)	X	0.145	=	pounds/sq in
TORQUE								
pound-inches	X	0.11298	=	newton-meters (N-m)	X	8.851	=	pound-inches
pound-feet	X	1.3558	=	newton-meters (N-m)	X	0.7376	=	pound-feet
pound-inches	X	0.0115	=	kilogram-meters (Kg-M)	X	87	=	pound-feet
pound-feet	X	0.138	=	kilogram-meters (Kg-M)	X	7.25	=	pound-feet

COMMON METRIC PREFIXES

mega	(M)	=	1,000,000	or	10^6	centi	(c)	=	0.01	or	10^{-2}
kilo	(k)	=	1,000	or	10^3	milli	(m)	=	0.001	or	10^{-3}
hecto	(h)	=	100	or	10^2	micro	(μ)	=	0.000,001	or	10^{-6}

Conversion Chart courtesy Ford Motor Company.